COMPARATIVE TOXICOGENOMICS

Other volumes in **Advances in Experimental Biology**

Volume 1   Nitric Oxide, by B. Tota and B. Trimmer (Eds.) – 2007,
            ISBN 978-0-444-53119-3

COMPARATIVE TOXICOGENOMICS

Edited by

Christer Hogstrand

King's College London
Nutritional Sciences Division
Franklin-Wilkins Building
150 Stamford Street
London SE1 9NH, UK

and

Peter Kille

Cardiff School of Biosciences
Biomedical Sciences Building
Cardiff University
Cardiff CF10 3US, UK

ELSEVIER

Amsterdam – Boston – Heidelberg – London – New York – Oxford
Paris – San Diego – San Francisco – Singapore – Sydney – Tokyo

Elsevier
Radarweg 29, PO Box 211, 1000 AE Amsterdam, The Netherlands
Linacre House, Jordan Hill, Oxford OX2 8DP, UK

First edition 2008

Notice
No responsibility is assumed by the publisher for any injury and/or damage to persons
or property as a matter of products liability, negligence or otherwise, or from any use
or operation of any methods, products, instructions or ideas contained in the material
herein. Because of rapid advances in the medical sciences, in particular, independent
verification of diagnoses and drug dosages should be made

Library of Congress Cataloging-in-Publication Data
A catalog record for this book is available from the Library of Congress

British Library Cataloguing in Publication Data
A catalogue record for this book is available from the British Library

ISBN: 978-0-444-53274-9
ISSN: 1872-2423

For information on all Elsevier publications
visit our website at books.elsevier.com

Printed and bound in the United Kingdom

Transferred to Digital Print 2011

Working together to grow
libraries in developing countries

www.elsevier.com | www.bookaid.org | www.sabre.org

ELSEVIER   BOOK AID
             International   Sabre Foundation

Information about the Society for Experimental Biology (SEB)

The Society for Experimental Biology (SEB) is Europe's leading, not-for-profit organisation embracing all disciplines of experimental biology. Through its large membership and passion for science the Society supports and promotes experimental biology, from molecular to ecological, to benefit both the scientific community and the general public. The Society was founded in 1923 at Birkbeck College and is now well established with a current worldwide membership of over 1,900 biological researchers, teachers and students. The SEB is somewhat unusual in catering for both plant and animal biologists.

The Animal Section of the SEB launched this serial "Advances in Experimental Biology" (AEB) in 2006, with the first volume published in 2007. The aim of this new serial is to provide state-of-the-art review volumes on timely issues that have international topicality within the field of Comparative and Integrative Biology. The series as a whole will therefore cover a rather broad range of topics from the role of individual molecules (e.g. nitric oxide) to system-level approaches applied to a particular research discipline (e.g. toxicogenomics).

Each volume will contain approximately ten chapters each providing a detailed review of current understanding within a sub-topic of the volume title. In some volumes the chapters will be organised according to the level at which the research is focused (e.g. molecular, biochemical, physiological, behavioural, ecological); in others the chapters may be organised on a more taxonomic basis. Each chapter is written by leading authorities in science that have been invited, based on their international reputation, to provide their perspective on the current status and recent developments within the field. All chapters are peer-reviewed by at least two independent referees prior to acceptance for publication.

The series thus aims to provide an excellent, up-to-date resource for a global research audience within each of the volume topics.

ix

Information about the Series Editors

Professor Mike Thorndyke

Professor Mike Thorndyke is the Director and Chair of Experimental Marine Biology at the Royal Swedish Academy of Sciences, Kristineberg Marine Research Station, Sweden. His laboratory specialises in understanding the diversity of cellular, genetic processes in marine organisms, and their importance in terms of evolution, adaptation and ecology, with particular reference to development and adult regeneration. Professor Thorndyke has over 25 years experience working with marine invertebrates. His laboratory has been the focus for the characterisation of neural complexity in invertebrate deuterostomes including echinoderms, tunicates and the Xenoturbellids, recently confirmed as a new Deuterostome phylum. Most recently, Professor Thorndyke has been one of the leaders of the neural group in the sea urchin genome annotation consortium.

Dr Rod Wilson

Dr Rod Wilson began his academic career reading Biological Sciences at the University of Birmingham, where he subsequently completed a Ph.D. in Fish Physiology and Ecotoxicology. After completing postdoctoral training at the University of Birmingham, McMaster University in Canada and the University of Manchester, Dr Wilson moved to the University of Exeter where he is currently a Senior Lecturer. His area of expertise covers comparative and integrative physiology, ranging from studies on water absorption in the mammalian kidney to the behaviour and physiology of fish in the wild. Dr Wilson's work currently focuses upon fish (marine and freshwater) and he believes it is important to approach research from a multidisciplinary angle in order to guarantee a holistic understanding of homeostasis in animals. He is particularly interested in how multiple physiological systems (*e.g.*, respiratory/cardiovascular, osmoregulation, acid–base balance, nitrogenous waste excretion) respond in an integrated manner to maintain whole animal homeostasis in the face of environmental changes (both natural and anthropogenic). His work utilises studies at the molecular, cellular, tissue and whole animal levels. Furthermore, he is increasingly integrating this approach with behavioural studies that help link physiological mechanisms with social behaviour in fish, both in the laboratory and in the wild.

Information about the Volume Editors

Professor Christer Hogstrand

Professor Christer Hogstrand's expertise lies in the biology and toxicology of minerals, with a particular focus on how zinc controls biological processes. Professor Hogstrand completed his Ph.D. at the University of Göteborg before undertaking a postdoctoral research position with Professor Chris Wood at McMaster University, Canada. During this time, he pioneered research on silver toxicity to fish and discovered uptake pathways for zinc across the gill. Professor Hogstrand moved to King's College London in 2001 after developing his research as Assistant and Associate Professor at the University of Kentucky and the University of Miami. He was promoted to Professor in 2006 where he is a leading figure on using post-genomic and proteomic technologies as tools for class prediction and mechanistically relating negative effects to affected networks.

Dr Peter Kille

Dr Peter Kille's areas of expertise include the biochemistry of heavy metals, molecular biology, molecular ecotoxicology, environmental biomonitoring and metal binding proteins. Specifically, his research harnesses genomic, proteomic and metabolomic techniques in order to study the mechanisms by which biological systems handle heavy metals and other xenobiotics. Dr Kille began and developed his research career at the University of Wales College of Cardiff. He is currently a Senior Lecturer at the School of Biosciences, University of Wales College of Cardiff, a position he has held since 2001.

List of Contributors

Ronny van Aerle, Environmental and Molecular Fish Biology, School of Biosciences, The Hatherly Laboratories, University of Exeter, Prince of Wales Road, Exeter, Devon, UK.

Jonathan Ball, Environmental and Molecular Fish Biology, School of Biosciences, The Hatherly Laboratories, University of Exeter, Prince of Wales Road, Exeter, Devon, UK.

Angela Brown, School of Civil Engineering and Geosciences, Newcastle University, Newcastle upon Tyne, UK.

Amanda Callaghan, School of Biological Sciences, University of Reading, Reading, UK.

John K. Colbourne, The Center for Genomics and Bioinformatics, Indiana University, Bloomington, Indiana, USA.

Isabelle Colson, Zoologisches Institut, Universität Basel, Biozentrum/Pharmazentrum, Klingelbergstrasse 50, 4056 Basel, Switzerland.

Brian D. Eads, Department of Biology, Indiana University, Bloomington, Indiana, USA.

Amy L. Filby, Environmental and Molecular Fish Biology, School of Biosciences, The Hatherly Laboratories, University of Exeter, Prince of Wales Road, Exeter, Devon, UK.

Donald Gilbert, Department of Biology, Indiana University, Bloomington, Indiana, USA.

Eugene P. Halligan, Molecular Diagnostics, Pathology SDU, St. Thomas' Hospital, Lambeth Palace Road, London, UK.

Richard D. Handy, School of Biological Sciences, University of Plymouth, Drake Circus, Plymouth, UK.

Colin R. Harwood, Institute for Cell and Molecular Biosciences, Newcastle University, Newcastle upon Tyne, UK.

Ian M. Head, School of Civil Engineering and Geosciences, Newcastle University, Newcastle upon Tyne, UK.

Bastiaan Jansen, Laboratory of Aquatic Ecology, Katholieke Universiteit Leuven, Ch. De Beriostraat 32, B-3000, Leuven, Belgium.

Peter Kille, Cardiff School of Biosciences, BIOSI 1, University of Cardiff, Cardiff, UK.

Rebecca Klaper, Great Lakes WATER Institute, University of Wisconsin-Milwaukee, 600 East Greenfield Ave, Milwaukee, Wisconsin, USA.

Anke Lange, Environmental and Molecular Fish Biology, School of Biosciences, The Hatherly Laboratories, University of Exeter, Prince of Wales Road, Exeter, Devon, UK.

Joseph Lunec, Cranfield Health, Cranfield University, Barton Road, Silsoe, Bedfordshire, UK.

A. John Morgan, Cardiff School of Biosciences, BIOSI 1, University of Cardiff, Cardiff, UK.

Michael E. Pfrender, Department of Biology, Utah State University, 5305 Old Main Hill Road, Logan, Utah, USA.

Helen C. Poynton, Department of Nutritional Sciences and Toxicology, University of California, Berkeley, California, USA.

Eduarda M. Santos, Environmental and Molecular Fish Biology, School of Biosciences, The Hatherly Laboratories, University of Exeter, Prince of Wales Road, Exeter, Devon, UK.

Joseph R. Shaw, The School of Public and Environmental Affairs, Bloomington, Indiana, USA.

Richard M. Sibly, School of Biological Sciences, University of Reading, Reading, UK.

Jason R. Snape, AstraZeneca, Brixham Environmental Laboratory, Freshwater Quarry Brixham, Devon, UK.

David J. Spurgeon, Centre for Ecology and Hydrology, Monks Wood, Abbots Ripton, Huntingdon, Cambridgeshire, UK.

Charles R. Tyler, Environmental and Molecular Fish Biology, School of Biosciences, The Hatherly Laboratories, University of Exeter, Prince of Wales Road, Exeter, Devon, UK.

Chris D. Vulpe, Berkeley Institute of the Environment, University of California, Berkeley, California, USA.

Henri Wintz, Department of Nutritional Sciences and Toxicology, University of California, Berkeley, California, USA.

Preface

Functional genomics has come of age. No longer is it an adventure for the avant-garde scientist, but it has become an increasingly standardised mainstream tool accessible to any modern biological laboratory. Toxicogenomics studies are now generating an avalanche of data which, with the aid of established informatic methodology, is being translated into biologically meaningful information. This is enabling us to start harvesting the benefits from years of investment in terms of technology, time, and (of course) money. It is therefore timely to bring together leading toxicologists with a wide variety of scientific aims to demonstrate how microarray technology can be successfully applied to different research areas.

There are a number of challenges in editing any volume whose remit encompasses the 'genomics' arena including jargon busting, demystifying technology or representing the sheer scope of the disciplines embracing this scientific approach. Our aim was to address the latter objective and bringing together a series of papers that represent the breadth of the toxicogenomic landscape. To achieve this we had to consider a myriad of related themes each representing a spectrum of interconnected applications. The major continuum revolves around the applied endpoint of the investigation whether the toxicological investigation is: directly applied to human health (*i.e.*, pharmalogical in nature); a surrogate for human response; or a measure of environmental risk (ecotoxicogenomics). Furthermore, intertwined with this is the level of genetic and functional knowledge that exists for the organisms exploited in the studies ranging from the so-called *model organism* (such as mouse and zebrafish) to those genetically underrepresented taxa where the small amount of genome information is overshadowed by our ignorance of their functional biology. The nature of the application is also reflected by the bio-material exploited with ethical drivers promoting the use of *in vitro* cell lines whilst tissue specific investigation are employed where there is a known mode-of-action (MOA) to whole organisms and population based analysis where drivers include size, unknown MOA or a desire to relate molecular responses to population level effects. These aspects are overlaid by the growing maturation of the field which is seeing a transfer from the fundamental science forums to the legislative arena. Thus, this book transects biology from bacteria to human, from ecologically relevant sentinel organisms to well characterised model species, and the toxicogenomics arena from exploratory 'blue sky' science to prospects of incorporation into regulatory frameworks.

In many ways the implications for toxicology of human genetic variation, as revealed by whole genome sequence analyses, is substantively less controversial than the disclosure of pre-disposure to either conditions that can be mitigated or are untreatable. As is discussed by Halligan and Lunec (Chapter 1) the potential to predict the differential responses of specific haplotypes to chemical exposure unlocks the Pandora's box which is personalised medicines, whilst providing opportunities to predict groups which are at increased risk from non-specific exposure to environmental or industrial chemicals. The impact of the prospective deployment of genomics within pharmaceutical, agrochemical and chemical legislative testing which is touched open in Chapter 1 is expanded on by Poynton *et al.* (Chapter 2) who explore the integration of 'omics' approaches into the present tier testing framework considering their predictive abilities against established life history endpoints. They continue the theme of applying genomics within risk assessment by touching on its application in environmental monitoring, an issue which is given increased emphasis by Tyler *et al.* (Chapter 3), Spurgeon *et al.* (Chapter 4), and Shaw *et al.* (Chapter 5). These Chapters explore the transition from laboratory-based ecotoxicology into 'real world' or field-based studies raising the thorny issue of the strengths and weaknesses of the so-called *genomic models* against '*environmental sentinels*' culminating in a detailed exploration of the recognised freshwater sentinel, *Daphnia*, whose genome is complete and where there are significant transcriptome studies (Shaw *et al.*, Chapter 5). Whether directed towards a defined endpoint, such as endocrine disruption (Tyler *et al.*, Chapter 3), or characterising the complexities posed by the overlay of geo-chemistry with complex pollution scenarios such as found in terrestrial systems (Spurgeon *et al.*, Chapter 4) genomics will have a significant role to play. We present a case study of the application of functional genomics to explore microbial toxicology (Brown *et al.*, Chapter 6) as an illustration of the power of genomics to unravel the mechanistic pathways underlying biological effects and the universality of the approach from bacterial to man. In recognition that the full potential of the genomics approaches cannot be realised without considering the biological or environmental context in which they are made, we have included a horizon scanning paper exploring the possibilities inherent in 'Systems toxicology' (Handy *et al.*, Chapter 7). Intriguingly, the corollary of the papers is to demonstrate the commonality of approach throughout the toxicological community. This has been driven by an alignment of the platform technologies; international agreement on quality/data standards; and a consolidation of methods for data interpretation. Whatever is the subject or objective of the study, the approach is similar, with the exception that scientists working on non-model species must put considerable effort in generating the primary sequence data, while biomedical scientists working on genetically well-defined organisms have genomic sequence at the click of a mouse-button.

Dr Craig Venter heralds the advent of a '*DNA-driven world*' and through toxicogenomics the toxicology community is embracing this mantra. Although there is significant sequence knowledge gaps which need to be closed, the emergence of massively parallel sequencing platforms (*e.g.*, pyro-sequencing) will ensure that the generation of both primary sequence together with the identification of functionally important sequence variations will be rapidly filled. These will exponentially accelerate the SNP-based population associations in human whilst provide an opportunity to explore the implication of natural variation within wild animal population. The extraordinary advance in technology for both sequence generation and transcript analysis is posing two questions; how do we better design are experiments? and, how we transfer the toxicogenomic knowledge to understanding?

In its infancy, the vision of toxicogenomics revolved around the generation of a global transcript profile that could be deconvoluted to reveal; the individual compounds to which the biological system had been exposed; and the combination of current and predicted impacts exposure on the molecular functionality and biological processes of the target biological system. Initially, however, microarray-based toxicogenomics studies often produced little beyond gene lists and investigators were pleased enough to see that genes they recognised from traditional methods were involved in the studied process were indeed picked up by microarray analysis as being differentially expressed (*e.g.*, CYP1A1, G6PD, MT). Although we still have not realised the full potential power of toxicogenomics, the time has passed when microarray analysis ended at the generation of flat gene lists. Over the last few years, a number of studies have shown that the analysis of genes placed in context of their functional pathways and classes generates subsets of tens of genes which may accurately predict specific MOA such as geno or nephridial toxicity. Furthermore, as these approaches are applied to a wider base of organisms the differences and conservation of pathways become apparent providing insight into differential species sensitivity and, as importantly, underlying conserved mechanisms. We must embrace intelligent experimental design to allow us to dissect dynamic responses to multifaceted exposures to better predict real-world experience. Harnessing the power of multivariant statistics may allow us to determine where mixture toxicity causes unexpected MOA and to characterise specific synergistic or antagonistic interactions between chemicals.

Measuring global genomic responses to the chemical perturbation of a biological system is analogous to a stone being dropped into a body of water. The initial interaction creates a wave or response that moves outwards impacting on pathways and networks throughout the cell. Furthermore, if the initial interaction or subsequent response affects key nodes, such as transcription factors, signalling molecules, or flux points, the impact of the chemical response is amplified. If biological systems are disrupted at multiple points (multiple toxicants or MOAs) the key, it is essential to determine the interference pattern establishing where the waves interact and reinforcement of effect will be felt or whether the effects cancel each other. Again the key is defining the effect on nodal molecules. Like a moving wave, gene expression patterns are not necessarily static, but may change over time as the cell and the organism experiences disruption of homeostasis. As a consequence if we consider the response a single time point it is extremely difficult to predict backwards to the initial perturbation. There is a deficiency in temporal data sets together with the tools required if we are to trace response back to the source. Furthermore, the lack of functional annotation within some key species makes us reliant on implied pathway/network architecture from mouse, man, fly, nematode, or yeast.

We would like to thank all of the authors who have contributed to this volume together with Dr John Morgan who have aided in constructing the ethos underlying this volume. Lastly we would like to recognise Suzanne Brockhouse at the Society for Experimental Biology for her forbearance and organisation skill in ensuring the timely construction of this volume.

Christer Hogstrand and Peter Kille

Contents

1

Toxicogenomics: Unlocking the potential of the human genome

Eugene P. Halligan[1],* and Joseph Lunec[2]

[1]*Molecular Diagnostics, Pathology SDU, 5th Floor North Wing, Room 211, St. Thomas' Hospital, Lambeth Palace Road, London, SE1 7EH, UK*
[2]*Cranfield Health, Cranfield University, Barton Road, Silsoe, Bedfordshire, MK45 4DT, UK*

Abstract. Toxicogenomics merges genomics with toxicology and observes the genome-wide effects of toxicants. The field is really only taking the first steps to establish the 'ground rules', nomenclature and standards by which it will develop and a standardised approach to toxicogenomic evaluation is still to be agreed. There is a great need to address toxicogenomics to iatrogenic morbidity, environmental health and safety and diet. In the case of iatrogenic morbidity, 0.5% of the UK hospital population is affected and toxicogenomics could lead to personalised medicine, *i.e.*, to define a drug dosage tailored to each patient's unique genetic make-up and medical condition that is beneficial, inadequate or toxic. Before this can become a reality, toxicogenomic profiles need to be generated for a host of commonly prescribed drugs and shown to be robust in cross-centre, cross-platform comparisons; the magnitude of the work needed will be vast and needs to be nationally coordinated and funded. The potential of predictive toxicogenomic and all tri-nomic methodologies is far greater than its current usefulness. The sequencing of genomes alone is not a panacea. Rather, genomic, tri-nomic and pharmacogenetic databases must be integrated with a comprehensive toxicant class database with validated tri-nomic profiles linked to traditional toxicity endpoints; this should be carried out as exhaustively as the sequencing effort itself. This must be undertaken in order to exploit the vast potential of this new field to provide personalised medicine, sensitive and quick environmental health and safety surveillance and accurate and scientifically supported dietary advice.

Keywords: toxicology; genomics; transcriptome; tri-nomics; National Health Service (NHS); statins; diesel particles; peroxides; red meat; lipid peroxidation; environment; pollution; bioinformatics; pharmacogenetics; proteomics; metabonomics; personalised medicine; study design; pharmaceutical industry.

Introduction

In day-to-day life we can be exposed to a vast number of toxic chemicals from drugs, food additives, industrial chemicals, environmental pollutants and natural toxins. The assessment of their subsequent impact on human health and disease comprises a hugely complex area of biology known as

Corresponding author: Tel.: 44(0)20 7188 1257.
E-mail: Eugene.Halligan@gstt.nhs.uk (E.P. Halligan).

© 2008 ELSEVIER B.V.
ALL RIGHTS RESERVED

'toxicology'. Toxicology has come to mean the 'study of poisons' but it can be more accurately defined as the study of the effects of chemicals on gene, cell, tissue and organ pathology and physiology, which can impact negatively on the health of living organisms. In the field of toxicology, one of our main aims is to elucidate the toxic properties of chemicals by understanding their mode of action specific mechanism contributing to their toxicity and subsequent disease progression. The function of the toxicologist is to evaluate the hazards of chemicals to organisms in relation to the concentration of these substances in the environment, offer risk estimation and advice on measures to control and prevent the harmful effects of chemicals. Toxicology is a broad church, as almost anything can be considered toxic given sufficient dose and exposure. It is prudent to discriminate 'all compounds' and narrow the focus to those that cross the line between tolerable and intolerable risk when encountered commonly in the environment (unintentional/accidental exposure), or by intentional ingestion (food, drugs, alcohol, and cigarettes). This delineation is arbitrary but can be more objectively specified by the magnitude of the dose–toxic response relationship and likelihood of that dose being encountered during a specific type of exposure.

Molecular toxicology focuses on harmful effects of chemicals at the cellular and molecular level in living organisms. It has greater potential to quickly determine mechanisms of toxicity, comprehensive specific biomarkers, accurate risk estimation and effective hazard control than conventional toxicology. This potential is due in large part to the improved sensitivity, accuracy and speed of molecular techniques such as microarray analysis and the fundamental wealth of biological knowledge contained in the human genome project. A host of new advances in high-throughput screening technologies will exploit this new data resource. Molecular toxicology has been used to study a relative handful of toxicants so far, but undoubtedly a wealth of toxic mechanistic information on many chemicals will be derived from this new field and will ultimately redefine the study of toxicology.

The current state of play in toxicogenomics

In the field of toxicology, a new subdiscipline termed toxicogenomics has emerged which promises to identify and characterise the molecular mechanisms that lead to toxicity. Gene expression profiling, through the use of microarray technology, is rapidly becoming a standard method of analysis in toxicology studies, and has the potential to play a pivotal role in all stages of drug safety evaluation (Waring and Halbert, 2002).

Toxicogenomics merges genomics with toxicology by observing the genome-wide effects of a toxic compound on an organism. The new molecular technology of microarray-based genomics have revolutionised the field of toxicology making it possible to study the effects of toxic substances on entire genomes thus leading to a better understanding of gene–environment interactions. Toxicogenomics, however, is not the endpoint of a study but rather a good starting point from which to develop experiments designed to gain a mechanistic insight into drug toxicities. These mechanisms must be then ideally confirmed using conventional biochemical, toxicological and pathological approaches (Lord et al., 2006).

A process is underway in the USA and well ahead of any similar initiative in the UK or Europe (Pennie et al., 2004), driven by agencies such as the National Institute of Environmental Health Sciences (NIEHS) and their allies in the National Toxicology Program, to 'molecularise' toxicology by fostering the emergence of a new discipline: toxicogenomics (Chhabra et al., 2003; Waters et al., 2003). Establishing toxicogenomics as a new discipline requires (1) the development of communication across a range of disciplines and integration with existing toxicology data; (2) integration of current and new devices, standards and practices; (3) grounding toxicogenomics in traditional toxicological standards and work practices; and (4) identification and stabilisation of roles for toxicogenomic knowledge in fields such as environmental health risk assessment and regulation (Shostak, 2005). Both governments and commercial organisations are building robust toxicogenomic databases to characterise hazard. The great challenge is how to link patterns of gene expression and gene clustering as well traditional toxicology biomarkers and endpoints to specific adverse effects of toxicants or classes of toxicants (Chan and Theilade, 2005). Indeed, the field is really only taking the first steps to establish the 'ground rules', nomenclature and standards by which it will develop as can be seen in comments made by some of the main players in the emergent field:

'The microarray is fairly new so, right now, researchers are using a lot of different methods and protocols in microarray experiments. That makes it hard for researchers to compare their results to results from other labs', 'When scientists start using the same methods, equipment and reagents, data can be compared across the entire field of medicine and scientific advances will come more quickly' said Kenneth Olden, PhD, Director of NIEHS. Researchers at the NIEHS, in a study initiated in 2001, have systematically examined the processes involved in most microarray or gene expression studies, and found that using a standardised process led to

4

more consistent results when they looked at causes of variation in gene expression experiments within and between laboratories, as well as within and between microarray platforms. The broad consensus of NIEHS researchers was that using commercially manufactured microarrays produced the best results that could be more easily replicated than in-house arrays from individual laboratories which gave less-consistent results (Bammler *et al.*, 2005; Shi *et al.*, 2006). Gene expression data was shown to be very useful in understanding diseases and biological processes but it was concluded that protocols must be rigidly standardised in order to inform diagnostics in clinical practice or objective risk assessment in environmental toxicology. The Toxicology Research Consortium or TRC is a consortium of seven research centres including the NIEHS Microarray Group of the National Center for Toxicogenomics, Duke University, Fred Hutchinson Cancer Research Center/University of Washington, Massachusetts Institute of Technology, Oregon Health and Sciences University and the University of North Carolina at Chapel Hill; Icoria Inc. is also a research partner. The TRC coordinates researchers who use microarrays to study disease processes to agreed standards in order to minimise variation along the lines of the many initiatives that are being evaluated and proposed (Bao *et al.*, 2005). The central aim is to ensure that microarrays are used optimally in order to be effective in clinical diagnostics to aid in the design of patient-tailored therapies. It is widely agreed that no single initiative or individual organisation, or laboratory, could afford to do this without global agreement and national/ international collaboration. For instance, the US Environmental Protection Agency (EPA) is currently evaluating high-throughput screening and toxicogenomics to forecast toxicity based on bioactivity profiling to predict and prioritise limited testing resources toward chemicals that likely represent the greatest hazard to human health and the environment under a research programme called 'ToxCast' (Dix *et al.*, 2007). Initially ToxCast will focus on several hundred well-characterised reference chemicals which represent numerous structural classes and phenotypic outcomes, including tumorigens, developmental and reproductive toxicants, neurotoxicants and immunotoxicants in cell-based assays (Dix *et al.*, 2007).

A standardised approach to toxicogenomic evaluation has still to be agreed and formulated in this new and exciting area of toxicology. Work continues apace providing a significant body of proof through numerous studies to demonstrate an agreed statistical/mathematical algorithm to provide a truly predictive toxicology. A large number of genes in any given microarray analysis dataset of differential expression will be irrelevant to the analysis and objective selection of discriminatory genes critical to predictive toxicity. In a study by Tsai *et al.* (2005) of a toxicogenomic dataset with nine treatments (a control and eight metals, As, Cd, Ni, Cr, Sb, Pb, Cu and AsV with a total of 55 samples), the authors were able to group the 55 samples into one of the nine treatments 85% of the time using hierarchical and partition (*k*-means) methods. While this gives great encouragement, the size of the task to be overcome in order to provide global objective risk assessment and true predictive toxicity requires a comprehensive powerful bioinformatics approach such as that used by Tsai *et al.* (2005), to systematically evaluate all biological toxicity data, not just selected heavy metal toxins in one specific case as cited here. One such global interpretive system and database proposed is dbZach (http:// dbzach.fst.msu.edu), a modular relational database that manages traditional toxicology and complementary toxicogenomic data to facilitate comprehensive data integration and analysis across a range of matrices, species and methods (Burgoon *et al.*, 2006). The toxicogenomic data which would be integrated with standard biometric measures of toxicity on such a database are to be derived from a number of studies. In the main these studies favour class prediction methods using the gene expression profiles of known toxins from representative toxicological classes in order to predict the toxicological effect of an uncharacterised compound based on the similarities between its gene expression profile and that of a thoroughly characterised toxicological class (Maggioli *et al.*, 2006). However, this resource is not currently available to toxicologists and is at present mainly theoretical conjecture (Lord *et al.*, 2006). It is encouraging however, that recent studies such as the MicroArray Quality Control (MAQC) project, using a biologically relevant toxicogenomics dataset of 36 RNA samples from rats treated with three chemicals (aristolochic acid, riddelliine and comfrey), was hybridised across four microarray platforms and was reported to show high concordance in intersite and cross-platform comparisons (Guo *et al.*, 2006).

Toxicogenomics and personalised medical treatment

Iatrogenic morbidity is an important and increasing problem to the NHS (UK) which affects an estimated 5–6.5% of the UK in-patient hospital population (Davies *et al.*, 2006; Wiffen *et al.*, 2002). There is a clinical need to better define the fine line between a drug dosage that is beneficial, has an inadequate therapeutic effect or produces toxicity. Providing the clinician with comprehensive diagnostic information about the genetic elements of the patient that may affect drug efficacy will eventually allow for the determination of individualised dosages and fewer adverse effects.

This will facilitate an often-stated government aim of introducing personalised medical treatment tailored to each patient's unique genetic make-up and medical condition (Ginsburg and McCarthy, 2001). Diagnosticians require from toxicogenomics highly consistent and reliable results of the same standards as those used in all aspects of current diagnostics testing. However, there is an often high degree of data artefact in typical microarray datasets which is currently contributed to by large intra-laboratory, intra-methodology and inter-patient sample variation of the microarray analysis on gene chips. These artefacts must be allowed for and overcome as they are unacceptable in a clinical setting.

Toxicogenomics is among the emerging disciplines that can provide relevant biomarkers for individual susceptibility in patient biomonitoring of drug efficacy and toxicity and can be used to build a comprehensively accessible pharmacogenomics and toxicogenomic knowledge base to aid drug treatment in the NHS. Improving and standardising toxicology microarray experiments, building a comprehensive integrated knowledge base and agreeing a universal interpretative algorithm of this data are the principal challenges toxicogenomics must overcome in order to be incorporated into routine clinical use.

It would be logical and highly desirable to establish a programme in the UK, similar to that in the USA, for the dedicated application of toxicogenomics to the risk assessment needs of the UK-based population. Such a government-sponsored project could be tasked with building a knowledge base of toxicogenomic, proteomic, metabonomic and pharmacogenomic data for directing individualised medicine in the NHS, and thus provide government with objective measures of risk in environmental toxicity. In line with government guidelines it should concentrate on common drugs/toxicants. Determining differential gene expression can provide indicators/biomarkers of toxicity at earlier time-points and at lower doses than traditional toxicology parameters and provide both a polymorphic genotypic link to particular drug morphology and insights into the pathogenesis of the toxic insult. Chemical-specific patterns of gene expression can be revealed using microarrays. Transcript profiling can potentially provide discrimination between classes of toxicants, as well as genome-wide insight into mechanism(s) of toxicity and iatrogenic disease progression. For example, a study which looked at chemical-specific patterns of gene expression using cDNA microarrays of two toxicants, a barbiturate and a peroxisome proliferator, discriminated between these classes of toxicants, and provided genome-wide insight into the mechanism of toxicity and disease progression (Amin et al., 2002). Ballatori (2003) noted that 'A holy grail of high-throughput screening "omics" is the so-called tri-nomics, a field of study that could integrate clinical observations/phenotype with, genomic, proteomic and metabonomic data and indeed pharmacogenomic information into a unified framework for understanding the biochemical and genetic basis for various diseases'. Thus tri-nomics could be employed to supply information to support and improve existing prescribing practices. The goal is to unify all the new molecular high-throughput resources *in vivo* or *in vitro* to categorise chemicals and mixtures of chemicals at various dose levels (Battershill, 2005; Oberemm *et al.*, 2005). A UK-based toxicogenomic consortium should aim to combine information from studies of genomic-scale mRNA profiling, cell-wide or tissue-wide protein profiling (proteomics), pharmacogenetic variability and metabolite profiles. This would lead to an understanding of the roles of drug–gene interactions and their impact on health and would help to identify (1) biomarkers of exposure and effect; (2) genomic global approaches to the study of toxic mechanisms to identify multiple-gene biomarkers likely to be more sensitive and discriminating than single genes/protein/metabolite biomarkers which have been the standard in pathology; and (3) the toxicant 'signature' of gene/protein/metabolite profiling and predictive toxicology.

The preceding sections have repeatedly emphasised the progress that could be made in the application of toxicogenomics to clinical decision, environmental health and risk assessment. This strategy would also include conventional biochemical, histological and toxicology endpoints, together with other high-throughput screening methods. There must also be an agreed evidence-supported consensus of standards and interpretative algorithms to establish the quality of the data produced. So at this stage of development in the field we are really entering a discussion of the rules of 'grammar' for toxicogenomics. However, what of the 'vocabulary'? This would be the application of the technology to specific toxicology problems. It may be obvious from the preceding text that no clear consensus and easily applied system exists for the application of toxicogenomics to specific toxicology assessments. The field to be integrated is obviously vast, but it is not realistic for researchers, diagnosticians, government and industry to wait until all classes, doses and combinations of toxicants are 'mapped' and '…omically' fingerprinted on all platforms, biological models, tissues and cellular matrices. The derived data should be integrated and universally interpreted to an agreed consensus before this exciting new area of toxicology can be exploited. The first steps are being taken and projects established that yet may be exemplar studies of the benefits of using a toxicogenomic (and tri-nomic) approach to individualising drug therapy. Ideally, such studies should focus on a small number of commonly prescribed drugs in the UK NHS with widely known iatrogenic morbidity and whose targeted optimised therapy would benefit a wide number of patients throughout the country. To illustrate both the potential value of toxicogenomics and the daunting scope of such an undertaking let us examine the toxicogenomics of a single commonly used class of drug widely used for cholesterol-lowering therapy, the statins.

Statin toxicity

Why statins? Statin therapy is generally well tolerated, but adverse events do occur. In August 2001, cerivastatin was removed from European and USA markets because of a higher risk of rhabdomyolysis associated with its use in comparison with other statins (Maggini *et al.*, 2004). Myotoxicity, subsequent myopathy or rhabdomyolysis and polyneuropathy occur in 1–7% of statin users (Banga, 2001; Moosmann and Behl, 2004; Thompson *et al.*, 2003). In rare instances, potentially fatal rhabdomyolysis can occur with myoglobinuria and renal failure. Myotoxicity, liver toxicity and rhabdomyolysis, can occur with statin monotherapy, and it has been demonstrated that all statins have a dose/toxicity relationship, with a marked increase in adverse effects when the dose is titrated from 40 to 80 mg, both common therapeutic doses (Davidson, 2004). However, adverse reactions are more common in patients receiving concomitant therapy with other drugs metabolised by, and inhibitors of, the cytochrome P450 enzyme system (Andreou and Ledger, 2003; Clark, 2003). Statin-associated complications can develop in susceptible patients with the use of medications that impede the biodegradation of statins, for example, biotransformation via the cytochrome P450 system which may result in the plasma and tissue concentrations of statins, and their active metabolites, increasing to levels that are toxic to striated muscle (Banga, 2001). Therefore, due to the common use and possible serious incidents of toxicity associated with statins, they would be one of the ideal candidates to benefit from a toxicogenomic study to provide clinicians with better prescriptive biomarkers, since the pathogenic mechanism of statin-induced myopathy and rhabdomyolysis is poorly understood. Theories have been proposed, but are as yet unproven. One theory suggests that lipophilic statins lead to depletion of intermediates normally formed after cholesterol synthesis within myocytes (Jamal *et al.*, 2004). This is based on the intriguing observation that the pattern of side effects associated with statins resembles the pathology of selenium deficiency, and that the mechanism may involve a negative effect of statins on selenoprotein synthesis leading to statin-induced myopathy, in part by inhibiting the geranylgeranylation of proteins (Johnson *et al.*, 2004; Moosmann and Behl, 2004). Oxidative stress has been implicated in the pathogenesis of several muscle diseases (Olive *et al.*, 2004). Myopathy during statin treatment has been associated with *in vivo* oxidation injury as assessed by increased isoprostane levels (Sinzinger *et al.*, 2000) but the cause is not known. Increased lipid peroxidation has been observed in patients with creatine kinase elevation and muscle pain during statin therapy (Sinzinger *et al.*, 2002) but the mechanism is again not understood. Thus the precise mechanism of statin-associated muscle toxicity remains unclear and is potentially related to genetically mediated muscle enzyme defects, drug interactions, intracellular depletion of metabolic intermediates and intrinsic properties of the statins *per se* (Farmer, 2003).

It is therefore of interest to develop diagnostic test systems, which would allow the identification of patients at increased risk of adverse drug reactions (Schmitz and Drobnik, 2003). Applying toxicogenomics may be an exemplar of the approach that a future NHS consortium may want to take. There has been relatively little work carried out in this area using the powerful technique of high-throughput genomic molecular toxicology.

In the work that has been undertaken, microarray study has been confined to muscle myopathy in general, *e.g.*, muscle biopsies taken from patients with polymyositis and dermatomyositis: altered gene expression involved in immune regulation and myofibrillar proteins disregulation was prominent in both features of myopathy (Zhou *et al.*, 2004). The molecular toxicology of statins is poorly understood and it is one of a number of commonly prescribed classes of drug in need of this kind of investigation.

Study design

So now that we have selected our hypothetical toxicogenomic candidate and justified its selection based on clinical grounds how would we conduct such a study? A research ethics committee would be unwilling to approve a study or a clinical governance review or approve an assay in routine clinical usage based on muscle or liver biopsy gene expression. A peripheral blood monocyte surrogate screen is much more likely to be of general application. The precedent of using microarray gene expression profiling to investigate the molecular pathology in blood as a surrogate is well established (Tang *et al.*, 2001). Microarray gene expression profiling has been successfully used to assess the impact of a variety of drugs and toxins on gene expression. These include among many others, statins on peripheral blood mononuclear cells (PBMCs) in coronary arterial disease

(CAD) (Waehre *et al.*, 2004), selective serotonin reuptake inhibitor citalopram in Alzheimer's disease (Palotas *et al.*, 2004), amphotericin B responsive genes (Rogers *et al.*, 2002) and acute smoke exposure (Ryder *et al.*, 2004). A study would need to involve the recruitment of patients from a single or group of NHS lipid clinics. Patient biometric, biochemical, drug pharmacokinetics and bioavailability data would be collected and correlated to determine subgroups of patients who have had beneficial, inadequate or adverse therapeutic statin therapy (as distinct from myopathy). Myopathy is defined by raised creatinine kinase (CK) ($>10,000$ IU) and clinical symptoms (skeletal-muscle pain/weakness) matched to statin-treated patients with no myopathy or clinical symptoms and good therapeutic response to the treatment.

In this way we could look at the therapeutic response to statin therapy and determine whether we can use our hypothetical approach to determine expression biomarkers for beneficial, inadequate and adverse drug therapy in general. The PBMC expression profiles, while distinct biomarkers in their own right, can then be exploited to identify suitable gene candidates for polymorphic study. A wide variety of drug metabolising enzymes are known to contribute to pharmacogenetic variability. Pharmacogenetics is the study of how genetic variations affect drug response. Polymorphisms in genes that affect drug metabolism affect toxicity and efficacy.

It is now readily apparent that almost every human gene is polymorphic, and the technical task of measuring this scope of pharmacogenetic variation is beyond conventional genotyping assays. A high-throughput system such as provided by oligonucleotide arrays would be eminently suitable and can be used to genotype in parallel vast numbers of specific alleles in individuals simultaneously and cheaply.

A combination of both toxicogenomics and pharmacogenetics can elucidate how the entire genome is involved in biological responses of organisms exposed to environmental toxicants/stressors. The use of polymorphisms for diagnosis and risk assessment *i.e.*, pharmacogenetics will not be discussed in this chapter but it would be remiss not to mention it in this context as the two disciplines are likely to be used in any 'omic' diagnostic evaluation of drug toxicity.

Bioinformatics

No discussion of mass parallel molecular screening would be complete without consideration of its analysis (data mining). Bioinformatics is the use of statistics, biology, genetics, scientific computing, information sciences, systems and technologies for management, analysis, interpretation and presentation of sequence genomics, functional genomics, transcriptomics, proteomics, metabonomics, and other large-scale biological information and high-throughput data acquisition methods. The volume of data that will be produced by gene expression, proteomic, metabonomic and high-throughput sequence analysis data in the hypothetical project demands massive bioinformatics support and a new series of bespoke integrative and interpretive algorithms (Tong *et al.*, 2003, 2004; Whittaker, 2003). Statistical analysis of microarray data must take account of both inherent biological and experimental error as well as the huge multiplicity of high-throughput analysis (typically of tens of thousands of simultaneous measurements) which will inevitably result in a significant false discovery rate (FDR). Most data analysis methods that have been used to analyse microarrays do not assess the degree to which significant changes in gene expression occurred by chance.

A method to calculate FDR in genomic data is vitally important to reduce these large datasets and focus on changes in biologically relevant gene expression. Our experimental design needs to ensure adequate power to validate expression profiles and sequence, proteomic and metabonomic biomarkers. We would need to create and maintain a database for management and analysis of gene expression, proteomic, metabonomic and sequence data, updating it, and make it publicly accessible to an agreed format and set of protocols for pharmacogenomic and toxicogenomic diagnostic resources.

Now we must consider dose interaction with other co-prescribed drugs, gender, race, age, co-pathology to make this resource-generating study truly comprehensive. For illustrative purposes it is now readily apparent that to produce a toxicogenomic profile of one drug in common usage would require a significant amount of work. Profiles would have to be shown to be robust in cross-centre, cross-platform comparisons and so on. While undoubtedly review after review optimistically forecasts, and we too, justifiably predict that toxicogenomics and indeed all 'omic' applications will prove a very widely used and powerful tool in toxicology (Chin and Kong, 2002; Marchant, 2002; Waring and Halbert, 2002; Waters and Fostel, 2004), the magnitude of work needed for one common compound is hopefully apparent and the lack of consensus on application plain.

No 'plug-in and play' unsupervised methodology currently exists for using toxicogenomic/proteomic/metabonomic screening, that is to say, a purely objective analysis algorithm of high-throughput toxicology screening data without recourse to specific candidate selection. This is not to discount the current and future usefulness of the 'incompletely developed' and 'evolving' field of toxicogenomics. Limiting the use of toxicogenomics to interpretation of more 'supervised' or guided toxico-genomic profiles linked to existing and established toxicology profiles is a valid and often used approach. Sometimes this approach is called 'phenotypic anchoring', which defines the relationships between chemically induced changes in gene expression and alterations in conventional toxicological parameters such as clinical chemistry and histopathology (Moggs *et al.*, 2004).

Environmental pollution and toxicogenomics

Human activity provides another vast source of toxic chemicals that we can be exposed to in our environment. These are produced during industrial activity ranging from factory flume exhausted gases, solid waste and water discharged effluent containing chemical by-products of industrial production and by large-scale plant or with internal engine combustion of fossil fuels products. Environmental pollution, such as airborne particulate matter from diesel exhaust fumes, typical in British urban areas, has been shown to adversely affect health outcomes inducing lung inflammation in both humans and rodents. A recent study showed a pulmonary toxicological response in male Sprague Dawley rats following the intra-tracheal instillation of airborne particulates. The rats showed mild but significant changes in lung permeability and swollen lung capillaries post-instillation. Significant changes in differential gene expression linked to the different histopathological events within the lung were observed when the lung was treated with airborne particulates (Wise *et al.*, 2006), showing the potential efficacy of forecasting lung toxicity based on bioactivity and toxicogenomic profiling.

Aquatic toxicogenomics has the potential to investigate diagnostic biomarkers, establish toxicant stress-specific signatures and molecular pathways hallmarking the adaptation to water-based toxic pollutants (Ju *et al.*, 2006). The xenoestrogen group of endocrine disruptors are thought to cause fertility and gender developmental pathology through stimulation or disruption of sex steroid nuclear receptor signalling pathways (Naciff and Daston, 2004). A better understanding of how they interact with biological systems at the molecular level will provide a mechanistic basis for improved safety assessment. The application of toxicogenomics to xenoestrogen-induced changes in gene expression is related to clear conventional physiological and toxicological endpoints (Currie *et al.*, 2005; Moggs, 2005). Another example of an environmental contaminant that has been toxicogenomically profiled is 2,3,7,8-tetrachlorodibenzo-*p*-dioxin (TCDD), a potent cardiovascular teratogen. When the effects of TCDD on gene expression during murine cardiovascular development were investigated it was found that TCDD significantly altered expression of a number of genes involved in xenobiotic metabolism, cardiac homeostasis, extracellular matrix production/remodelling and cell-cycle regulation (Thackaberry *et al.*, 2005).

Food is also a major source of potentially toxic, environmentally encountered chemicals (Dybing *et al.*, 2005; Sawa *et al.*, 1998; Sesink *et al.*, 2000; Turesky, 2007). A potential list of toxic compounds that may be in our food might contain microbiological contaminants both in food and the water supply, a complex assortment of phytochemicals found naturally in our plant food products, flavour enhancers, emulsifiers, preservatives, colourants and other food processing additives, crop insecticides and pesticides, intentionally ingested nutri-pharmaceuticals dietary supplements and others. Indeed the overall calorific content (Vega *et al.*, 2004), the proportion of fat, carbohydrate and protein as well as the method of cooking are known to render a specific diet 'toxic' (Sawa *et al.*, 1998; Sesink *et al.*, 2000).

For instance, intake of dietary components such as fat and red meat, alcohol and fried food are thought to increase oxidative stress. A study which examined a high-fat diet showed that there was increased urinary excretion of 8oxodG, hence increased oxidative DNA damage but that the degree of saturation of the fat in that study did not influence levels of oxidative damage; rather the energy intake of the high-fat diet appears to be the major determinant of the rate of oxidative modification (Sorensen *et al.*, 2003).

Unsaturated fish oil protected the intestine colonic cells against oxidative stress when compared to a saturated corn oil-supplemented diet. Cells had lower levels of 8oxodG and increased apoptotic cell death in the upper (crypt) region of the intestine surface cells (Bancroft *et al.*, 2003). A typical Western diet, with a high amount of red meat is associated with a high risk of colon cancer. Haem (or heme), the iron carrier in red meat, is thought to be involved in red meat diet-induced colonic epithelial damage, resulting in increased cytotoxic epithelial proliferation and a subsequently greater risk of colonic cancer (Sesink *et al.*, 1999, 2000). Investigators have demonstrated that a diet rich in lipid peroxides and heme components generated peroxyl radical species which exert DNA cleaving activity (Sesink *et al.*, 2000). The lipid peroxyl radicals (ROS species) generated, would also be expected in routine dietary components such as fat and red meat, and may contribute, at least in part, to the high incidence of colon cancer attributed to a red meat diet (Sawa *et al.*, 1998).

We have observed that oxidation of unsaturated lipid in the diet has the potential to be genotoxic and hence carcinogenic. Such carcinogenic processes originate within stem cells of the colon. These cells appear to be predisposed and particularly vulnerable to the carcinogenic process as the undifferentiated cells could mount an effective recombinational repair/ transcriptional coupled repair response to an endogenous peroxidative DNA damage insult, but not to an external exogenous peroxidative insult as would be encountered from a dietary source. This phenotypically anchored (in this case immunohistochemistry of DNA damage lesions *in situ*) toxicogenomic study suggest that defects in such specific DNA repair may play a role in tumour development in undifferentiated colonocytes exposed to diet-derived lipid peroxides.

All these discreet toxicology endpoint-'anchored' genomic studies of food and environmental toxicants are proving to be useful mechanistic tools for examining toxicity. The empirical selection of 'toxic' profiles and manually supervising the candidate genes based on those already validated toxic biomarkers reduces the potential power of truly unsupervised 'omic' systems biology.

Toxicogenomics in the pharmaceutical industry: Drug candidate stratification

There is usually a fine line between a pharmacological agent which when brought forward into stage 1, 2 and 3 trials shows clearly beneficial, inadequate therapeutic effect or produces toxicity. However the cost of this difference is financially calamitous. Providing comprehensive informa-tion about the likely efficacy or toxicity of a compound will give powerful predictive input to stratify proprietal compounds into those most likely to be excellent animal/clinical trial drug candidates. Toxicogenomic studies have shown that gene expression is predictive for drug candidate phenotype. Toxicogenomics can elucidate how the entire genome is involved in the biological responses of organisms exposed to toxicants/ stressors. Determining differential gene expression can provide indicators/ biomarkers of toxicity at earlier time-points and at lower doses than traditional toxicology parameters and provide insights into the pathogene-sis of the toxic insult.

Major pharma companies are investing heavily in the technology and expertise to study genomic-scale mRNA profiling, protein profiling (proteomics) and metabolic pathology profiles in order to achieve an understanding of the roles of drug candidate–gene interactions in human and animal species. The aim of industry is principally to develop new pharmacological chemicals but to avoid the financially ruinous consequences of running hugely expensive and dangerous failed stage 2 and 3 clinical trials.

The commercially sensitive nature of intellectual property derived during in-house toxicogenomic studies will mean a relatively slow availability of this information and these commercial toxicogenomic tools to the public sector. This emphasises the need for a national toxicogenomic approach to the health and environmental concerns prescient to the UK. Indeed the clinicians/diagnosticians and environmental safety agencies of government are more concerned with the predictive toxicity of commonly prescribed drugs and commonly encountered toxic compounds in our environment. The pharmaceutical industry, in contrast, is more concerned with discovering novel pharmacologically active compounds which they can exploit commercially and will only go as far in finding toxicity endpoints as current legislation requires. A new field of toxicology like toxicogenomics would need a new set of minimum toxicity biomarker standards and legislation would have to be drafted to enforce these standards for public safety. Indeed pharmaceutical companies are obviously less than keen to find new potential toxic properties of existing compounds from which they are already deriving commercial benefit. Therefore, this and other potential conflicts of commercial/public interest show that it is imperative that an independent body be tasked with gathering, coordinating and evaluating a toxicogenomic/tri-nomic programme.

Conclusion

The field of toxicogenomics is a subdiscipline that merges genomics with toxicology, and examines the expression of thousands of genes simultaneously in response to chemical exposure to better understand the underlying mechanisms of chemical toxicity. While toxicogenomic and all 'omic' technologies have great potential for investigating questions in biology it is worth remembering that all tri-nomic technology really does is to take a large list, in these cases the entire genome, proteome and metabonome, and generate a smaller list from them. Among toxicologists there needs to be a combined effort, nationally coordinated, of 'joined-up-thinking' to generate a universally agreed format of experimental design, data mining and the interpretive tools to use them. The goal should ultimately be the integration and benchmarking of standard toxicology endpoints and tri-nomic data that will lead to the delivery of high-throughput biological-systems toxicology. This will be crucial to providing truly predictive toxicology which is often heralded in current scientific literature that will deliver a very powerful and accurate tool for the health and environmental risk assessment of chemical species. While this is undoubtedly possible, this review has highlighted both the need and the effort required to make this a reality. The potential of predictive toxicogenomic and all tri-nomic methodologies is far greater than its current usefulness. The sequencing of genomes is not the panacea that will answer every biological question in isolation. Genomic, tri-nomic and pharmacogenetic databases are an excellent start, but must be integrated with a comprehensive toxicant class database with validated tri-nomic profiles linked to traditional toxicity endpoints which should be every bit as exhaustive as the sequencing effort itself.

References

Amin, R. P., Hamadeh, H. K., Bushel, P. R., Bennett, L., Afshari, C. A. and Paules, R. S. (2002). Genomic interrogation of mechanism(s) underlying cellular responses to toxicants. *Toxicology* 181–182,555–563.

Andreou, E. R. and Ledger, S. (2003). Potential drug interaction between simvastatin and danazol causing rhabdomyolysis. *Can. J. Clin. Pharmacol.* 10,172–174.

Ballatori, N., Boyer, J. L. and Rockett, J. C. (2003). Exploiting genome data to understand the function, regulation, and evolutionary origins of toxicologically relevant genes. *EHP Toxicogenomics* 111(1T),61–65.

Bammler, T., Beyer, R. P., Bhattacharya, S., Boorman, G. A., Boyles, A., Bradford, B. U., Bumgarner, R. E., Bushel, P. R., Chaturvedi, K., Choi, D., Cunningham, M. L., Deng, S., Dressman, H. K., Fannin, R. D., Farin, F. M., Freedman, J. H., Fry, R. C., Harper, A., Humble, M. C., Hurban, P., Kavanagh, T. J., Kaufmann, W. K., Kerr, K. F., Jing, L., Lapidus, J. A., Lasarev, M. R., Li, J., Li, Y. J., Lobenhofer, E. K., Lu, X., Malek, R. L., Milton, S., Nagalla, S. R., O'Malley, J. P., Palmer, V. S., Pattee, P., Paules, R. S., Perou, C. M., Phillips, K., Qin, L. X., Qiu, Y., Quigley, S. D., Rodland, M., Rusyn, I., Samson, L. D., Schwartz, D. A., Shi, Y., Shin, J. L., Sieber, S. O., Slifer, S., Speer, M. C., Spencer, P. S., Sproles, D. I., Swenberg, J. A., Suk, W. A., Sullivan, R. C., Tian, R., Tennant, R. W., Todd, S. A., Tucker, C. J., Van Houten, B., Weis, B. K., Xuan, S. and Zarbl, H. (2005). Standardizing global gene expression analysis between laboratories and across platforms. *Nat. Methods* 2,351–356.

Bancroft, L. K., Lupton, J. R., Davidson, L. A., Taddeo, S. S., Murphy, M. E., Carroll, R. J. and Chapkin, R. S. (2003). Dietary fish oil reduces oxidative DNA damage in rat colonocytes. *Free Radic. Biol. Med.* 35,149–159.

Banga, J. D. (2001). Myotoxicity and rhabdomyolisis due to statins. *Ned. Tijdschr. Geneeskd.* 145,2371–2376.

Bao, W., Schmid, J. E., Goetz, A. K., Ren, H. and Dix, D. J. (2005). A database for tracking toxicogenomic samples and procedures. *Reprod. Toxicol.* 19,411–419.

Battershill, J. M. (2005). Toxicogenomics: Regulatory perspective on current position. *Hum. Exp. Toxicol.* 24,35–40.

Burgoon, L. D., Boutros, P. C., Dere, E. and Zacharewski, T. R. (2006). dbZach: A MIAME-compliant toxicogenomic supportive relational database. *Toxicol. Sci.* 90,558–568.

Chan, V. S. and Theilade, M. D. (2005). The use of toxicogenomic data in risk assessment: A regulatory perspective. *Clin. Toxicol. (Phila.)* 43,121–126.

Chhabra, R. S., Bucher, J. R., Wolfe, M. and Portier, C. (2003). Toxicity characterization of environmental chemicals by the US National Toxicology Program: An overview. *Int. J. Hyg. Environ. Health* 206,437–445.

Chin, K. V. and Kong, A. N. (2002). Application of DNA microarrays in pharmacogenomics and toxicogenomics. *Pharm. Res.* 19,1773–1778.

Clark, L. T. (2003). Treating dyslipidemia with statins: The risk–benefit profile. *Am. Heart J.* 145,387–396.

Currie, R. A., Orphanides, G. and Moggs, J. G. (2005). Mapping molecular responses to xenoestrogens through gene ontology and pathway analysis of toxicogenomic data. *Reprod. Toxicol.* 20,433–440.

Davidson, M. H. (2004). Rosuvastatin safety: Lessons from the FDA review and post-approval surveillance. *Expert Opin. Drug Saf.* 3,547–557.

Davies, E. C., Green, C. F., Mottram, D. R. and Pirmohamed, M. (2006). Adverse drug reactions in hospital in-patients: A pilot study. *J. Clin. Pharm. Ther.* 31,335–341.

Dix, D. J., Houck, K. A., Martin, M. T., Richard, A. M., Setzer, R. W. and Kavlock, R. J. (2007). The ToxCast program for prioritizing toxicity testing of environmental chemicals. *Toxicol. Sci.* 95,5–12.

Dybing, E., Farmer, P. B., Andersen, M., Fennell, T. R., Lalljie, S. P., Muller, D. J., Olin, S., Petersen, B. J., Schlatter, J., Scholz, G., Scimeca, J. A., Slimani, N., Tornqvist, M., Tuijtelaars, S. and Verger, P. (2005). Human exposure and internal dose assessments of acrylamide in food. *Food Chem. Toxicol.* 43,365–410.

Farmer, J. A. (2003). Statins and myotoxicity. *Curr. Atheroscler. Rep.* 5,96–100.

Ginsburg, G. S. and McCarthy, J. J. (2001). Personalized medicine: Revolutionizing drug discovery and patient care. *Trends Biotechnol.* 19,491–496.

Guo, L., Lobenhofer, E. K., Wang, C., Shippy, R., Harris, S. C., Zhang, L., Mei, N., Chen, T., Herman, D., Goodsaid, F. M., Hurban, P., Phillips, K. L., Xu, J., Deng, X., Sun, Y. A., Tong, W., Dragan, Y. P. and Shi, L. (2006). Rat toxicogenomic study reveals analytical consistency across microarray platforms. *Nat. Biotechnol.* 24, 1162–1169.

Jamal, S. M., Eisenberg, M. J. and Christopoulos, S. (2004). Rhabdomyolysis associated with hydroxymethylglutaryl-coenzyme A reductase inhibitors. *Am. Heart J.* 147, 956–965.

Johnson, T. E., Zhang, X., Bleicher, K. B., Dysart, G., Loughlin, A. F., Schaefer, W. H. and Umbenhauer, D. R. (2004). Statins induce apoptosis in rat and human myotube cultures by inhibiting protein geranylgeranylation but not ubiquinone. *Toxicol. Appl. Pharmacol.* 200,237–250.

18

Ju, Z., Wells, M. C. and Walter, R. B. (2006). DNA microarray technology in toxicogenomics of aquatic models: Methods and applications. *Comp. Biochem. Physiol. C Toxicol. Pharmacol.* 145,5–14.

Lord, P. G., Nie, A. and McMillian, M. (2006). Application of genomics in preclinical drug safety evaluation. *Basic Clin. Pharmacol. Toxicol.* 98,537–546.

Maggini, M., Raschetti, R., Traversa, G., Bianchi, C., Caffari, B., Da Cas, R. and Panei, P. (2004). The cerivastatin withdrawal crisis: A "post-mortem" analysis. *Health Policy* 69,151–157.

Maggioli, J., Hoover, A. and Weng, L. (2006). Toxicogenomic analysis methods for predictive toxicology. *J. Pharmacol. Toxicol. Methods* 53,31–37.

Marchant, G. E. (2002). Toxicogenomics and toxic torts. *Trends Biotechnol.* 20, 329–332.

Moggs, J. G. (2005). Molecular responses to xenoestrogens: Mechanistic insights from toxicogenomics. *Toxicology* 213,177–193.

Moggs, J. G., Tinwell, H., Spurway, T., Chang, H. S., Pate, I., Lim, F. L., Moore, D. J., Soames, A., Stuckey, R., Currie, R., Zhu, T., Kimber, I., Ashby, J. and Orphanides, G. (2004). Phenotypic anchoring of gene expression changes during estrogen-induced uterine growth. *Environ. Health Perspect.* 112,1589–1606.

Moosmann, B. and Behl, C. (2004). Selenoprotein synthesis and side-effects of statins. *Lancet* 363,892–894.

Naciff, J. M. and Daston, G. P. (2004). Toxicogenomic approach to endocrine disrupters: Identification of a transcript profile characteristic of chemicals with estrogenic activity. *Toxicol. Pathol.* 32(Suppl. 2),59–70.

Oberemm, A., Onyon, L. and Gundert-Remy, U. (2005). How can toxicogenomics inform risk assessment?. *Toxicol. Appl. Pharmacol.* 207,592–598.

Olive, M., Unzeta, M., Moreno, D. and Ferrer, I. (2004). Overexpression of semicarbazide-sensitive amine oxidase in human myopathies. *Muscle Nerve* 29, 261–266.

Palotas, A., Puskas, L. G., Kitajka, K., Palotas, M., Molnar, J., Pakaski, M., Janka, Z., Penke, B. and Kalman, J. (2004). The effect of citalopram on gene expression profile of Alzheimer lymphocytes. *Neurochem. Res.* 29,1563–1570.

Pennie, W., Pettit, S. D. and Lord, P. G. (2004). Toxicogenomics in risk assessment: An overview of an HESI collaborative research program. *Environ. Health Perspect.* 112,417–419.

Rogers, P. D., Pearson, M. M., Cleary, J. D., Sullivan, D. C. and Chapman, S. W. (2002). Differential expression of genes encoding immunomodulatory proteins in response to amphotericin B in human mononuclear cells identified by cDNA microarray analysis. *J. Antimicrob. Chemother.* 50,811–817.

Ryder, M. I., Hyun, W., Loomer, P. and Haqq, C. (2004). Alteration of gene expression profiles of peripheral mononuclear blood cells by tobacco smoke: Implications for periodontal diseases. *Oral Microbiol. Immunol.* 19,39–49.

Sawa, T., Akaike, T., Kida, K., Fukushima, Y., Takagi, K. and Maeda, H. (1998). Lipid peroxyl radicals from oxidized oils and heme-iron: Implication of a high-fat diet in colon carcinogenesis. *Cancer Epidemiol. Biomarkers Prev.* 7,1007–1012.

Schmitz, G. and Drobnik, W. (2003). Pharmacogenomics and pharmacogenetics of cholesterol-lowering therapy. *Clin. Chem. Lab. Med.* 41,581–589.

Sesink, A. L., Termont, D. S., Kleibeuker, J. H. and Van der Meer, R. (1999). Red meat and colon cancer: The cytotoxic and hyperproliferative effects of dietary heme. *Cancer Res.* 59,5704–5709.

Sesink, A. L., Termont, D. S., Kleibeuker, J. H. and Van Der Meer, R. (2000). Red meat and colon cancer: Dietary haem, but not fat, has cytotoxic and hyperproliferative effects on rat colonic epithelium. *Carcinogenesis* 21,1909–1915.

Shi, L., Reid, L. H., Jones, W. D., Shippy, R., Warrington, J. A., Baker, S. C., Collins, P. J., de Longueville, F., Kawasaki, E. S., Lee, K. Y., Luo, Y., Sun, Y. A., Willey, J. C., Setterquist, R. A., Fischer, G. M., Tong, W., Dragan, Y. P., Dix, D. J., Frueh, F. W., Goodsaid, F. M., Herman, D., Jensen, R. V., Johnson, C. D., Lobenhofer, E. K., Puri, R. K., Schrf, U., Thierry-Mieg, J., Wang, C., Wilson, M., Wolber, P. K., Zhang, L., Amur, S., Bao, W., Barbacioru, C. C., Lucas, A. B., Bertholet, V., Boysen, C., Bromley, B., Brown, D., Brunner, A., Canales, R., Cao, X. M., Cebula, T. A., Chen, J. J., Cheng, J., Chu, T. M., Chudin, E., Corson, J., Corton, J. C., Croner, L. J., Davies, C., Davison, T. S., Delenstarr, G., Deng, X., Dorris, D., Eklund, A. C., Fan, X. H., Fang, H., Fulmer-Smentek, S., Fuscoe, J. C., Gallagher, K., Ge, W., Guo, L., Guo, X., Hager, J., Haje, P. K., Han, J., Han, T., Harbottle, H. C., Harris, S. C., Hatchwell, E., Hauser, C. A., Hester, S., Hong, H., Hurban, P., Jackson, S. A., Ji, H., Knight, C. R., Kuo, W. P., LeClerc, J. E., Levy, S., Li, Q. Z., Liu, C., Liu, Y., Lombardi, M. J., Ma, Y., Magnuson, S. R., Maqsodi, B., McDaniel, T., Mei, N., Myklebost, O., Ning, B., Novoradovskaya, N., Orr, M. S., Osborn, T. W., Papallo, A., Patterson, T. A., Perkins, R. G., Peters, E. H., Peterson, R., Philips, K. L., Pine, P. S., Pusztai, L., Qian, F., Ren, H., Rosen, M., Rosenzweig, B. A., Samaha, R. R., Schena, M., Schroth, G. P., Shchegrova, S., Smith, D. D., Staedtler, F., Su, Z., Sun, H., Szallasi, Z., Tezak, Z., Thierry-Mieg, D., Thompson, K. L., Tikhonova, I., Turpaz, Y., Vallanat, B., Van, C., Walker, S. J., Wang, S. J., Wang, Y., Wolfinger, R., Wong, A., Wu, J., Xiao, C., Xie, Q., Xu, J., Yang, W., Zhang, L., Zhong, S., Zong, Y. and Slikker, W., Jr. (2006). The MicroArray Quality Control (MAQC) project shows inter- and intraplatform reproducibility of gene expression measurements. *Nat. Biotechnol.* 24,1151–1161.

Shostak, S. (2005). The emergence of toxicogenomics: A case study of molecularization. *Soc. Stud. Sci.* 35,367–403.

Sinzinger, H., Lupattelli, G. and Chehne, F. (2000). Increased lipid peroxidation in a patient with CK-elevation and muscle pain during statin therapy. *Atherosclerosis* 153, 255–256.

Sinzinger, H., Chehne, F. and Lupattelli, G. (2002). Oxidation injury in patients receiving HMG-CoA reductase inhibitors: Occurrence in patients without enzyme elevation or myopathy. *Drug Saf.* 25,877–883.

Sorensen, M., Autrup, H., Moller, P., Hertel, O., Jensen, S. S., Vinzents, P., Knudsen, L. E. and Loft, S. (2003). Linking exposure to environmental pollutants with biological effects. *Mutat. Res.* 544,255–271.

Tang, Y., Lu, A., Aronow, B. J. and Sharp, F. R. (2001). Blood genomic responses differ after stroke, seizures, hypoglycemia, and hypoxia: Blood genomic fingerprints of disease. *Ann. Neurol.* 50,699–707.

Thackaberry, E. A., Jiang, Z., Johnson, C. D., Ramos, K. S. and Walker, M. K. (2005). Toxicogenomic profile of 2,3,7,8-tetrachlorodibenzo-p-dioxin in the murine fetal heart: Modulation of cell cycle and extracellular matrix genes. *Toxicol. Sci.* 88,231–241.

Thompson, P. D., Clarkson, P. and Karas, R. H. (2003). Statin-associated myopathy. *JAMA* 289,1681–1690.

Tong, W., Cao, X., Harris, S., Sun, H., Fang, H., Fuscoe, J., Harris, A., Hong, H., Xie, Q., Perkins, R., Shi, L. and Casciano, D. (2003). ArrayTrack – supporting toxicogenomic research at the U.S. Food and Drug Administration National Center for Toxicological Research. *Environ. Health Perspect.* 111,1819–1826.

Tong, W., Harris, S., Cao, X., Fang, H., Shi, L., Sun, H., Fuscoe, J., Harris, A., Hong, H., Xie, Q., Perkins, R. and Casciano, D. (2004). Development of public toxicogenomics software for microarray data management and analysis. *Mutat. Res.* 549,241–253.

Tsai, C. A., Lee, T. C., Ho, I. C., Yang, U. C., Chen, C. H. and Chen, J. J. (2005). Multi-class clustering and prediction in the analysis of microarray data. *Math. Biosci.* 193,79–100.

Turesky, R. J. (2007). Formation and biochemistry of carcinogenic heterocyclic aromatic amines in cooked meats. *Toxicol. Lett.* 168,219–227.

Vega, V. L., De Cabo, R. and De Maio, A. (2004). Age and caloric restriction diets are confounding factors that modify the response to lipopolysaccharide by peritoneal macrophages in C57BL/6 mice. *Shock* 22,248–253.

Waehre, T., Yndestad, A., Smith, C., Haug, T., Tunheim, S. H., Gullestad, L., Froland, S. S., Semb, A. G., Aukrust, P. and Damas, J. K. (2004). Increased expression of interleukin-1 in coronary artery disease with downregulatory effects of HMG-CoA reductase inhibitors. *Circulation* 109,1966–1972.

Waring, J. F. and Halbert, D. N. (2002). The promise of toxicogenomics. *Curr. Opin. Mol. Ther.* 4,229–235.

Waters, M. D. and Fostel, J. M. (2004). Toxicogenomics and systems toxicology: Aims and prospects. *Nat. Rev. Genet.* 5,936–948.

Waters, M. D., Olden, K. and Tennant, R. W. (2003). Toxicogenomic approach for assessing toxicant-related disease. *Mutat. Res.* 544,415–424.

Whittaker, P. A. (2003). What is the relevance of bioinformatics to pharmacology? *Trends Pharmacol. Sci.* 24,434–439.

Wiffen, P., Gill, M., Edwards, J. and Moore, A. (2002). Adverse drug reactions in hospital patients: A systematic review of the prospective and retrospective studies. *Bandolier Extra* 1–15.

Wise, H., Balharry, D., Reynolds, L. J., Sexton, K. and Richards, R. J. (2006). Conventional and toxicogenomic assessment of the acute pulmonary damage induced by the instillation of Cardiff PM10 into the rat lung. *Sci. Total Environ.* 360,60–67.

Zhou, X., Dimachkie, M. M., Xiong, M., Tan, F. K. and Arnett, F. C. (2004). cDNA microarrays reveal distinct gene expression clusters in idiopathic inflammatory myopathies. *Med. Sci. Monit.* 10,BR191–BR197.

Progress in ecotoxicogenomics for environmental monitoring, mode of action, and toxicant identification

Helen C. Poynton[1], Henri Wintz[1] and Chris D. Vulpe[1,2,*]

[1]Department of Nutritional Sciences and Toxicology, University of California, Berkeley, CA 94720, USA

[2]Berkeley Institute of the Environment, University of California, Berkeley, CA 94720, USA

Abstract. The holistic tools developed through genomic technologies are becoming rapidly integrated into many biological fields including ecotoxicology. Ecotoxicogenomics encompasses the incorporation of genomic technologies, including transcriptomics, proteomics, and metabolomics, into ecological studies. Like biomarkers, ecotoxicogenomic techniques may be applied to many areas of ecotoxicology, and offer increased sensitivity and specificity, and may be more informative than traditional toxicity endpoints. There are several potential applications for ecotoxicogenomics, including chemical screening, environmental monitoring, and risk assessment. In each of these areas, recent studies are laying the foundations for the field and establishing proof-of-principle for ecotoxicogenomics. However, many challenges remain. Ecosystem complexity, limited or non-existing sequence data of relevant organisms, the need for bioinformatics tools and the cost of the technology are currently delaying the growth of ecotoxicogenomics and making interpretation of the results difficult. Consortiums could play a large role in propelling the field forward by facilitating the development of standardised genomic tools and protocols.

Keywords: ecotoxicology; genomics; microarrays; 2D-gel electrophoresis; nuclear magnetic resonance; transcriptomics; proteomics; metabolomics; gene expression; protein expression; biomarkers; risk assessment; environmental monitoring; chemical screening; ecotoxicity; environment; toxicant identification; expression profiling; molecular indicators; phenotypic anchoring.

Abbreviations

BAC	bacterial artificial chromosomes
B(a)P	benzo(a)pyrene
BDE-47	BDE-47, 2,2′,4,4′-tetrabromodiphenyl ether
BFR	brominated flame retardants
BPA	bisphenol A
CEBS	chemical effects in biological systems
COMET	Consortium for Metabonomics Technology

Corresponding author: Tel.: +510-642-1834. Fax: +510-642-0535.
E-mail: vulpe@berkeley.edu (C.D. Vulpe).

ADVANCES IN EXPERIMENTAL BIOLOGY
VOLUME 02   ISSN 1872-2423
DOI: 10.1016/S1872-2423(08)00002-1

CTD	Comparative, Toxicogenomics Database
CYP	cytochrome P450
DD	differential display
p,p'-DDE	1,1-dichloro-2,2-bis(*p*-chlorophenyl)ethylene
DDT	dichloro-diphenyl-trichloroethane
DGC	*Daphnia* Genome Consortium
DIGE	2-D Fluorescence Differential Gel Electrophoresis
DIM	3,3'-diinolymethane
DHT	dihydrotestosterone
2,4-DNT	2,4-dinitrotoluene
2D-PAGE	two-dimensional polyacrylamide gel electrophoresis
E_2	17β-estradiol
EDC	endocrine-disrupting chemical
EE_2	17α-ethynylestradiol
ESI	electrospray ionisation
EST	expressed sequence tags
FIFRA	Federal Insecticide, Fungicide, and Rodenticide Act
FT-ICR MS	Fourier transform ion cyclotron resonance mass spectrometry
GEO	Gene Expression Omnibus
GRASP	Genomic Research on All Salmon Project
HPV	high production volume
I3C	indole-3-carbinol
ILSI	International Life Sciences Institute
11-KT	11-ketotestosterone
LC	liquid chromatography
L-FABP	L-fatty acid-binding protein
MALDI-TOF	matrix-assisted laser desorption/ionisation
MOA	mode of action
MS	mass spectrometry
MT	metallothionein
NCT	National Center for Toxicogenomics
β-NF	beta naphtoflavone
NIEHS	National Institute of Environmental Health Sciences
NMR	nuclear magnetic resonance
NOTEL	no observable transcriptional effect level
4NP	4-nonylphenol
NPDES	National Pollutant Discharge and Elimination System
OP	organophosphate
OPPT	Office of Pollution Prevention and Toxics
PAH	polyaromatic hydrocarbon

PCA	principal component analysis
PCB	polychlorinated biphenyls
PCP	pentachlorophenol
PDBE	polybrominated diphenylethers
PES	protein expression signature
POP	persistent organic pollutant
PPARα	peroxisome proliferators regulator alpha
ppb	parts per billion
ppt	parts per trillion
QSAR	quantitative structure activity relation
REACH	Registration, Evaluation and Authorisation of Chemicals
SAGE	serial analysis of gene expression
SELDI	surface-enhanced laser desorption ionisation
SOP	standard operating protocol
SSH	suppressive subtractive hybridisation
TBDE	2,2,4,4-tetrabromo-diphenylether
TCDD	2,3,7,8,-tetrachlorodibenzo-p-dioxin
TIE	toxicity identification evaluation
TSCA	Toxic Substances Control Act
VTG	vitellogenin
USEPA	US Environmental Protection Agency

Introduction

The first eukaryotic genome was sequenced 10 years ago, and the continuing and accelerating pace of genome sequencing has fostered the development of holistic approaches to understanding the complex interaction of organisms with their environment. Transcriptomics, proteomics and metabolomics, also referred to by the generic name of 'omics', are disciplines that were born in the wake of genome sequencing projects to examine the fluctuating state of a cell, a tissue or an organism by measuring gene expression, protein levels and metabolite fluxes. The application of these tools has permeated every field of biology, radically altering experimental approaches and giving investigators insight into questions only possible through a whole-genome approach.

The development of 'omics' technologies has dramatically changed the way toxicological problems are investigated (Waters and Fostel, 2004). For instance, the data generated by microarray-based toxicity assays has provided valuable insight into the mode of toxicity of many xenobiotics and has been used extensively in drug discovery and assessment programmes (Pennie *et al.*, 2004). Although slow to adopt the genomics revolution, ecotoxicology has in recent years embraced the emerging genomic technologies to create the rapidly growing field of ecotoxicogenomics (Snape *et al.*, 2004). The past year has seen an acceleration of articles utilising ecotoxicogenomics techniques. We therefore found this to be an important time to highlight the accomplishments of the new field. In this chapter we will review the current state of the field of ecotoxicogenomics, its promise for ecotoxicology, and the considerable challenges that hamper its widespread utilisation.

'Omics' as biomarkers

In ecotoxicology, the risk presented by potentially harmful chemicals is measured at several different levels of organisation. Although effects are most dramatic when seen at the level of populations of ecosystems, these endpoints are also the most difficult to measure and relate to a particular toxicant. Therefore, biomarkers were developed as physiological indicators of toxicant exposure or adverse effects to individuals within a population (Hugget *et al.*, 1992). Biomarkers comprise a broad variety of endpoints ranging from pathological changes in histology to molecular alterations in protein or gene expression levels. In general, they are measured with quick, simple assays, and provide biological evidence of exposure or effect (Walker *et al.*, 2006). Although their reliability and relationship to population or community effects is sometimes questioned (Forbes *et al.*, 2006), they are an important component of the ecotoxicology toolbox that can be used alongside analytical measurements, toxicity assays and bioassessments to strengthen monitoring programmes and risk assessments (Handy *et al.*, 2003).

As genomic technologies target the molecular responses an organism experiences in reaction to a pollutant, they provide a comprehensive picture of the toxic effects experienced by the organism, and compensatory mechanisms the organism has mobilised in its defence. Because a large number of genes, proteins, and metabolites may be affected in response to a stressor, expression patterns acquired through genomic methods, collectively referred to as genomic profiles, innately represent a suite of biomarkers, which will potentially be more informative, more specific and more sensitive than traditional toxicity endpoints. Figure 1 illustrates how genomic technologies produce genomic profiles that may be used as biomarkers of exposure, effect and susceptibility. However, before ecotoxicogenomics can realise its potential goals, conclusive linkages must be made between genomic

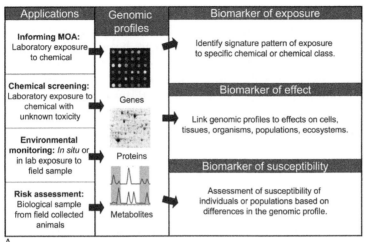

A.

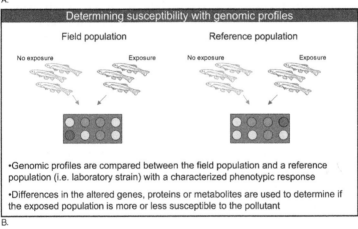

B.

Fig. 1. The potential applications of ecotoxicogenomics. (A) The pattern of gene expression, protein expression, or metabolite levels provides genomic profiles that can act as biomarkers of exposure, effect, and susceptibility. These genomic biomarkers can be applied to the following applications: informing mode of action (MOA), chemical screening, environmental monitoring, and risk assessment. (B) Illustration of how genomic profiles can be used to determine the susceptibility of a wild field population. Both a field population and a reference population are exposed to the pollutant of interest; controls that are not exposed are also used in both populations. Genomic profiles are established for both populations using a variety of genomic methods (microarrays are shown here as an example technique). The differences between the two profiles are investigated to determine if any of the differences correspond to biomarkers of effect that might determine whether the populations are being affected in the same way. (See color figure 2.1 in color plate section).

profiles and tissue concentrations of specific chemicals or adverse outcomes on the individual or population level (Ankley *et al.*, 2006; Miracle and Ankley, 2005). Few biomarkers have been able to establish such a relationship (Forbes *et al.*, 2006), and those that have, have undergone a stringent validation process (Mayer *et al.*, 1992). However, with the combined power of measuring thousands of endpoints in a single experiment, genomic technologies may prove to be optimal biomarkers.

'Omics' methods

In this chapter, genomic technologies include three targets for investigation: the transcriptome, or gene expression; the proteome, or protein expression; and the metabolome (also referred to as the metabonome), or metabolite levels. The most common technologies utilised to probe gene expression, protein expression and metabolite levels are listed in Table 1. Since these technologies have been described in detail elsewhere (Ankley *et al.*, 2006; Lettieri, 2006; Snape *et al.*, 2004; Snell *et al.*, 2003), we will only include a short discussion of these methods in this chapter.

Transcriptomics
Alterations in mRNA levels between two populations can be assessed using high-throughput techniques. DNA microarrays, serial analysis of gene expression (SAGE) and differential display (DD) are methods that compare the steady-state levels of mRNAs at a given time in the cell and identify genes that are affected by a condition (see Table 1).

Proteomics
Changes in gene expression can lead to changes in protein levels. In addition, protein levels can be modulated independently of mRNA levels. Separation technologies such as two-dimensional polyacrylamide gel electrophoresis (2D-PAGE) or other chromatographic approaches such as high-performance liquid chromatography (HPLC) coupled to mass spectrometry (MS) are utilised to quantify changes at the protein level and to identify the proteins (Table 1).

Metabolomics (or metabonomics)
Metabolomics is the systematic study of the unique chemical fingerprints (metabolome) that result from cellular processes. The

Table 1. Commonly utilised techniques in genomics.

Technique	Reference
Transcriptomics	
cDNA and oligonucleotide microarray	Lockhart *et al.* (1996) and Schena *et al.* (1995)
Differential display	Liang and Pardee (1992)
Suppressive subtractive hybridization (SSH)	Diatchenko *et al.* (1996)
Serial analysis of gene expression (SAGE)	Yamamoto *et al.* (2001)
Proteomics[a]	
Protein separation and fractionation techniques	
2D gel electrophoresis (PAGE)	Gevaert and Vandekerckhove (2000)
DIGE	Unlu *et al.* (1997)
Liquid chromatography (LC)	Aebersold and Mann (2003)
Protein identification by mass spectrometry	
MALDI-TOF	Aebersold and Mann (2003)
LC-ESI-MS	Aebersold and Mann (2003)
SELDI-MS	Hutchens and Yip (1993)
SELDI (Protein chips)-MS	Fung *et al.* (2001)
Metabolomics[b]	
H^1 – Nuclear magnetic resonance	Nicholson *et al.* (1999)
FT-ICR MS	Brown *et al.* (2005)
ESI-MS	Plumb *et al.* (2002)

[a]DIGE, 2-D Fluorescence Differential Gel Electrophoresis; MALDI-TOF, matrix-assisted laser desorption/ionization-time of flight; ESI, electrospray ionization; MS, mass spectrometry; PAGE, polyacrylamide gel electrophoresis; SELDI-MS, surface enhanced laser desorption ionization.
[b]FT-ICR MS, fourier-transform ion cyclotron resonance mass spectrometry.

metabolome refers to the complete set of small-molecule metabolites (such as metabolic intermediates, hormones, lipids and other signalling molecules, and secondary metabolites) found in the cell at a given time. Metabolites levels can be measured using various separation techniques and either MS methods or nuclear magnetic resonance (NMR) to identify the metabolites (Table 1). Although metabolomics is quite promising, a large number of metabolites have not yet been identified and referenced, which can make interpretation difficult (Robertson, 2005; Viant *et al.*, 2003a).

Advantages of the 'omics' approach

Sensitivity
A potential advantage of 'omic' techniques is that they may provide enhanced sensitivity over more traditional toxicity endpoints. Molecular alterations often occur at lower concentrations than the tissue or whole organism changes normally assessed. This is supported by toxicogenomic studies. A proteomic study using 2D-PAGE (Shepard and Bradley, 2000) investigated a copper dose–response (0–80 parts per billion (ppb) for 24 h) and correlated the results with a measure of lysosomal damage in the gill tissue of *Mytilus edulis* (Linnaeus). A protein-expression pattern distinct from the control was seen at the low-dose exposures (20 and 40 ppb) prior to any discernable lysosomal damage. A recent study in rats (*Rattus norvegicus*, Berkenhout) attempted to test the sensitivity of gene-expression profiling for detecting acetaminophen-induced toxicity and link the gene expression changes to phenotypic responses. It was found that gene expression changes suggestive of oxidative damage were apparent at concentrations lower than those that cause toxicity measured by traditional measures including clinical chemistry and histopathology (Heinloth *et al.*, 2004). Further investigation showed that the gene-expression changes are correlated with acetaminophen's mechanism of action, glutathione depletion and DNA adducts (Powell *et al.*, 2006). These studies together illustrate that investigating gene- or protein-expression patterns provides a sensitive and early indicator of toxicity.

It may also be possible to predict the chronic toxicity of chemicals based on an acute exposure, because gene-expression changes that occur after a few hours of exposure may result in a phenotypic response that is not observable for several days (Steiner and Anderson, 2000). Two studies in the water flea, *Daphnia magna* (Straus), suggest that this may be possible. Exposure to concentrations of fenarimol and zinc (Zn^{2+}) previously shown to cause a disruption of moulting and reproduction, caused a downregulation of a number of genes involved in the moulting process (Poynton *et al.*, 2007; Soetaert *et al.*, 2007). These exposures were much shorter (24 h or 4 days) than the classic chronic toxicity bioassay performed on *D. magna* (21 days). Therefore, a short-term exposure caused changes in genes that predicted the long-term effect on reproduction.

Specificity
One criticism of traditional biomarkers of exposure is their inability to distinguish between specific compounds; they are usually only indicative

of general stress or a class of compounds. With genomics, the power to screen thousands of genes, proteins or metabolites in one experiment increases the specificity of the response caused by each pollutant. Therefore, many expect that 'omic' technologies will have the capacity to differentiate similar toxicants even within the same chemical class, due to slight differences in the molecular alterations they cause or secondary modes of action. One example is shown by comparing gene-expression profiling to the biomarker vitellogenin (VTG) to study oestrogenic endocrine disruptors. VTG induction in male or juvenile fish has proven to be a good predictor of oestrogenic endocrine disruption in many fish species (Heppell *et al.*, 1995; Sumpter and Jobling, 1995); however, it is unable to discern the specific compound. Using a small 132 cDNA microarray, Larkin *et al.* (2003a, 2003b) investigated gene-expression changes in large mouth bass (*Micropterus salmoides*, Lacepede) in response to 17β-estradiol (E_2), 4-nonylphenol (4NP), and 1,1-dichloro-2, 2-bis(p-chlorophenyl)ethylene (p,p'-DDE). VTGs were induced by all three chemicals, but there were other genes that distinguished the three toxicants; in particular, 4NP induced genes suggesting a secondary mode of toxicity independent of the oestrogen receptor (Larkin *et al.*, 2003a, 2003b). Using a genomic approach, these investigators demonstrated that gene-expression profiling provides more specificity than the use of a traditional biomarker, such as VTG.

A more informative approach
Ecotoxicology, at both the scientific and the regulatory levels, has relied on the use of descriptive methods to establish the effect of chemicals on the environment. Standard toxicological methods rely on physiological endpoints, such as death and reproduction. The advent of genome-based technologies that enable the collection and analysis of large genomic data sets can bring a new dimension to environmental studies and offer a more detailed investigation of the mode of action (MOA) of toxicants. Transcriptomic, proteomic and metabolomic approaches provide powerful tools that can help generate hypotheses on the MOA and lead to a better understanding of the underlying molecular mechanisms of toxicity. Furthermore, they will help classify chemicals based on their MOA and help establish a link between MOA and population fitness.

A mechanistic approach to toxicology is well under way in human/ mammalian toxicology (Aardema and MacGregor, 2002; Amin *et al.*, 2002; Fielden and Zacharewski, 2001; Hamadeh *et al.*, 2001; Tennant, 2002; Waters and Fostel, 2004), and many examples can be found in the literature since the first report where the mechanism of action was inferred by comparing the gene-expression pattern in exposed livers with expression profiles of known hepatotoxins (Waring *et al.*, 2002). In ecotoxicology, genomics-based predictions of modes of toxicity are facing a number of challenges, one of which is the lack of sequence data for environmentally relevant organisms. Understanding the MOA of toxicants based on genomic profiles is only possible if the data can be compared to known molecular, biochemical, and physiological pathways, and this is difficult with limited sequence data and annotation of the genomes of interest. This is possible in human and mammalian toxicology because of the wealth of information available and the supportive bioinformatics tools to mine the data (Fielden and Zacharewski, 2001). Until this level of molecular understanding and the tools are developed for the different organisms used in ecotoxicology, the MOA of toxicants in environmental species can only be predicted by extrapolating from our knowledge in better-studied model organisms. However, with the decreasing cost of sequencing projects, the genomes of environmentally relevant organisms are slowly receiving their deserved attention as evidenced by the recent sequencing and ongoing annotation of the *Daphnia pulex* (De Greer) genome (http:// daphnia.cgb.indiana.edu).

Despite the limitations, several recent studies have successfully inferred MOAs using a toxicogenomic approach in ecologically important organisms, many of which are represented in Table 2. In our work on the toxicity of 2,4-dinitrotoluene (2,4-DNT) in the fathead minnow (*Pimephales promelas*, Rafinesque), we established that lipid metabolism and respiration (oxygen transport) were affected by 2,4-DNT. Furthermore, we were able to map the genes affected on a metabolic pathway based on the known mammalian pathways, as shown in Fig. 2 (Wintz *et al.*, 2006). In other instances, classification of genes by gene ontology may point to general functions that are affected by the toxicant. For example, expression profiles in kidney and liver of exposed trout revealed that two compounds (pyrene and carbon tetrachloride, CCl_4) had potentially different mode of actions (Krasnov *et al.*, 2005a). While pyrene toxicity targeted essentially the genetic apparatus, the immune response, glycolysis and metabolism of iron, CCl_4 appears to induce a response to cellular stress and protein folding as well as a modification of steroid metabolism. Similarly van der Ven *et al.* (2005) have shown that chlorpromazine, a neuropharmaceutical, triggers an oestrous cycle-related response in zebrafish brain, and correlations could be made with data from mammalian systems. Two additional studies have helped to provide interesting hypotheses on how metals exhibit toxicity to

Table 2. Organisms currently used in ecotoxicogenomics studies. This table gives an overview of the organisms that are currently being used in ecotoxicogenomics studies. Specific genomics technologies used and toxicants studied are indicated for each organism. Refer to Table 1 for techniques descriptions.

Organism	Technique	Stress/treatment	References
Algae (*Nannochloropsis oculata*, Hibberd)	Proteomics	Cd^{2+}	Kim *et al.* (2005)
Invertebrates			
American oyster (*Crassostrea virginica*, Gmelin)	cDNA microarray	Cd^{2+}	Jenny *et al.* (2004)
Blue mussel (*Mytilus galloprovincialis*, Lamarck)	cDNA microarray	Metal and organics	Marsano *et al.* (2004)
Clams (*Chamaelea gallina*, Linnaeus)	2D-PAGE and MALDI-TOF	Aroclor 1254, copper(II), tributyltin, and arsenic(III)	Rodriguez-Ortega *et al.* (2003)
Coral (*Montastraea faveolata*, Ellis and Solander), (*Acropora cervicornis*, Lamarck)	cDNA microarray	Sediment and metals Pesticides and PAH	Edge *et al.* (2005) Morgan *et al.* (2005)
Eastern oyster (*Crassostrea virginica*, Gmelin)	2D-PAGE	Zn^{2+}	Meiller and Bradley (2002)
Grooved carpet shell (*Tapes decussatus*, Linnaeus)	2D-PAGE	*p-p′*-dichlorodiphenyl-dichloroethylene	Dowling *et al.* (2006)
Marine amphipod (*Eurythenes gryllus*, Lichenstein)	Proteomics	Cu^{2+}, Hg^{2+}, and benzo(a)pyrene	Pampanin *et al.* (2006)
Mitten crab (*Eriocheir sinensis*, H Milne-Edwards)	2D-PAGE	Cd^{2+}	Silvestre *et al.* (2006)
Mussel (*Mytilus edulis*, Linnaeus)	Protein chip/SELDI-TOF Proteomics/SELDI-MS Proteomics/SELDI-MS	Alkylphenols and PAH Heavy metals and PAHs Crude oil spiked alkylphenols and 4-nonylphenol	Bjornstad *et al.* (2006) Knigge *et al.* (2004) Gomiero *et al.* (2006)
Shore crabs (*Carcinus maeanas*, Linnaeus)			
Pacific oyster (*Crassostrea gigas*, Thunberg)	SSH	Hydrocarbons	Boutet *et al.* (2004)

Table 2. (*Continued*)

Organism	Technique	Stress/treatment	References
Peppery furrow shell (*Scrobicularia plana*, da Costa)	2D-PAGE	Field pesticide contamination	Lopez-Barea and Gomez-Ariza (2006)
Red abalone (*Haliotis rufescens*, Swainson)	Metabolomics/NMR	Withering syndrome (biotic stress)	Viant *et al.* (2003a)
Red Crab (*Procamarus clarkia*, Girard)	2D-PAGE	Field pesticide contamination	Lopez-Barea and Gomez-Ariza (2006)
Sea squirt (*Ciona intestinalis*, Linnaeus)	cDNA microarray	Tributyl tin	Azumi *et al.* (2004)
Water flea (*Daphnia magna*, Straus)	cDNA microarray, SSH cDNA microarray Oligonucleotide array Differential display	Propiconazole Cu^{2+}, Cd^{2+}, and Zn^{2+} Cu^{2+}, H_2O_2, PCP, and β-NF	Soetaert *et al.* (2006) Poynton *et al.* (2007) Watanabe *et al.* (2007) Diener *et al.* (2004)
Zebra mussels (*Dreissena polymorpha*, Pallas)	SSH, differential display	Chemicals	Bultelle *et al.* (2004)
Earthworm (*Eisenia fetida*, Savigny)	Metabolomics, NMR	Anilines Metals and chemical warfare agents	Bundy *et al.* (2002a, 2002b) Bundy *et al.* (2004)
Flatworm (*Caenorhabditis elegans*, Maupas)	Proteomics cDNA microarray	Mustard gas and nerve agent EDCs	Kuperman *et al.* (2003) Custodia *et al.* (2001) and Reichert and Menzel (2005)
Midge (*Chironomus riparius*, Meigen), (*Chironomus tentans*, Fabricius)	Proteomic/MALDI-TOF *Differential display*	Cd^{2+} PAH and metals	Lee *et al.* (2006) Perkins *et al.* (2004)
Vertebrates			
Artic Charr (*Salvelinus alpinus*, Linnaeus)	cDNA microarray	PCBs	Vijayan *et al.* (2005)
Atlantic salmon (*Salmo salar*, Linnaeus)	cDNA microarray	Saxitoxin	Bard *et al.* (2005)

33

Species	Method	Stressor	Reference
Atlantic Salmon Fry – Baltic Sea	cDNA microarray	M74 (unknown stress)	Vuori et al. (2006)
Bastard halibut (*Paralichthys olivaceus*, Temminck and Schlegel)	2D-PAGE	Cd^{2+}	Zhu et al. (2006)
Chinook salmon (*Oncorhynchus tshawytscha*, Walbaum)	Metabolomics/H-1 NMR	Inoseb, diazinon, and and esfenvalerate	Viant et al. (2006a)
Cod (*Gadus morhua*, Linnaeus)	Proteomics/SELDI-MS	Nonylphenol and bisphenol A	Larsen et al. (2006)
Coho salmon (*Oncorhynchus kisutch*, Walbaum)	cDNA microarray	Zn^{2+} and estrogens	Tremblay et al. (2005)
European flounder (*Platichthys flesus*, Linnaeus)	SSH	B(a)P and Cd^{2+}	Sheader et al. (2004)
	cDNA microarray	UK estuaries	Williams et al. (2002, 2003)
	cDNA microarray	Cd^{2+}	Sheader et al. (2006)
	cDNA microarray	2,4-Dinitrotoluene	Wintz et al. (2006)
Fathead minnow (*Pimephales promelas*, Rafinesque)	cDNA microarray, SSH	DHT and 11-KT	Blum et al. (2004)
	cDNA microarray	EDCs	Larkin et al. (2003b)
	cDNA array	EDCs	Denslow et al. (2002)
	Differential display, cDNA microarray	Papermill effluent	Denslow et al. (2004a)
Largemouth bass (*Micropterus salmoides*, Lacepede)	DIGE/MS	Diallyl phthalate, PBDE-47, and bisphenol-A	Apraiz et al. (2006)
	2D-PAGE/MS	Exposure to a polluted environment	Mi et al. (2005)
	Proteomics/2D-PAGE	PCBs, Cu, and salinity stress	Shepard et al. (2000)
	Proteomics/2D-PAGE	Cu^{2+}, diallylphthalate, and crude oil	Shepard and Bradley (2000)
	Proteomics/2D-PAGE	Alkylphenols and PAH	Jonsson et al. (2006)
	Proteomics/2D-PAGE	Polluted and reference site	McDonagh et al. (2005)
	Proteomics/2D-PAGE	Extracted Baltic sediments	Olsson et al. (2004)
Medaka (*Oryzias latipes*, Temminck and Schlegel)	cDNA microarray	4NP	Kim et al. (2006)
	SSH	TCDD	Volz et al. (2005)
	cDNA microarray	TCDD	Volz et al. (2006)
	NMR metabolomics	Dinoseb	Viant et al. (2006b)

Table 2. (Continued)

Organism	Technique	Stress/treatment	References
Mummichog/Killifish (*Fundulus heteroclitus*, Linnaeus)	Differential display	Cr(III)	Maples and Bain (2004) and Peterson and Bain (2004)
	Differential display	Anthracene	Roling et al. (2004)
	Differential display	Pyrene, creosote, and Superfund site	Meyer et al. (2005)
Plaice (*Pleuronectes platessa*, Linnaeus)	cDNA microarray	Arsenic	Gonzalez et al. (2006)
	SSH	Ethynylestradiol	Brown et al. (2004)
Rainbow trout (*Oncorhynchus mykiss*, Walbaum)	cDNA microarray	EE_2, BDE-47, Cr(VI), and B(a)P	Hook et al. (2006b)
	cDNA microarray	β-NF, Cd^{2+}, CCl_4, and pyrene	Koskinen et al. (2004)
	cDNA microarray	CCl_4 and pyrene	Krasnov et al. (2005b)
	Oligonucleotide microarray	AFB	Tilton et al. (2006)
	Oligonucleotide microarray	I3C, E_2, and β-NF	Hook et al. (2006a)
	cDNA microarrays	E_2	Samuelsson et al. (2006)
	Metabolomics/NMR	E_2	Viant et al. (2003b)
	cDNA array and SELDI	Zn^{2+}	Bradley et al. (2002)
Sheepshead minnow (*Cyprinodon variegates*, Lacepede)	cDNA mciroarray	EDC	Denslow et al. (2004b) and Larkin et al. (2003a)
		E_2	Larkin et al. (2002)
Winter flounder (*Pseudopleuronectes americanus*, Walbaum)	Subtractive library	Cr(IV)	Chapman et al. 2004)

Species	Technique	Chemical/Contaminant	Reference
Zebra fish (*Danio rerio*, Hamilton)	cDNA microarray	4NP	Hoyt et al. (2003)
	Gene expression	BFRs	Huyskens et al. (2004)
	Brain cDNA microarray	Chlorpromazine	van der Ven et al. (2005)
	Proteomics	E$_2$ and 4NP	Shrader et al. (2003)
	Oligonucleotide microarray	Mianserin (drug)	van der Ven et al. (2006)
	Oligonulceotide microarray	Arsenic	Lam et al. (2006)
	cDNA microarray	TCDD	Carney et al. (2006)
Frog (*Xenopus laevis*, Daudin)	Differential display	Methyl-Hg	Monetti et al. (2002)
		Platinum	Monetti et al. (2003)
Wild mouse (*Mus spretus*, Lataste)	2D-PAGE	Field pesticide contamination	Lopez-Barea and Gomez-Ariza (2006)
Common cormorant (*Phalacrocorax carbo*, Linnaeus)	Oligonucleotide microarray	Organochlorines	Nakayama et al. (2006)

Notes: B(a)P, benzo(a)pyrene; BDE-47, 2,2′,4,4′-tetrabromodiphenyl ether; BFR, brominated flame retardants; DHT, dihydrotestosterone; DIGE, fluorescence 2D difference in gel electrophoresis; EDC, endocrine-disrupting chemical; E$_2$, 17β-estradiol; EE$_2$, 17α-estradiol; 11-KT, 11-ketotestosterone; β-NF, beta naphtoflavone; MALDI-TOF, matrix-assisted laser desorption/ionization; MS, mass spectrometry; NMR, nuclear magnetic resonance; 4NP, 4-nonylphenol; PAGE, polyacrylamide gel electrophoresis; PAH, polyaromatic hydrocarbon; PBDE, polybrominated diphenylethers; PCB, polychlorinated biphenyls; PCP, pentachlorophenol; SELDI, surface-enhanced laser desorption ionization; SSH, suppressive subtractive hybridization; TCDD, 2,3,7,8,-tetrachlorodibenzo-*p*-dioxin; and TOF, time of flight.

36

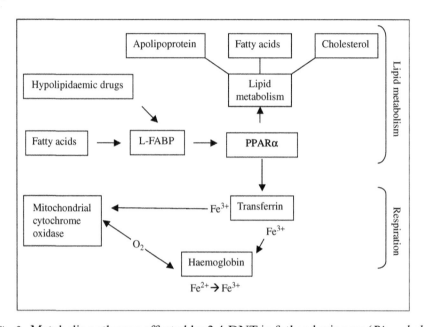

Fig. 2. Metabolic pathways affected by 2,4-DNT in fathead minnow (*Pimephales promelas*, Rafinesque) liver. Genes affected by 2,4-DNT and how they relate to each other within known pathways are represented. Fatty acids or hypolipidemic drugs signals are relayed to the nucleus via the fatty acid-binding protein (L-FABP) where it activates peroxisome proliferator activator receptor (PPARα), which controls expression of lipid metabolism genes (apolipoproteins and fatty acid metabolism genes) as well as transferrin gene (*Tf*) expression. Transferrin carries iron which is an essential cofactor of haemoglobin and of the mitochondrial respiratory chain. Oxygen is a substrate of cytochrome oxidase and haemoglobin. 2,4-DNT is known to affect oxygen transport by oxidising haemoglobin ferrous iron (Fe^{2+}) to its ferric state (Fe^{3+}). Reprinted from Wintz *et al.* (2006) with permission from Oxford University Press.

invertebrates. A study conducted in our laboratory with *D. magna* revealed that Zn^{2+} causes a decrease in chitinase gene expression resulting in decreased enzymatic activity and may explain the chronic effects exerted by exposure to Zn^{2+}. This study also revealed other potential MOAs for metals. Copper (Cu^{2+}), cadmium (Cd^{2+}), and Zn^{2+} all caused a downregulation of digestion-related genes, suggesting effects on digestion or feeding behaviour. By causing a downregulation of genes involved in the innate immune response, Cu^{2+} may increase the susceptibility of certain aquatic invertebrates to disease. This previously unknown mechanism may have adverse ecological consequences. Although the *D. magna* genome has not been sequenced, homology searches predicted the functions of the downregulated genes and helped to establish these novel modes of action (Poynton *et al.*, 2007). In other example, Lee *et al.* (2006) used a proteomic approach to discern a MOA for Cd^{2+} in the midge (*Chironomus riparius*, Meigen). They found several proteins involved in glutathione biosynthesis were affected by Cd^{2+} exposure, and mapped these proteins onto a pathway map, illustrating how Cd^{2+} exposure may cause an increase in glutathione production and present a possible mode of detoxification. Finally, Iguchi *et al.* (2006) are using genomic approaches to study endocrine disruption in a number of diverse, ecologically important species and are gaining insight into the important genes and processes involved in both vertebrates and invertebrates.

Ultimately, a goal of ecotoxicogenomic approaches would be to use genomic data to develop MOA-based expert systems and quantitative structure–activity relationship (QSAR) predictive algorithms, which could provide more accurate and cost-effective predictive tools for ecotoxicity. Currently, QSAR predictions are hampered by a lack of mechanistic data and basic endpoints such as acute and chronic toxicity, which may not reflect mechanistic relationships. Genomics data could be useful to inform QSAR databases by providing additional, more sensitive toxicity endpoints (*e.g.*, differential gene expression at subacute toxicity levels), which could help identify mechanistically related compounds. Such insight into MOAs can allow the identification or selection of physicochemical or molecular (structural) descriptors of chemicals that are more mechanistically related than commonly used descriptors such as hydrophobicity or electrophilicity. Incorporation of these MOA-based endpoints and descriptors could help improve current QSAR models. Similarly, the MOA and pathway information could assist in prediction of chemical interactions including synergistic or antagonistic effects.

Cross-species extrapolation
In human toxicity risk assessment, in cases when human data is absent, data from a limited number of model laboratory animals, such as the mouse and rat are extrapolated to humans to assess the potential health risk to humans exposed to harmful chemicals (Dixon, 1976; Mueller *et al.*, 1985; Reichsman and Calabrese, 1979; Waters and Fostel, 2004; Winneke and Lilienthal, 1992). As mentioned earlier, ecotoxicologists are interested in the fitness of populations of the multitude of species that make up an ecosystem. Because it is not practical to undertake large-scale genomic projects for every species that may be affected in an

38

ecosystem, extrapolation from one species to another is essential. In addition, the goal is not simply to predict the effect on one species, humans, but to predict the effect on multiple organisms of diverse evolutionary origin (Fig. 3). However, the conservation of sequences between organisms of individual genes and cellular, metabolic, and physiological pathways may make interspecies comparisons and extrapolations possible. It is possible that similar changes in the expression of homologous and presumed orthologous sets of genes in different species may produce similar physiological or biochemical endpoints. Genomic tools will allow us to address this issue, and we

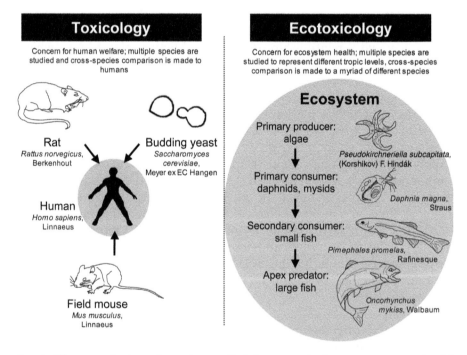

Fig. 3. The complexity of species studied in ecotoxicology as compared with toxicology. Model species studied in toxicology and ecotoxicology are shown. Because the diversity of ecosystems cannot be expressed in a single figure, the aquatic ecosystem was selected as a representative ecosystem. Toxicology and ecotoxicology utilise model species for opposite approaches. In toxicology the effects seen in many species are focused inward to explain what may occur in one species, humans. In ecotoxicology the effects seen in many species representing different trophic levels are extrapolated outward to explain what may occur in hundreds of other species.

suggest that the degree of similarity in gene-expression pattern between different species can provide a new tool to help determine whether results in one experimental species are relevant to another. In this perspective, fish such as the laboratory model zebrafish (*Danio rerio,* Hamilton) or the sentinel fish fathead minnow (*P. promela*) have been proposed to be used as models for toxicity assessment (Ankley *et al.,* 2006; Hill *et al.,* 2005). We have found a correlation between toxicity symptoms described in mammals with gene-expression profiles in the fathead minnow (Wintz *et al.,* 2006). Development of databases such as the Comparative Toxicogenomics Database (CTD; http://ctd.mdibl. org/) will make interspecies extrapolation for risk assessment more feasible (Mattingly *et al.,* 2006). Cross-species extrapolation based on a mechanistic understanding of toxicity achieved through genomics could allow invertebrate species to be used as surrogates for testing chemicals known to act through similar pathways across species boundaries. Studies that have correlated toxicity values between invertebrate and vertebrate species provide reasonable assurance that this may be possible (Botsford, 2002; Guilhermino *et al.,* 2000). However, we should be alert that there can be species-specific responses, even between closely related species. Such is the case for the differential regulation of hepatic genes by the peroxisome proliferator regulator alpha (PPARα) in humans and rodents (Lambe *et al.,* 1999; Lawrence *et al.,* 2001).

Potential applications of ecotoxicogenomics

Having reviewed the potential benefits that genomics offers to ecotoxicology, the question remains as to how ecotoxicogenomics will be integrated into the current field of ecotoxicology. We will consider several areas where genomics may be applied including: basic research to define mechanisms of action (as described earlier), screening of novel chemicals, environmental monitoring and risk assessment. These applications, first described by Nuwaysir *et al.* (1999), have been further explored and articulated recently (Ankley *et al.,* 2006; Genomics Task Force Workgroup, 2004). Figure 1 describes the potential applications of ecotoxicogenomics and illustrates how a genomic profile may be used as a biomarker of exposure and effect in each application. The following sections of this article will explore in depth each of these applications, the current research in these areas, and the research necessary to fully integrate genomics into ecotoxicology.

New chemical screening and prioritising

For the vast majority of chemicals produced, there is little or no toxicity information. The burden of proof in general lies upon environmental regulatory agencies to demonstrate toxicity to the ecosystem or people (www.epa.gov/opptintr/chemtest/pubs/sct4main.htm). In the United States, under the Toxic Substances Control Act (TSCA) and the Federal Insecticide, Fungicide, and Rodenticide Act (FIFRA) certain new chemicals must undergo rigorous testing, including toxicity bioassays in several species, before they can be released into the environment. Comprehensive toxicity testing of a pesticide, for example, currently can take several years and cost $20 million (Freeman, 2004). Because of the expense, it is not routine to test other chemicals for environmental toxicity during their development, resulting in an enormous number of compounds being released into the environment with unknown environmental toxicity. In the last decade, regulatory agencies have began to tackle the challenge of chemicals without toxicity data, through volunteer initiatives promoted by the US Environmental Protection Agency (USEPA) and the International Council of Chemical Associations (ICCA) to obtain toxicity screening information for high production volume (HPV) chemicals (www.epa.gov/chemrtk/; www.cefic.org/activities/hse/mgt/hpv/hpvinit.htm). In addition, USEPA Office of Pollution Prevention and Toxics (OPPT) has encouraged the development of 'safer chemicals through a combination of regulatory and voluntary efforts' (www.epa.gov/opptintr/) through the Green Chemistry Program (www.epa.gov/greenchemistry). In the European Union, the Registration, Evaluation and Authorisation of Chemicals (REACH) regulatory framework may fundamentally alter the regulatory situation in Europe by requiring testing of chemical entities for potential environmental toxicity (Walker *et al.*, 2006). Given the enormous backlog of chemicals and the significant costs associated with standard multi-species ecotoxicity testing, it is clear that these recent initiatives require strategies for prioritising chemicals for testing.

One potential use of ecotoxicogenomics would be in this process of prioritising chemicals for follow-up toxicity screening. The mechanistic information provided by genomics could guide the selection of appropriate follow-up studies. For example, a chemical with a genomic profile similar to that of known endocrine disruptors could be evaluated for endocrine effect with additional assays. Ecotoxicogenomics could also provide information on sublethal effects not provided by standard ecotoxicity testing, such as carcinogenicity, endocrine or immune system effects. Combining traditional approaches to water quality assessment with a genomic approach could provide a novel and cost-effective approach to predict the potential toxicity of a new chemical. Of course, for such a screen to be effective, an extensive database of chemicals with documented effects needs to be established at a range of concentrations.

Another interesting concept in toxicity testing has recently been suggested. Many chemicals may be environmentally benign at likely environmental contaminant levels. The No Observable Transcriptional Effect Level (NOTEL) has been suggested as a means to describe the dose of chemical that results in no significant changes in gene expression (Ankley et al., 2006; Lobenhofer et al., 2004).

The determination of the NOTEL for many chemicals could provide a simple means to prioritise the potency of environmental contaminants. If the NOTEL was greater than the likely environmental contaminant level, the chemical would be a lower priority or excluded from further testing. The key issue then is to determine the likely environmental contaminant level. To maximise throughput, the highest likely environmental concentration could be used, and if this concentration produces no observable expression level change then no further analysis would be needed. Ideally, this would be a short exposure that could predict whether chronic effects are expected. However, important considerations in developing a NOTEL for a chemical are the duration and the route of exposure. The exposure time must be adequate to allow absorption of the chemical depending on the method used for administering the chemical. Possible bioaccumulation must also be taken into account. However, coupled with chemical structure analysis to predict absorption rates and the likelihood of bioaccumulation, NOTEL may become an important tool in chemical screening.

Using the genomic screens described earlier, chemicals likely to have damaging effects on the environment can be identified early in their development. Microarray-based expression profiles could help expand the USEPA's Green Chemistry Program and similar initiatives by providing the chemical industry with a much-needed tool for predicting the toxicity of candidate chemicals.

In order for genomic techniques to be utilised in chemical screening, a number of proof-of-principle studies are needed to illustrate (1) chemical-specific profiles, (2) similar expression profiles among toxicants with similar MOAs, and (3) sensitivity of the techniques and the existence of a NOTEL. The initial studies in ecotoxicogenomics focused on a single contaminant with the goal of understanding the MOA or developing biomarkers. More recently, studies have explored the issue of a chemical-specific expression profiles (Table 2), providing evidence of unique expression profiles for chemicals and suggesting that like mammalian systems, ecologically relevant organisms exhibit a robust and varied response to environmental contaminants. In the following sections, we will explore these studies in depth to illustrate that expression profiles vary over dose, time, and age, and are specific for individual toxicants (also see Fig. 4).

Dose-, time-, and tissue-dependent responses
Gene, protein, and metabolite profiling have demonstrated dose-, time-, and tissue-dependent responses to individual toxicants. In the *M. edulis* proteomic study mentioned earlier, investigators performed a copper (Cu^{2+}) dose response (0–80 ppb for 24 h) and found 13 proteins consistently altered across the range of concentrations tested. However, each Cu^{2+} concentration also resulted in unique patterns of protein spots (Shepard and Bradley, 2000). Two earthworm (*Eisenia veneta*, Savigny) studies (Bundy *et al.*, 2001; Warne *et al.*, 2000), and three studies in rainbow trout (*Oncorhynchus mykiss*, Walbaum), Chinook salmon (*Oncorhynchus tshawytscha*, Walbaum), and medaka (*Oryzias latipes*, Temmick and Schlegel) used ^{1}H NMR to study metabolomic changes in response to different toxicant concentrations (Samuelsson *et al.*, 2006; Viant *et al.*, 2006a, 2006b). In these studies the authors used principal component analysis (PCA) and other computational methods to illustrate that different concentrations of toxicants produce distinct metabolic profiles. Dose-specific effects can also be observed at the level of gene expression. For example, a dose–response study (10, 50, or $100 \, \text{ng} \, \text{l}^{-1}$ EE_2) in the rainbow trout (*O. mykiss*) using cDNA microarrays, found significant changes in gene expression at all three doses (Hook *et al.*, 2006a). Many genes had consistent patterns of expression at the different doses; however, it is clear that each dose induces a unique expression profile illustrating the different sensitivities of individual genes (Fig. 4B). Our study on the effect of 2,4-DNT on the expression profiles in the fathead minnow (*P. promela*) has also shown dose-specific responses (Wintz *et al.*, 2006). The amplitude of the response of specific genes varied in response to the different concentrations tested and there was also variability in the identity and number of the genes affected.

Similar to dose–response, the duration of exposure to a fixed dose can influence the observed expression profiles. In a study of zebrafish exposed to one concentration of arsenic (As (V)) ($\sim 1/2 \, LC_{50}$) for four periods of time between 8 h and 4 days, there were 285 gene probes in common between early and late response while 465 were differentially expressed

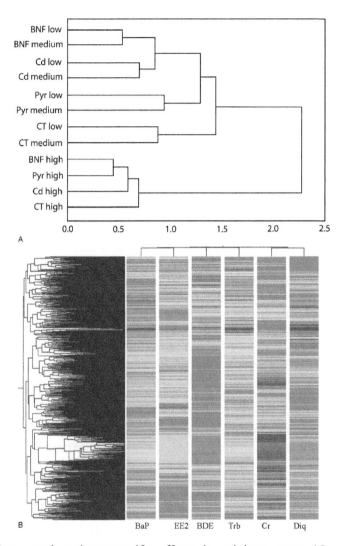

Fig. 4. Dose- and toxicant-specific effects in rainbow trout (*Oncorhynchus mykiss*, Walbaum) toxicogenomics studies. (A) Pearson correlation clustering of expression profiles in this study of exposures of trout fry show specific profiles at low doses of chemicals, while overlapping expression profiles at higher doses suggest non-specific effects. β-naphthoflavone (BNF), cadmium (Cd^{2+}), carbon tetrachloride (CT), pyrene (Pyr). Adapted from Fig. 1 of Koskinen *et al.* (2004) and reprinted with permission from Elsevier. (B) Pearson correlation clustering of the expression profile of differentially expressed genes in rainbow trout exposed to model contaminants (*y* axis) vs. contaminants (*x* axis) show contaminant-specific profiles. The gene tree is coloured as a gradient with respect to expression level, with red denoting five-fold induction, yellow denoting no change (fold change of 1), and green denoting five-fold reduction in expression levels. Brominated diphenylether-47 (BDE), benzopyrene (BaP), chromium (Cr), Diquat (Diq), ethinylestradiol (EE2), and trenbolone (Trb). Reprinted from Fig. 8 of Hook *et al.* (2006b), with permission from Elsevier. (See color figure 2.4 in color plate section).

only in the two early time points, and 1,417 only in the later two time points (Lam *et al.*, 2006). Similarly, progressively different gene-expression patterns over a 12-hr period were demonstrated in cardiac tissue of zebrafish after exposure to a dose of 2,3,7,8,-tetrachlorodibenzo-*p*-dioxin (TCDD) that was previously demonstrated to result in cardiac abnormalities (Carney *et al.*, 2006).

Several studies also indicate tissue-specific responses to toxicants. Segment-specific ^{1}H-NMR spectra was demonstrated in two earthworm species (*Elsinoe veneta* (Jenkins) and *Lumbricus terrestris*, Linnaeus) exposed to cold and starvation (4°C for 5 days) (Lenz *et al.*, 2005). A study of juvenile trout exposed to CCl_4 and pyrene found clear differences in the response of kidney and liver to the same dose of toxicant, which highlights the importance of the selection of a consistent tissue for comparison between toxicants (Krasnov *et al.*, 2005b). Overall, these studies demonstrate tissue-, concentration-, and time-dependent genomic responses to individual toxicants in ecotoxicity-relevant organisms.

Toxicant-specific responses
Multiple ecotoxicogenomic studies also indicate a robust and varied response to different toxicants. In an initial proteomics study, the term 'protein-expression signatures' (PES) was coined to describe 'sets of proteins within the proteome that are specific to stressors, physical, chemical, or biological'. Mussels (*M. edulus*) were exposed for 7 days to Aroclor 1,248 (1 ppb), copper (70 ppb), and low salinity (3 parts per trillion, ppt). Using 2D-PAGE, the authors identified PES composed of approximately 20 spots, which were uniquely induced or repressed by each condition. The proteins were not identified and the relative toxicity of these stressors was not stated, but this study demonstrated the potential to identify stressor-specific protein-expression patterns (Shepard *et al.*, 2000). In another study, embryonic zebrafish exposed to E_2 or 4NP to two dose regimens showed $\sim 30\%$ 'specific' proteins analysed by 2D-PAGE for each compound at each dose. The ability of metabolomics to distinguish between different toxicant exposures was shown in two studies of two different species. PCA analysis of the NMR spectra of extracts of whole worms (*E. veneta*) exposed to three aniline compounds differentiated exposed from control worms. In addition, exposure to two of the compounds resulted in similar metabolic profiles distinct from the profile elicited by the third compound (Bundy *et al.*, 2002a). Similarly, in the study of pesticide exposure described earlier, PCA of ^{1}H NMR spectra also allowed differentiation of the different exposures (Viant *et al.*, 2006a).

More recently, gene-expression studies have also provided evidence of toxicant-specific responses. Different toxicants could be distinguished from each other by the expression patterns observed in *Chironomus tentans* (Fabricius) larvae exposed at one-tenth LC_{50} to dichloro-diphenyl-trichloroethane (DDT), phenanthrene, fluoranthene, Cd^{2+}, Cu^{2+}, and Zn^{2+} (Perkins *et al.*, 2004). Four studies in rainbow trout (*O. mykiss*) exposed to a variety of toxicants under different experimental conditions provide further evidence of specific expression profiles in response to toxicants. In one study, 10-day-old rainbow trout fry were exposed to four model toxicants, β-naphthoflavone (β-NF), Cd^{2+}, CCl_4, and pyrene, at three doses (low, medium, and high for each compound but it is not clear how these doses compare in toxicity between compounds) for 4 days. They used a 1,380 clone cDNA microarray and found a unique expression profile for each compound at low and medium doses. However at high doses, the expression profiles of the different compounds were similar to each other, which could represent non-specific toxic effects at high doses (Fig. 4A; Koskinen *et al.*, 2004). This illustrates the importance of assessing gene-expression effects at doses below acutely toxic doses. The same group compared exposure of juvenile trout ($\sim$ 5 weeks old) to CCl_4 and pyrene exposure, and similarly found overlapping but distinct responses using cDNA arrays (Krasnov *et al.*, 2005b). An additional study of juvenile trout (12–18 months) using GRASP (Genomic Research on All Salmon Project) arrays compared liver-expression profiles in response to indole-3-carbinol (I3C), 3,3'-dinolymethane (DIM – active metabolite of I3C), E_2 and β-NF. They compared two doses of I3C and DIM to doses of E_2 and β-NF designed to maximally induce VTG or cytochrome P450 gene 1A1 (CYP1A1). Interestingly, they found that the expression profiles of I3C, DIM, and E_2 were highly correlated with each other while the β-NF profile was distinct. Based on this correlation, the authors suggest that I3C and DIM could have oestrogenic properties in rainbow trout (Tilton *et al.*, 2006). Finally, gene expression was assessed after exposure of 5–7-month-old isogenic trout to 6 model toxicants (EE_2, trenbolone, 2,2,4,4-tetrabromodiphenyl ether (TBDE), benzo(a)pyrene (B(a)P), diquat, and chromium VI) using a GRASP array. The majority of the differentially expressed genes were altered in only one treatment (Hook *et al.*, 2006b). As shown in Fig. 4B, 2D clustering of the gene-expression changes clearly illustrates that each toxicant has a specific expression profile.

Several reports this year describe further proteomics approaches to identifying specific patterns of protein expression in response to toxicant exposure in diverse species. Several studies looked at the effect on different species as part of a large mesocosm study designed to model environmental exposure of North Sea organisms to effluent from offshore oil instillations (Larsen et al., 2006). By performing proteomics using SELDI (surface-enhanced laser desorption ionisation) in several crab species and mussels, these studies showed a clear separation of the different treatment groups (Bjornstad et al., 2006; Gomiero et al., 2006). PCA of DIGE (2-D Fluorescence Differential Gel Electrophoresis, a variant of 2D-PAGE) of peroxisome-enriched fractions of the digestive gland of M. edulis also identified a unique pattern for several endocrine disruptors (Apraiz et al., 2006). In another study, plasma protein profiles using SELDI of juvenile fish of two species, turbot (Scophthalmus maximus, Linnaeus) and cod (Gadus morhua, Linnaeus) exposed to 4NP and bisphenol A (BPA) identified species-, sex-, and exposure-specific responses, although complex SELDI patterns made interpretation difficult (Larsen et al., 2006).

The examples described in the previous sections, help lay a foundation for ecotoxicogenomics in chemical screening, by providing proof of chemical-specific expression profiles. By investigating dose- and time-dependent changes, they also emphasise the areas where further research is needed. However, there is considerable variability in the platform, the strain of the test species used, exposure conditions, dose, age of the animals and bioinformatics methods, making it difficult to compare across studies. This highlights the pressing need for the development of consistent protocols to enable effective cataloguing and comparison of toxicant responses generated by multiple groups.

Environmental monitoring

Genomic technologies can be applied to several different environmental monitoring situations including general monitoring of the distribution of contamination and pollution, regulatory monitoring of effluents from point sources and monitoring of clean-up efforts at toxic sites. The objectives of general monitoring programmes is to gain a overview of the health of water bodies and are best exemplified in the United States by Clean Water Act monitoring for compiling 305(d) (water quality reports) and 303(d) lists (impaired waters list) (Cooter, 2004), and marine tissue banking programmes such as Mussel Watch (Becker et al., 1997). Genomic technologies could contribute to these programmes, providing holistic screening tools to predict contaminants present and potential adverse effects, which could guide more extensive investigations and chemical analysis. Point source effluent monitoring could also benefit by including genomic tools in discharge permits. In a diagnostic setting, genomics could aid in identifying the causal agents responsible for causing an observed toxicity (Miracle and Ankley, 2005). Finally, in environmental samples, contaminants are often present in mixtures, and determining the causal agents in these samples has always been a challenge in ecotoxicology. Toxicity Identification Evaluations (TIEs) have been developed for some pollutant classes to aid in determining which agents are responsible for the observed toxicity, but the process is not straightforward and TIE strategies rarely single out individual chemicals. Used in conjunction with TIE, genomic techniques could help identify causal agents by providing an expression fingerprint that is distinct for the chemicals exerting an adverse effect.

The use of genomic technologies for environmental monitoring is still in its infancy, with the current research emphasising the development of genomic tools using controlled laboratory exposures. Many of the studies listed in Table 2 are laboratory exposures carried out to establish expression profiles of environmental toxicants to be used later in environmental monitoring. Early studies in proteomics (Shepard and Bradley, 2000; Shepard et al., 2000), gene-expression profiling (Custodia et al., 2001), and metabolomics (Bundy et al., 2002a), attempted to assess the practicality of using genomic technologies, and showed initial potential for genomics in environmental monitoring. The studies that have followed explore in depth the ability of expression profiles to distinguish toxicants. Since these studies were described in detail in the previous section, our discussion will shift to recent studies that have used genomic techniques to distinguish and identify pollutants in the field.

Field studies of exposure
A few studies have ventured out to the field providing a preview to the potential applications and obstacles that genomics will face outside of a controlled laboratory setting. Williams *et al.* led a pilot study to determine the feasibility of toxicogenomics study in a field situation. They collected wild European flounder (*Platichthys flesus*, Linnaeus) from the Alde estuary, a relatively clean site, and the Tyne estuary, a site that is highly polluted with polyaromatic hydrocarbons (PAHs) and heavy metals, and used a 160 cDNA microarray to determine if gene expression could discriminate between the two field sites. Although, they observed high individual variability in gene expression especially in the female fish, they successfully identified a signature profile consisting of 11 genes that were differentially expressed in the male fish from the

contaminated site. This study provided initial confidence that gene expression is a robust enough measure to overcome individual variation in wild populations (Williams *et al.*, 2003). A similar study was done on three species of terrestrial worms using metabolomics. The study provided some potential for the metabolomic technique by identifying three biomarkers that correlated with the contaminated sites (Bundy *et al.*, 2004). However, because these studies did not measure concentrations of the pollutants in the organisms' tissues, they were unable to link any particular biomarker response to a specific pollutant.

Three recent studies investigating gene expression changes in the mummichog or killifish (*Fundulus heteroclitus*, Linnaeus) have been more closely linked to specific toxicant exposures. Maples and Bain (2004) compared gene-expression changes following laboratory exposures of trivalent chromium with the expression profile of wild fish collected from a creek contaminated with trivalent chromium. A second study attempted to validate biomarkers of exposure identified from laboratory exposure to pyrene with wild mummichog from a PAH-contaminated creek (Roling *et al.*, 2004). In both of these studies, individual variability in gene expression proved to be a major limitation. A final study used mummichog in an investigation of the PAH resistance in a wild population of fish inhabiting a Superfund site highly polluted with PAHs. Using a 60 cDNA macroarray, the investigators found several biomarkers that correlated robustly with PAH resistance (Meyer *et al.*, 2005).

Proteomics have been applied to study the effects of pollution on mussels. McDonagh *et al.* explored the differences in protein expression in mussels (*M. edulis*) collected from a clean site vs. those collected from a site contaminated with heavy metals and PAHs, known inducers of oxidative stress. Though the expression profiles examined by 2D-PAGE were similar, many of the proteins from the polluted site were carbonylated and glutathionylated. A laboratory exposure to hydrogen peroxide confirmed that the observed effects were probably due to oxidative stress (McDonagh *et al.*, 2005). Another study investigated the peroxisome proteome of the mussel, *Mytitus galloprovincialis* (Lamarck). They observed several changes in protein expression in mussels gathered from a polluted site (Mi *et al.*, 2005). Based on the identity of the proteins affected, the authors suggested that peroxisome proliferators may be responsible; however, without chemical analysis of the mussel tissues or sediment, it is difficult to speculate on the causal agents.

A final noteworthy example of toxicogenomics being applied to a field situation was shown in a study of the common cormorant, a fish-eating bird. Nakayama *et al.* (2006) correlated gene-expression changes in wild cormorants (*Phalacrocorax carbo*, Linnaeus) in response to measured tissue concentrations of several persistent organic pollutants (POPs). The use of wild birds with mixtures of contaminants in their tissues inhibited the authors from differentiating which compound or mixture of compounds caused the differential expression. However, many of the genes had high correlation coefficients indicating that a dose–response relationship exists for many of the genes. Further validation of the differentially expressed genes by single contaminant exposures will help to link the gene expression changes to specific pollutants. Overall the study identified genes responsive to general POP stress, which may help monitor the general health of these birds (Nakayama *et al.*, 2006).

While the previous examples illustrate the feasibility of applying genomics to the field setting, the bridge connecting a genomic response to an environmental exposure has not yet been built. The work must begin in the lab and be followed by carefully designed field validation. Though the mummichog studies described earlier attempted to correlate changes in gene expression observed in laboratory fish with wild fish exposed to the same contaminant, several aspects of the laboratory exposures, including the exposure time and differences in accumulated toxicants in tissues, prevented them from making a direct comparison between the lab exposures and the field-exposed fish. In order to identify reliable biomarkers of exposure in laboratory exposures for use in the field, laboratory exposures must (1) reflect the time period that wild animals may be exposed to a toxicant and (2) cover a range of concentrations that include environmentally relevant concentrations and the acute lethal to chronic sublethal toxicity of the pollutant to the organism being studied. The field sites selected for validation should contain concentrations that have been included in the laboratory exposures and reflect a similar exposure time as the lab exposures. Is the organism exposed to the pollutant over its entire life span, or it is a pulse of contamination resulting from pesticide application or effluent discharge? Caged organisms or laboratory exposures to collected field samples could be used to shorten exposure time when chronic exposures are not feasible in laboratory settings. If genomic-expression profiles are to be used as biomarkers of exposure in environmental monitoring, they will have to adhere to stringent criteria and validation processes explained earlier for biomarkers.

Field studies of effect
Several monitoring programmes do not just rely on exposure data, but also monitor for biological effects using toxicity assays. Bioassays to

indicate effects from unanticipated exposures are used in whole effect toxicity testing and also in monitoring water bodies for beneficial uses and constructing impairment lists (303(d) list). Genomic technologies could also benefit these programmes by providing genomic profiles indicative of an effect.

Two recent studies using different genomic approaches have shown potential for genomics in field monitoring of effects. A pilot study investigated proteomic and metabolomic responses in liver tumours of the wild flatfish (*Limanda limanda*, Linnaeus) collected from contaminated estuaries of the UK. The study revealed 12 proteomic features and three metabolites that were differentially present in tumours across many individuals. They suggested use of a multiple biomarker screen that can discriminate between disease and non-disease state (Stentiford *et al.*, 2005). Another important study investigated the effect of papermill effluent on gene expression in the large mouth bass (*M. salmoides*). The female bass exhibited a downregulation of many reproductively important genes, which correlated with decreased reproduction and decreased hormone levels (Denslow *et al.*, 2004a).

Risk assessment

In other areas of ecotoxicogenomics, researchers have looked to human toxicology to speculate on the role of genomics to ecotoxicology. For example, many investigators have used the pharmaceutical industry's use of toxicogenomics in screening novel chemicals for potency and toxicity as a model for how ecotoxicology could utilise similar genomic tools. However, the role of genomics in human risk assessment has yet to be defined (Boverhof and Zacharewski, 2006) and may be different for human and ecological risk assessments because the tolerated level of adverse effects is much higher in ecotoxicology.

Risk assessment is used in two distinct areas of ecotoxicology: (1) defining the risk posed by the controlled release of a chemical into the environment or prospective risk assessment and (2) defining the threat to the surrounding ecosystem at a contaminated site, or diagnostic risk assessment. The first scenario occurs during pesticide registration, outlined in the United States by FIFRA, or during the application for discharge permits occurring to NPDES (National Pollutant Discharge Elimination System). The second situation occurs during risk assessment at Superfund sites and the USEPA has published guidelines for risk assessors confronted with contaminated field sites (Prenger and Charters, 1997). A recent review has already outlined extensively the potential applications of genomic tools to risk assessment in both prospective and diagnostic assessment scenarios. Readers are therefore referred to Ankley *et al.* (2006) for further details on these applications.

Although many potential applications of genomics in risk assessment have been proposed by regulatory agencies including the USEPA (Genomics Task Force Workgroup, 2004), to date, genomic technologies have not been utilised in any risk-assessment scenario and this is not likely to occur soon. Validated expression profiles of exposure and effect are needed before they can be used in any risk-assessment strategy. In human risk assessment, detailed dose–response and time-course information is still needed to link long-term adverse effects with genomic profiles (Boverhof and Zacharewski, 2006). This process of 'phenotype anchoring' of genomic responses is vital before genomics can be used in risk assessment (Paules, 2003). In ecological risk assessment, expression profiles will have to be linked not only to individual effects, but also to population- and community-level effects where the most concern lies. This requirement presents challenges unique to ecotoxicology, but will have to be resolved before genomic technologies are integrated into ecological risk assessment.

Susceptibility

It is well established that some individuals are more susceptible to the effects of environmental toxicants, and Evenden and Depledge (1997) also note that certain populations within a species may be more susceptible to toxicants. Populations with differing susceptibilities will respond differently to the same toxicant and this will affect our ability to predict potential adverse effects. In addition, several microarray studies have demonstrated clear individual differences in gene expression within and between natural populations of fish (Gerwick *et al.*, 2006; Kurtz *et al.*, 2006; Oleksiak *et al.*, 2002; Whitehead and Crawford, 2005), and similar differences have been observed in fish responding to stress (Picard and Schulte, 2004; Place *et al.*, 2004; Scott and Schulte, 2005). Such intrinsic and stressor-specific variation needs to be considered in any risk-assessment approach. As proposed by Evenden and Depledge (1997), the development of biomarkers of susceptibility would aid in characterising the sensitivity of populations and better enable ecotoxicologists to perform risk assessments that are specific for localised population.

Genomic technologies could help investigators understand the sensitivity of different individuals and develop biomarkers of susceptibility. Oleksiak *et al.* (2005) recently showed that variation in a physiological parameter (cardiac metabolism) was partially explained by differences in gene-expression patterns of different individuals, illustrating an application of genomic technologies to studying physiological variation among individuals, and possibly populations. This approach could be extended to the variation of responses to toxicants and the susceptibility of different populations. Expression profiles of effect could be compared between the population of concern and a reference population. Using the known susceptibility of the reference population and an understanding of the effects predicted by the expression profile, investigators could predict whether the exposed population is more or less sensitive than the reference population. In Fig. 1B, there are two representative genes, which differ in their expression pattern from the reference and exposed population (of course, there could be many more). If the expression pattern of these genes has been linked to adverse effects, it can be used to determine the susceptibility of the exposed population. Understanding the susceptibility of the populations present at a contaminated site will greatly aid diagnostic risk assessment by helping to predict the risk to that particular population.

Challenges

Ecosystem complexity

Diversity and interaction between many species considered
The application of genomics to ecotoxicology provides some unique challenges as compared to its application to mammalian toxicology, where the emphasis is on one species, humans. Whereas a variety of model organisms may be used in toxicogenomics to evaluate the potential effects of environmental contaminants, the use of different model systems is to provide different perspectives on toxicity. The final goal is always to relate the effects to only one species. In contrast, the goal of ecotoxicology is to monitor the health of many diverse species, all interacting in communities, and spread out over the different layers (strata) of an ecosystem. Ecotoxicity studies and ecotoxicogenomics must therefore consider the effects on multiple different organisms of diverse phylogenetic origin. The different approaches utilised by toxicology and ecotoxicology are illustrated in Fig. 3. Whereas there may be commonalities in the mechanism of action in different organisms, the sensitivity of organisms can be dramatically different and ecosystems can be as sensitive as the most sensitive species. One approach to deal with this complex issue is to utilise a variety of indicator organisms representative of the different strata and metabolic systems present in

an ecosystem. Many regulatory agencies, such as the USEPA have adopted this approach as part of their risk-assessment process. Standard test organisms for each type of ecosystem (*e.g.*, freshwater aquatic) are utilised to assess the potential effects. Within a stratum, model organisms are selected to represent the components of the food web such as a primary producer, primary consumer and an apex predator. An extensive database of toxicity information is available for these model species (http://cfpub.epa.gov/ecotox/). For example, in the aquatic ecosystem, a freshwater alga, *Pseudokirchneriella subcapitata*, (Korshikov) F. Hindák, is used as an example of a primary producer in the water column, water fleas (*Ceriodaphnia dubia* (Jurine), *D. pulex* or *D. magna*) are used as examples of primary consumers, fathead minnows (*P. promelas*) as intermediate consumers, and trout (rainbow or brook) as the top predator. The sediment organisms are represented by the midge larvae (*C. tentans*) or amphipod (*Hyallea azteca*, Saussure). Different ecosystems require different suites of organisms that are representative of each system. So ecotoxicology, and by extension ecotoxicogenomics, must consider the effects on multiple organisms to evaluate possible toxicity to an ecosystem. While the indicator species are widely utilised, this method is not site specific and the actual species present at a particular area of concern may not include these indicator species. In some cases, different sets of indicator species, specific to a region are required by regulatory agencies (*e.g.*, CalEPA www.oehha.ca.gov/ecotox/documents/index.html recommends a particular suite of organisms for ecotoxicity studies within California). Finally, the evaluation of toxicity in some cases can entail utilising organisms from field sites. As a result, a major challenge of ecotoxicogenomics is the multitude of organisms of concern.

Confounding factors
Genomic profiles can be indiscriminate and reveal both effects that are related to exposure to a toxicant as well as the effect of other confounding factors. This issue is particularly problematic when analysing field organisms or water samples collected in the field. There are many factors such as nutrients, temperature, salinity, dissolved organic carbon, and water hardness (for aquatic environments), which will probably modulate the response of an organism to any toxicant and may induce their own expression changes. In addition, the multiple possible exposure routes in each environment (*e.g.*, diet, suspended sediment, or water column for aquatic systems) further complicate the possible effects. It is clear that the route of exposure can affect toxicity and is likely to similarly affect how the organism responds on a molecular level, as shown by the

recent study of the effects of Cd^{2+} after waterborne vs. dietary exposure in carp (Reynders et al., 2006). Long-standing, pre-existing contamination resulting in chronic exposure could result in pre-adapted organisms, which respond to environmental insults differently than naive populations (Bossuyt and Janssen, 2003). To minimise the uncertain effects of confounding factors, we need to study and understand the effects of these variables and how they alter toxicity at the molecular level, and well-designed controls should be utilised to compensate for their effects.

A number of studies have begun to investigate how other environmental factors affect genomic profiles and have found that confounding factors may themselves have distinct expression patterns. For example, recent studies of the effect of cold on carp (*Cyprinus carpio*, Linnaeus) found a robust adaptive response reflected in myriad gene expression changes (Chang et al., 2005; Cossins et al., 2006). Hypoxia, as well, has been shown to result in series of expression changes in zebrafish (van der Meer et al., 2005). A number of metabolomic studies have looked at a variety of stressors besides toxicants including starvation in the earthworm (*E. veneta* and *L. terrestris*) (Warne et al., 2001), infection in the abalone (*Haliotis rufescens*, Swainson) (Viant et al., 2003a), temperature in juvenile trout (Viant et al., 2003b), and hypoxia in medaka (Pincetich et al., 2005). In general, data-reduction techniques such as principal component analysis allowed separation of the stressed groups from the controls. By determining these stressor-specific expression profiles, the studies described earlier are helping to create an understanding of how confounding variables will influence the molecular profiles of toxicants. Their work also highlights the importance of developing comprehensive exposure databases that include non-toxicant stressors to effectively distinguish between toxicant-associated alterations in genomic profiles.

Mixtures
In field situations, organisms are exposed to not just one compound but a mélange of contaminants, which can interact within the environment and individual organisms. The next major hurdle to incorporating genomics into ecotoxicology is to understand the effects of mixtures on genomic profiles. The major questions are whether genomic profiles of multiple chemicals will reflect the combined expression profiles of the individual chemicals, or will create a completely unique pattern.

Recent studies have used gene-expression analysis to better understand how chemical mixtures will influence genomic expression profiles. In a study of the effects of a mixture of chromium and benzo(a)pyrene (B(a)P) in a hepatoma cell line, Wei et al. (2004) found that the gene profile of the mixture was distinct from the profiles of the individual contaminants. However, these two known carcinogens target similar transcriptional pathways in opposing directions, causing an antagonistic response. Therefore, the suppression of many of the (B(a)P)-induced genes in the mixture was predicted. Two recent studies in trout have also addressed the issue of mixtures. In the first study, the authors investigated a mixture of four unique toxicants at one concentration. They found that only a small subset of the genes overlapped in differential expression in the mixtures and single chemical exposures. Overall, the chemical-specific gene-expression profiles were not identifiable in the mixture profile (Finne et al., 2007). However, these authors only examined a single concentration of the chemicals and they attempted to compare the expression profile of the mixture to the single chemicals at the same concentration. It is likely that the organism is experiencing increased stress due the combination of toxicants, and the expression profile may resemble the profile of a single chemical at a higher concentration. In a broader study of three model toxicants, Cd^{2+}, CCl_4, and pyrene, Krasnov et al. (2007) showed that the expression profiles were additive at low-level exposures, and that the Cd^{2+} and pyrene expression profiles could be dissected from the mixture profile. It is known that different chemicals interact differently in mixtures, causing different outcomes to survival and reproduction (Walter et al., 2002). It is likely that different combinations of chemicals which have different effects on the gene-expression profiles of that organism. Chemicals that cause an additive response in acute or chronic bioassays may also have an additive expression profile. Chemicals that show synergistic or antagonist effects in standard bioassays may have distinctive expression profiles, not resembling the expression profiles of the single chemicals. It is apparent that more studies are needed to understand the effects of mixtures on gene and protein expression, and it also appears unlikely that there will be a single pattern that explains the behaviour of all mixtures. The knowledge gained from acute and chronic toxicity bioassays and other standard ecotoxicological tests should be employed to help predict the likely genomic responses to particular mixtures.

Limiting sequence data and annotation

It is important to discuss the difficulty of applying genomic technologies to the species commonly studied in ecotoxicology and to explore some of the ways researchers are working with these limitations. The species commonly studied in ecotoxicology are not the standard laboratory species, and therefore little to no sequence information is available (Snape *et al.*, 2004). While gene-expression profiling and proteomics require sequence data for meaningful results, one alternative is metabolomics, which is not limited by the absence of sequence data. However, many investigators have found strategies for constructing cDNA microarrays with limited sequence information (Snell *et al.*, 2003). One method is to create subtracted libraries enhanced for transcripts that respond to stress or represent a particular developmental stage of interest (Bultelle *et al.*, 2002; Snell *et al.*, 2003; Soetaert *et al.*, 2006). Subtracted libraries decrease the amount of sequencing needed for microarray construction, by increasing the relevance of chosen transcripts. Another method explored in our lab is the use of anonymous microarrays. Microarrays are printed with unsequenced cDNAs from a normalised library. Only the transcripts that are responsive to a particular treatment are sequenced, thus drastically reducing the cDNAs sequenced without reducing the number of transcripts investigated for differential expression (Poynton *et al.*, 2007; Wintz *et al.*, 2006). There are a few species where large sequencing efforts have led to sequenced genomes (*D. pulex,* zebrafish) or databases of several thousand cDNAs (*D. magna,* fathead minnow, rainbow trout). However, even for these species, a general lack of annotation has left many investigators puzzled about the results of gene-expression data. While homology searches can sometimes provide fruitful results, often the closest homologues are proteins of unknown function. The need for better annotation of these genomes cannot be stressed enough.

Bioinformatics for predicative ecotoxicogenomics

Since biomarkers measure only a single, or a few endpoints, linking the observed effect to an exposure or effect is usually straightforward. With genomic technologies, patterns in the expression of genes, protein, or metabolites are used to determine exposure or effects. Nuwaysir *et al.* (1999) and Snape *et al.* (2004) illustrated how expression patterns could be used to find matches indicative of an exposure to particular class of compounds. If expression profiles of effect are also established, it is conceivable to use a similar model to predict adverse effects. The simple illustrations in these review articles, however, do not adequately depict the complications that arise when you attempt to find patterns in very large data sets. To achieve this, computational methods are utilised to establish the similarities in the expression patterns (Maggioli *et al.*, 2006).

To enable the prediction of chemical effects or exposure, there are two main bioinformatics needs: databases composed of expression data, and sophisticated bioinformatics tools to screen large data sets for expression profile matches. In toxicogenomics, several databases are currently under development, including the Chemical Effects in Biological Systems (CEBS) knowledge base (http://cebs.niehs.nih.gov/) (Waters et al., 2003a) and the Comparative Toxicogenomics Database (CTD) (http://ctd.mdibl.org) (Mattingly et al., 2006). The CEBS database, developed by the National Institute for Environmental Health Sciences (NIEHS), not only incorporates toxicogenomics information, but also includes phenotypic responses such as blood chemistry and histopathology. The aim is to create an integrated view of the molecular responses measured by genomics and the pathological effects measured using traditional toxicology. The end result will allow phenotypic anchoring of molecular effects. Although the database is set up to accept data from a diverse range of species, it is unclear whether it is configured to accept ecological endpoints such as population decline. Additionally, it would be advantageous to integrate expression data with established ecotoxicology datasets such as ECOTOX (http://epa.gov/ecotox). It is therefore important for the ecotoxicogenomics community to maintain active dialogue with those developing tools for toxicogenomics, to ensure that ecotoxicology is not left out.

Once these databases are established, they must be populated with ecotoxicogenomics data. As stated by Ankley et al. (2006) proof-of-principle studies are still needed to establish the methods and determine the appropriate protocols. For the databases to have functional usage, expression data must be generated for many different chemical exposures over different doses and time points. Investigators must submit their data to these databases, perhaps as a requirement for publication, in a similar manner to the way many journals require submission of gene-expression data to the Gene Expression Omnibus (GEO) or ArrayExpress databases. In toxicogenomics, several national (NIEHS) and international (International Life Sciences Institute, ILSI) institutes have been instrumental in providing support for the proof-of-principle studies (Waters et al., 2003b). However, there are no equivalent organisations in ecotoxicology and it is not clear where support for this vital work will be found. National and international environmental organisations must come forward and establish ecotoxicogenomics task forces similar to the NIEHS National Center for Toxicogenomics (NCT) and the ILSI Health and Environmental Services Institute Genomics Committee.

For the genomic databases to have significant value to ecotoxicology, they must be accompanied by powerful bioinformatics tools. The proposed idea for the CEBS database is that investigators will be able to take expression data from an unknown and 'BLAST' the CEBS database and it will return 'hits' of expression data that match the data submitted (Waters *et al.*, 2003a). Ultimately, a user-friendly interface could be developed where users could input an expression profile and receive a list of candidate pollutants and effects without knowing details of molecular events. However, the discussion of this proposal is still superficial and it is not clear what kinds of bioinformatics software could tackle such a task. Supervised clustering techniques may hold the greatest potential for predicative toxicogenomics. These clustering methods use known data sets to 'train' an algorithm and design a mathematical model consisting of different classes. When presented with an unknown data set, the model assigns the unknown to the class it most closely resembles, thereby predicting its origin. If integrated with an expression database, the data sets present in the database would act as training sets, and the user would input the unknown expression profile. Recent studies have surveyed the potential of supervised clustering methods for predicting chemical classes (reviewed by Maggioli *et al.*, 2006). However, it is not clear whether supervised clustering methods will be able to differentiate between the expression profiles of vast numbers of chemicals. Perhaps only chemical class or generalised effect will be distinguishable. Only after large data sets comprising multiple chemical exposures are compiled will we be able to answer this important question, which we define the applicability of ecotoxicogenomics. Additionally, there are several different approaches to supervised clustering and no one method has proven to be the best for every application. In general, the methods need to be further developed and validated, and a standard method must be chosen for use alongside these databases (Maggioli *et al.*, 2006).

The cost of technology

Another barrier to the full adoption of genomics technologies in ecotoxicology is the cost. Microarray costs continue to decrease as the technology becomes routine but they can still be substantial. For ecotoxicology organisms there has been a lack of genomic resources, requiring most groups to develop their own microarrays, which requires a substantial outlay of resources and access to microarray printing equipment. While some traditional proteomics approaches using 2D gel electrophoresis are reasonably inexpensive, new approaches such as SELDI or DIGE or any mass spectrometry-based system require access to sophisticated and expensive equipment. Similarly, metabolomic approaches require chromatographic and spectrometric equipment that is usually beyond the capabilities of individual labs. Many universities and academic centres have developed genomic core facilities that can make these technologies more accessible, but the focus is usually on standard model organisms rather than ecotoxicology-relevant organisms. The development of consortia and shared genomic resources could significantly help reduce these initial outlays and help make genomic tools accessible to a wider ecotoxicity community.

The need for standardisation and the role of consortiums

Throughout this chapter, we have mentioned the need to standardise protocols and focus on model organisms. This final section outlines these needs (also summarised in Box 1) and suggests how collaborative efforts and consortiums will aid in bringing ecotoxicogenomics out of its infancy to where its techniques can be adopted for regulatory purposes.

Owing to the vast number of species represented in ecotoxicology, and the substantial resources needed for expressed sequence tags (EST) and genome-sequencing projects, model organisms should be chosen and the development of genomic tools should concentrate on these organisms. Table 2 illustrates the range of species currently pursued in ecotoxicogenomics. While the table includes species from many different trophic levels and habitats, there is also a lot of overlap. For example, there are 12 species of freshwater fish represented in Table 2, indicating that sequencing efforts have been spread out over 12 separate species. Instead these diffuse efforts could have concentrated on a couple species representing different trophic levels, enabling more-rapid development of genomic tools. Because of the myriad of stressors and conditions that

---

Box 1. Standardisation needs

- A focus on defined strains of a select model species representing different trophic levels
- Development of consistent protocols for organism care
- Standardised protocols for exposure and establishment of toxicity endpoints for equitoxic comparisons
- Shared genomic tools including data analysis algorithms to minimise data variability between laboratories
- Minimum reporting requirements which include water quality measurements and field site locations

need to be investigated using genomics, future efforts need to focus on a few model species. International committees should be established to identify these species and specific strains and establish consortiums to lead sequencing efforts.

In addition to selecting model species, standardised protocols must be developed to enable comparison of ecotoxicogenomics results across labs. The sensitivity of genomic techniques affords many advantages; however, there is potential for confounding variables to also influence genomics results to a greater extent than traditional ecotoxicology assays. Because the effects of culturing media, hardness, pH, temperature, and other confounding variables are well known in ecotoxicology, standard operating protocols (SOPs) that attempt to control these variables have been established for all assays used in a regulatory setting. Similarly, ecotoxicogenomics could adopt standardised culturing and treatment protocols to minimise the influence of confounding variables. To apply ecotoxicogenomics to the development of predicative tools for environmental monitoring, standard conditions for exposure experiments will be necessary, which also include defined exposure durations and toxicological definitions for establishing equitoxic concentrations, or concentrations of a similar level of toxicity. In addition to organism care and handling, genomic platforms, experimental protocols, and data-analysis tools should also be standardised. Recent studies have shown that cross-laboratory comparison of genomic data is possible if common protocols are followed (Baker et al., 2004; Chu et al., 2004; Thompson et al., 2004). Additionally, standardisation of techniques and platforms is needed to apply supervised clustering methods (Maggioli et al., 2006). There is an urgent need to develop standardisation protocols to ensure that the studies conducted today are useful for toxicant predictions in the future.

Because of the resources needed to develop genomic tools, and the urgent need for standardised methods, consortiums have a major role in propelling the field forward. For each model species designated for genomic studies, a consortium should be established to (1) design and construct shared genomic tools to increase access of genomic technologies and ensure that similar platforms are used among investigators studying a common organism; (2) establish standardised methods for organism care and study design; (3) delegate tasks and chemicals to speed development and discourage overlap; and (4) support and perform replicate experiments across labs to help define inter-laboratory variability of the techniques.

There are a few examples of consortiums created in toxicogenomics. These include the NCT, the ILSI Health and Environmental Services Institute Genomics Committee, and the Consortium for Metabonomics

Technology (COMET). These consortiums have played a role in developing genomic tools, are supporting proof-of-principle studies, are establishing databases of expression data, and are suggesting protocols for experimental design and data analysis (Waters et al., 2003b). In ecotoxicogenomics, one notable example is the Daphnia Genome Consortium (DGC). The cooperative effort of this consortium has established genomic tools for two daphnid species including cDNA microarrays and bacterial artificial chromosomes (BAC) libraries, and authored the successful white paper for the sequencing of D. pulex. The advantages of a collaborative approach to ecotoxicogenomics are obvious, and will be necessary to rapidly integrate genomics into ecotoxicology.

Conclusion

Ecotoxicogenomics holds great promise in ecotoxicology. The holistic nature of genomic methods offers increased sensitivity and specificity, is more informative than other toxicity endpoints, and may aid in cross-species extrapolation. Although several challenges remain, studies over the past few years have shown the potential of 'omic' methods in MOA studies, chemical screening, and environmental monitoring. Creating consortiums to establish standards for ecotoxicogenomics studies will aid in overcoming the challenges the field faces and will help facilitate the integration of genomics into risk assessment and regulation.

References

Aardema, M. J. and MacGregor, J. T. (2002). Toxicology and genetic toxicology in the new era of 'toxicogenomics': Impact of '-omics' technologies. Mutat. Res. 499,13–25.
Aebersold, R. and Mann, M. (2003). Mass spectrometry-based proteomics. Nature 422,198–207.
Amin, R. P., Hamadeh, H. K., Bushel, P. R., Bennett, L., Afshari, C. A. and Paules, R. S. (2002). Genomic interrogation of mechanism(s) underlying cellular responses to toxicants. Toxicology 181–182,555–563.
Ankley, G. T., Daston, G. P., Degitz, S. J., Denslow, N. D., Hoke, R. A., Kennedy, S. W., Miracle, A. L., Perkins, E. J., Snape, J., Tillitt, D. E., Tyler, C. R. and Versteeg, D. (2006). Toxicogenomics in regulatory ecotoxicology. Environ. Sci. Technol. 40,4055–4065.
Apraiz, I., Mi, J. and Cristobal, S. (2006). Identification of proteomic signatures of exposure to marine pollutants in mussels (Mytilus edulis). Mol. Cell. Proteomics 5,1274–1285.

62

Azumi, K., Fujie, M., Usami, T., Miki, Y. and Satoh, N. (2004). A cDNA microarray technique applied for analysis of global gene expression profiles in tributyltin-exposed ascidians. *Mar. Environ. Res.* 58,543–546.

Baker, V. A., Harries, H. M., Waring, J. F., Duggan, C. M., Ni, H. A., Jolly, R. A., Yoon, L. W., De Souza, A. T., Schmid, J. E., Brown, R. H., Ulrich, R. G. and Rockett, J. C. (2004). Clofibrate-induced gene expression changes in rat liver: A cross-laboratory analysis using membrane cDNA arrays. *Environ. Health Perspect.* 112,428–438.

Bard, S. M., Zucchi, S., Goralski, K., Gubbins, M., Williams, J., Richards, R., Eddy, B., Stagg, R. M., Gallacher, S., Sinal, C. J., Douglas, S. and Ewart, K. V. (2005). Differential gene expression in Atlantic salmon (*Salmo salar*) exposed to saxitoxin using cDNA microarray and qPCR analysis. *Can. Tech. Rep. Fish. Aquat. Sci.* 2617,70.

Becker, P. R., Wise, S. A., Thorsteinson, L., Koster, B. J. and Rowles, T. (1997). Specimen banking of marine organisms in the United States: Current status and long-term prospective. *Chemosphere* 34,1889–1906.

Bjornstad, A., Larsen, B. K., Skadsheim, A., Jones, M. B. and Andersen, O. K. (2006). The potential of ecotoxicoproteomics in environmental monitoring: Biomarker profiling in mussel plasma using proteinchip array technology. *J. Toxicol. Environ. Health Part A* 69,77–96.

Blum, J. L., Knoebl, I., Larkin, P., Kroll, K. J. and Denslow, N. D. (2004). Use of suppressive subtractive hybridization and cDNA arrays to discover patterns of altered gene expression in the liver of dihydrotestosterone and 11-ketotestosterone exposed adult male largemouth bass (*Micropterus salmoides*). *Mar. Environ. Res.* 58,565–569.

Bossuyt, B. T. and Janssen, C. R. (2003). Acclimation of *Daphnia magna* to environmentally realistic copper concentrations. *Comp. Biochem. Physiol. C, Comp. Pharmacol.* 136,253–264.

Botsford, J. L. (2002). A comparison of ecotoxicological tests. *Altern. Lab. Anim.* 30,539–550.

Boutet, I., Tanguy, A. and Moraga, D. (2004). Response of the Pacific oyster *Crassostrea gigas* to hydrocarbon contamination under experimental conditions. *Gene* 329,147–157.

Boverhof, D. R. and Zacharewski, T. R. (2006). Toxicogenomics in risk assessment: Applications and needs. *Toxicol. Sci.* 89,352–360.

Bradley, B. P., Shrader, E. A., Kimmel, D. G. and Meiller, J. C. (2002). Protein expression signatures: An application of proteomics. *Mar. Environ. Res.* 54,373–377.

Brown, M., Robinson, C., Davies, I. M., Moffat, C. F., Redshaw, J. and Craft, J. A. (2004). Temporal changes in gene expression in the liver of male plaice (*Pleuronectes platessa*) in response to exposure to ethynyl oestradiol analysed by macroarray and Real-Time PCR. *Mutat. Res.* 552,35–49.

Brown, S. C., Kruppa, G. and Dasseux, J. L. (2005). Metabolomics applications of FT-ICR mass spectrometry. *Mass Spectrom. Rev.* 24,223–231.

Bultelle, F., Panchout, M., Leboulenger, F. and Danger, J. M. (2002). Identification of differentially expressed genes in *Dreissena polymorpha* exposed to contaminants. *Mar. Environ. Res.* 54,385–389.

Bultelle, F., Panchout, M., Masson, R., Leboulenger, F. and Danger, J. M. (2004). Gene expression profiling in *Dreissena polymorpha* exposed to chemical contaminants. *Mar. Environ. Res.* 58,587.

Bundy, J. G., Osborn, D., Weeks, J. M., Lindon, J. C. and Nicholson, J. K. (2001). An NMR-based metabonomic approach to the investigation of coelomic fluid biochemistry in earthworms under toxic stress. *FEBS Lett.* 500,31–35.

Bundy, J. G., Lenz, E. M., Bailey, N. J., Gavaghan, C. L., Svendsen, C., Spurgeon, D., Hankard, P. K., Osborn, D., Weeks, J. M., Trauger, S. A., Speir, P., Sanders, I., Lindon, J. C., Nicholson, J. K. and Tang, H. (2002a). Metabonomic assessment of toxicity of 4-fluoroaniline, 3,5-difluoroaniline and 2-fluoro-4-methylaniline to the earthworm *Eisenia veneta* (Rosa): Identification of new endogenous biomarkers. *Environ. Toxicol. Chem.* 21,1966–1972.

Bundy, J. G., Lenz, E. M., Osborn, D., Weeks, J. M., Lindon, J. C. and Nicholson, J. K. (2002b). Metabolism of 4-fluoroaniline and 4-fluorobiphenyl in the earthworm *Eisenia veneta* characterized by high-resolution NMR spectroscopy with directly coupled HPLC-NMR and HPLC-MS. *Xenobiotica* 32,479–490.

Bundy, J. G., Spurgeon, D. J., Svendsen, C., Hankard, P. K., Weeks, J. M., Osborn, D., Lindon, J. C. and Nicholson, J. K. (2004). Environmental metabonomics: Applying combination biomarker analysis in earthworms at a metal contaminated site. *Ecotoxicology* 13,797–806.

Carney, S. A., Chen, J., Burns, C. G., Xiong, K. M., Peterson, R. E. and Heideman, W. (2006). Aryl hydrocarbon receptor activation produces heart-specific transcriptional and toxic responses in developing zebrafish. *Mol. Pharm.* 70,549–561.

Chang, L. W., Toth, G. P., Gordon, D. A., Graham, D. W., Meier, J. R., Knapp, C. W., Denoyelles, F. J., Jr., Campbell, S. and Lattier, D. L. (2005). Responses of molecular indicators of exposure in mesocosms: Common carp (*Cyprinus carpio*) exposed to the herbicides alachlor and atrazine. *Environ. Toxicol. Chem.* 24,190–197.

Chapman, L. M., Roling, J. A., Bingham, L. K., Herald, M. R. and Baldwin, W. S. (2004). Construction of a subtractive library from hexavalent chromium treated winter flounder (*Pseudopleuronectes americanus*) reveals alterations in non-selenium glutathione peroxidases. *Aquat. Toxicol.* 67,181–194.

Chu, T. M., Deng, S., Wolfinger, R., Paules, R. S. and Hamadeh, H. K. (2004). Cross-site comparison of gene expression data reveals high similarity. *Environ. Health Perspect.* 112,449–455.

Cooter, W. S. (2004). Clean Water Act assessment processes in relation to changing U.S. Environmental Protection Agency management strategies. *Environ. Sci. Technol.* 38,5265–5273.

Cossins, A., Fraser, J., Hughes, M. and Gracey, A. (2006). Post-genomic approaches to understanding the mechanisms of environmentally induced phenotypic plasticity. *J. Exp. Biol.* 209,2328–2336.

Custodia, N., Won, S. J., Novillo, A., Wieland, M., Li, C. and Callard, I. P. (2001). *Caenorhabditis elegans* as an environmental monitor using DNA microarray analysis. *Ann. N. Y. Acad. Sci.* 948,32–42.

Denslow, N. D., Bowman, C., Sabo, T. and Nelson, M. (2002). Changes in estrogen-regulated gene expression in largemouth bass as a biomarker of reproductive health and environmental contaminants. *Mar. Environ. Res.* 54,748.

Denslow, N. D., Kocerha, J., Sepulveda, M. S., Gross, T. and Holm, S. E. (2004a). Gene expression fingerprints of largemouth bass (*Micropterus salmoides*) exposed to pulp and paper mill effluents. *Mutat. Res.* 552,19–34.

64

Denslow, N. D., Larkin, P., Sabo-Attwood, T. L., Kocerha, J., Kroll, K. J., Hemmer, M. and Folmar, L. C. (2004b). Analysis of changes in gene expression patterns in fish exposed to natural, pharmaceutical and environmental estrogens using gene arrays. *Mar. Environ. Res.* 58,579–580.

Diatchenko, L., Lau, Y. F., Campbell, A. P., Chenchik, A., Moqadam, F., Huang, B., Lukyanov, S., Lukyanov, K., Gurskaya, N., Sverdlov, E. D. and Siebert, P. D. (1996). Suppression subtractive hybridization: A method for generating differentially regulated or tissue-specific cDNA probes and libraries. *Proc. Natl. Acad. Sci. USA* 93,6025–6030.

Diener, L. C., Schulte, P. M., Dixon, D. G. and Greenberg, B. M. (2004). Optimization of differential display polymerase chain reaction as a bioindicator for the cladoceran *Daphnia magna*. *Environ. Toxicol.* 19,179–190.

Dixon, R. L. (1976). Problems in extrapolating toxicity data for laboratory animals to man. *Environ. Health Perspect.* 13,43–50.

Dowling, V., Hoarau, P. C., Romeo, M., O'Halloran, J., Van Pelt, F., O'Brien, N. and Sheehan, D. (2006). Protein carbonylation and heat shock response in *Ruditapes decussatus* following *p,p'*-dichlorodiphenyldichloroethylene (DDE) exposure: A proteomic approach reveals that DDE causes oxidative stress. *Aquat. Toxicol.* 77,11–18.

Edge, S. E., Morgan, M. B., Gleason, D. F. and Snell, T. W. (2005). Development of a coral cDNA array to examine gene expression profiles in *Montastraea faveolata* exposed to environmental stress. *Mar. Pollut. Bull.* 51,507–523.

Evenden, A. J. and Depledge, M. H. (1997). Genetic susceptibility in ecosystems: The challenge for ecotoxicology. *Environ. Health Perspect.* 105(Suppl. 4),849–854.

Fielden, M. R. and Zacharewski, T. R. (2001). Challenges and limitations of gene expression profiling in mechanistic and predictive toxicology. *Toxicol. Sci.* 60,6–10.

Finne, E. F., Cooper, G. A., Koop, B. F., Hylland, K. and Tollefsen, K. E. (2007). Toxicogenomic responses in rainbow trout (*Oncorhynchus mykiss*) hepatocytes exposed to model chemicals and a synthetic mixture. *Aquat. Toxicol.* 81,293–303.

Forbes, V. E., Palmqvist, A. and Bach, L. (2006). The use and misuse of biomarkers in ecotoxicology. *Environ. Toxicol. Chem.* 25,272–280.

Freeman, K. (2004). Toxicogenomics data: The road to acceptance. *Environ. Health Perspect.* 112,A678–A685.

Fung, E. T., Thulasiraman, V., Weinberger, S. R. and Dalmasso, E. A. (2001). Protein biochips for differential profiling. *Curr. Opin. Biotech.* 12,65–69.

Genomics Task Force Workgroup. (2004). *Potential Implications of Genomics for Regulatory and Risk Assessment Application at EPA*. US Environmental Protection Agency, Washington, DC.

Gerwick, L., Corley-Smith, G. and Bayne, C. J. (2006). Gene transcript changes in individual rainbow trout livers following an inflammatory stimulus. *Fish Shellfish Immunol.* 22,157–171.

Gevaert, K. and Vandekerckhove, J. (2000). Protein identification methods in proteomics. *Electrophoresis* 21,1145–1154.

Gomiero, A., Pampanin, D. M., Bjornstad, A., Larsen, B. K., Provan, F., Lyng, E. and Andersen, O. K. (2006). An ecotoxicoproteomic approach (SELDI-TOF mass spectrometry) to biomarker discovery in crab exposed to pollutants under laboratory conditions. *Aquat. Toxicol.* 78,S34–S41.

Gonzalez, H. O., Roling, J. A., Baldwin, W. S. and Bain, L. J. (2006). Physiological changes and differential gene expression in mummichogs (*Fundulus heteroclitus*) exposed to arsenic. *Aquat. Toxicol.* 77,43–52.

Guilhermino, L., Diamantino, T., Silva, M. C. and Soares, A. M. (2000). Acute toxicity test with *Daphnia magna*: An alternative to mammals in the prescreening of chemical toxicity? *Ecotoxicol. Environ. Saf.* 46,357–362.

Hamadeh, H. K., Bushel, P., Paules, R. and Afshari, C. A. (2001). Discovery in toxicology: Mediation by gene expression array technology. *J. Biochem. Mol. Toxicol.* 15,231–242.

Handy, R. D., Galloway, T. S. and Depledge, M. H. (2003). A proposal for the use of biomarkers for the assessment of chronic pollution and in regulatory toxicology. *Ecotoxicology* 12,331–343.

Heinloth, A. N., Irwin, R. D., Boorman, G. A., Nettesheim, P., Fannin, R. D., Sieber, S. O., Snell, M. L., Tucker, C. J., Li, L., Travlos, G. S., Vansant, G., Blackshear, P. E., Tennant, R. W., Cunningham, M. L. and Paules, R. S. (2004). Gene expression profiling of rat livers reveals indicators of potential adverse effects. *Toxicol. Sci.* 80,193–202.

Heppell, S. A., Denslow, N. D., Folmar, L. C. and Sullivan, C. V. (1995). Universal assay of vitellogenin as a biomarker for environmental estrogens. *Environ. Health Perspect.* 103(Suppl. 7),9–15.

Hill, A. J., Teraoka, H., Heideman, W. and Peterson, R. E. (2005). Zebrafish as a model vertebrate for investigating chemical toxicity. *Toxicol. Sci.* 86,6–19.

Hook, S. E., Skillman, A. D., Small, J. A. and Schultz, I. R. (2006a). Dose-response relationships in gene expression profiles in rainbow trout, *Oncorhyncus mykiss*, exposed to ethynylestradiol. *Mar. Environ. Res.* 62(Suppl.),S151–S155.

Hook, S. E., Skillman, A. D., Small, J. A. and Schultz, I. R. (2006b). Gene expression patterns in rainbow trout, *Oncorhynchus mykiss*, exposed to a suite of model toxicants. *Aquat. Toxicol.* 77,372–385.

Hoyt, P. R., Doktycz, M. J., Beattie, K. L. and Greeley, M. S., Jr. (2003). DNA microarrays detect 4-nonylphenol-induced alterations in gene expression during zebrafish early development. *Ecotoxicology* 12,469–474.

Hugget, R. J., Kimerle, R. A., Mehrle, P. M., Bergman, H. L., Dickson, K. L., Fava, J. A., Mccarthy, J. F., Parrish, R., Dorn, P. B., Mcfarlan, V. and Lahvis, G. (1992). Introduction. In *Biomarkers Biochemical, Physiological, and Histological Markers of Anthropogenic Stress* (eds R. J. Hugget, R. A. Kimerle, P. M. Mehrle and H. L. Bergman), pp. 1–3, Lewis Publishers, Chelsea, MI.

Hutchens, T. W. and Yip, T.-T. (1993). New desorption strategies for the mass spectrometric analysis of macromolecules. *Rapid Commun. Mass. Spectrom.* 7,576–580.

Huyskens, G., Van Der Ven, K., Moens, L., Blust, R. and De Coen, W. (2004). Toxicogenomic analysis of the liver of zebrafish (*Danio rerio*) as a potential new tool for hazard identification of brominated flame retardants (BFRS). *Mar. Environ. Res.* 58,585–586.

Iguchi, T., Watanabe, H. and Katsu, Y. (2006). Application of ecotoxicogenomics for studying endocrine disruption in vertebrates and invertebrates. *Environ. Health Perspect.* 114(Suppl. 1),101–105.

Jenny, M. J., Warr, G. W., Ringwood, A. H., Gross, P. S., Robledo, J. A. F., Vasta, G. R. and Chapman, R. W. (2004). Ecogenomics: The use of gene expression

66

profiles for evaluating relationships between ecosystems and *Crassostrea virginica*. *Mar. Environ. Res.* 58,594–595.

Jonsson, H., Schiedek, D., Grosvik, B. E. and Goksoyr, A. (2006). Protein responses in blue mussels (*Mytilus edulis*) exposed to organic pollutants: A combined CYP-antibody/proteomic approach. *Aquat. Toxicol.* 78,S49–S56.

Kim, I. C., Lee, Y. M., Lee, C., Kim, H. M., Oda, S., Lee, Y. S., Mitani, H. and Lee, J. S. (2006). Expression profiles of 4-nonylphenol-exposed medaka (*Oryzias latipes*) analyzed with a 3.4 K microarray. *Mar. Environ. Res.* 62,S141–S146.

Kim, Y. K., Yoo, W. I., Lee, S. H. and Lee, M. Y. (2005). Proteomic analysis of cadmium-induced protein profile alterations from marine alga *Nannochloropsis oculata*. *Ecotoxicology* 14,589–596.

Knigge, T., Monsinjon, T. and Andersen, O. K. (2004). Surface-enhanced laser desorption/ionization-time of flight-mass spectrometry approach to biomarker discovery in blue mussels (*Mytilus edulis*) exposed to polyaromatic hydrocarbons and heavy metals under field conditions. *Proteomics* 4,2722–2727.

Koskinen, H., Pehkonen, P., Vehniainen, E., Krasnov, A., Rexroad, C., Afanasyev, S., Molsa, H. and Oikari, A. (2004). Response of rainbow trout transcriptome to model chemical contaminants. *Biochem. Biophys. Res. Commun.* 320,745–753.

Krasnov, A., Afanasyev, S. and Oikari, A. (2007). Hepatic responses of gene expression in juvenile brown trout (*Salmo trutta lacustris*) exposed to three model contaminants applied singly and in combination. *Environ. Toxicol. Chem.* 26,100–109.

Krasnov, A., Koskinen, H., Pehkonen, P., Rexroad, C. E., 3rd, Afanasyev, S. and Molsa, H. (2005a). Gene expression in the brain and kidney of rainbow trout in response to handling stress. *BMC Genomics* 6,3.

Krasnov, A., Koskinen, H., Rexroad, C., Afanasyev, S., Molsa, H. and Oikari, A. (2005b). Transcriptome responses to carbon tetrachloride and pyrene in the kidney and liver of juvenile rainbow trout (*Oncorhynchus mykiss*). *Aquat. Toxicol.* 74,70–81.

Kuperman, R. G., Checkai, R. T., Ruth, L. M., Henry, T., Simini, M., Kimmel, D. G., Phillips, C. T. and Bradley, B. P. (2003). A proteome-based assessment of the earthworm *Eisenia fetida*: Response to chemical warfare agents in a sandy loam soil. *Pedobiologia* 47,617–621.

Kurtz, J., Wegner, K. M., Kalbe, M., Reusch, T. B., Schaschl, H., Hasselquist, D. and Milinski, M. (2006). MHC genes and oxidative stress in sticklebacks: An immuno-ecological approach. *Proc. Natl. Acad. Sci. USA* 273,1407–1414.

Lam, S. H., Winata, C. L., Tong, Y., Korzh, S., Lim, W. S., Korzh, V., Spitsbergen, J., Mathavan, S., Miller, L. D., Liu, E. T. and Gong, Z. (2006). Trancriptome kinetics of arsenic-induced adaptive response in zebrafish liver. *Physiol. Genomics* 27, 351–361.

Lambe, K. G., Woodyatt, N. J., Macdonald, N., Chevalier, S. and Roberts, R. A. (1999). Species differences in sequence and activity of the peroxisome proliferator response element (PPRE) within the acyl CoA oxidase gene promoter. *Toxicol. Lett.* 110,119–127.

Larkin, P., Folmar, L. C., Hemmer, M. J., Poston, A. J., Lee, H. S. and Denslow, N. D. (2002). Array technology as a tool to monitor exposure of fish to xenoestrogens. *Mar. Environ. Res.* 54,395–399.

Larkin, P., Knoebl, I. and Denslow, N. D. (2003a). Differential gene expression analysis in fish exposed to endocrine disrupting compounds. *Comp. Biochem. Physiol., B* 136B,149–161.

Larkin, P., Sabo-Attwood, T., Kelso, J. and Denslow, N. D. (2003b). Analysis of gene expression profiles in largemouth bass exposed to 17-beta-estradiol and to anthropogenic contaminants that behave as estrogens. *Ecotoxicology* 12,463–468.

Larsen, B. K., Bjornstad, A., Sundt, R. C., Taban, I. C., Pampanin, D. M. and Andersen, O. K. (2006). Comparison of protein expression in plasma from nonylphenol and bisphenol A-exposed Atlantic cod (*Gadus morhua*) and turbot (*Scophthalmus maximus*) by use of SELDI-TOF. *Aquat. Toxicol.* 78,S25–S33.

Lawrence, J. W., Li, Y., Chen, S., Deluca, J. G., Berger, J. P., Umbenhauer, D. R., Moller, D. E. and Zhou, G. (2001). Differential gene regulation in human versus rodent hepatocytes by peroxisome proliferator-activated receptor (PPAR) alpha. PPAR alpha fails to induce peroxisome proliferation-associated genes in human cells independently of the level of receptor expresson. *J. Biol. Chem.* 276,31521–31527.

Lee, S.-E., Yoo, D.-H., Son, J. and Cho, K. (2006). Proteomic evaluation of cadmium toxicity on the midge *Chironomus riparius* Meigen larvae. *Proteomics* 6,945–957.

Lenz, E. M., Weeks, J. M., Lindon, J. C., Osborn, D. and Nicholson, J. K. (2005). Qualitative high field 1H-NMR spectroscopy for the characterization of endogenous metabolites in earthworms with biochemical biomarker potential. *Metabolomics* 1,123–136.

Lettieri, T. (2006). Recent applications of DNA microarray technology to toxicology and ecotoxicology. *Environ. Health Perspect.* 114,4–9.

Liang, P. and Pardee, A. B. (1992). Differential display of eukaryotic messenger RNA by means of the polymerase chain reaction. *Science* 257,967–971.

Lobenhofer, E. K., Cui, X., Bennett, L., Cable, P. L., Merrick, B. A., Churchill, G. A. and Afshari, C. A. (2004). Exploration of low-dose estrogen effects: Identification of No Observed Transcriptional Effect Level (NOTEL). *Toxicol. Pathol.* 32,482–492.

Lockhart, D. J., Dong, H., Byrne, M. C., Follettie, M. T., Gallo, M. V., Chee, M. S., Mittmann, M., Wang, C., Kobayashi, M., Horton, H. and Brown, E. L. (1996). Expression monitoring by hybridization to high-density oligonucleotide arrays. *Nat. Biotechnol.* 14,1675–1680.

Lopez-Barea, J. and Gomez-Ariza, J. L. (2006). Environmental proteomics and metallomics. *Proteomics* 26(Suppl. 1),S51–S62.

Maggioli, J., Hoover, A. and Weng, L. (2006). Toxicogenomic analysis methods for predictive toxicology. *J. Pharmacol. Toxicol.* 53,31–37.

Maples, N. L. and Bain, L. J. (2004). Trivalent chromium alters gene expression in the Mummichog (*Fundulus heteroclitus*). *Environ. Toxicol. Chem.* 23,626–631.

Marsano, F., De Pitta, C., Pallavicini, A., Vergani, L., Rebelo, M., Piacentini, L., Vitulo, N., Dal Monego, S., Venier, P., Dondero, F., Lanfranchi, G. and Viarengo, A. (2004). Gene transcription profiling in the mussel *Mytilus galloprovincialis* using low density and high density DNA microarrays. *Mar. Environ. Res.* 58,591–592.

Mattingly, C. J., Rosenstein, M. C., Davis, A. P., Colby, G. T., Forrest, J. N., Jr. and Boyer, J. L. (2006). The comparative toxicogenomics database: A cross-species resource for building chemical-gene interaction networks. *Toxicol. Sci.* 92,587–595.

68

Mayer, F. L., Versteeg, D. J., Mckee, M. J., Folmar, L. C., Graney, R. L., Mccume, D. C. and Rattner, B. A. (1992). Physiological and nonspecific biomarkers. In *Biochemical, Physiological, and Histological Markers of Anthropogenic Stress Biomarkers* (eds R. J. Hugget, R. A. Kimerle, P. M. Mehrle and H. L. Bergman), pp. 5–60, Lewis Publishers, Chelsea, MI.

McDonagh, B., Tyther, R. and Sheehan, D. (2005). Carbonylation and glutathionylation of proteins in the blue mussel *Mytilus edulis* detected by proteomic analysis and Western blotting: Actin as a target for oxidative stress. *Aquat. Toxicol.* 73,315–326.

Meiller, J. C. and Bradley, B. P. (2002). Zinc concentration effect at the organismal, cellular and subcellular levels in the eastern oyster. *Mar. Environ. Res.* 54,401–404.

Meyer, J. N., Volz, D. C., Freedman, J. H. and Di Giulio, R. T. (2005). Differential display of hepatic mRNA from killifish (*Fundulus heteroclitus*) inhabiting a Superfund estuary. *Aquat. Toxicol.* 73,327–341.

Mi, J., Orbea, A., Syme, N., Ahmed, M., Cajaraville, M. P. and Cristobal, S. (2005). Peroxisomal proteomics, a new tool for risk assessment of peroxisome proliferating pollutants in the marine environment. *Proteomics* 5,3954–3965.

Miracle, A. L. and Ankley, G. T. (2005). Ecotoxicogenomics: Linkages between exposure and effects in assessing risks of aquatic contaminants to fish. *Reprod. Toxicol.* 19, 321–326.

Monetti, C., Bernardini, G., Vigetti, D., Prati, M., Fortaner, S., Sabbioni, E. and Gornati, R. (2003). Platinum toxicity and gene expression in *Xenopus* embryos: Analysis by FETAX and differential display. *Altern. Lab. Anim.* 31,401–408.

Monetti, C., Vigetti, D., Prati, M., Sabbioni, E., Bernardini, G. and Gornati, R. (2002). Gene expression in *Xenopus* embryos after methylmercury exposure: A search for molecular biomarkers. *Environ. Toxicol. Chem.* 21,2731–2736.

Morgan, M. B., Edge, S. E. and Snell, T. W. (2005). Profiling differential gene expression of corals along a transect of waters adjacent to the Bermuda municipal dump. *Mar. Pollut. Bull.* 51,524–533.

Mueller, W. F., Coulston, F. and Korte, F. (1985). The role of the chimpanzee in the evaluation of the risk of foreign chemicals to man. *Regul. Toxicol. Pharmacol.* 5,182–189.

Nakayama, K., Iwata, H., Kim, E.-Y., Tashiro, K. and Tanabe, S. (2006). Gene expression profiling in common cormorant liver with an oligo array: Assessing the potential toxic effects of environmental contaminants. *Environ. Sci. Technol.* 40,1076–1083.

Nicholson, J. K., Lindon, J. C. and Holmes, E. (1999). 'Metabonomics': Understanding the metabolic responses of living systems to pathophysiological stimuli via multivariate statistical analysis of biological NMR spectroscopic data. *Xenobiotica* 29,1181–1189.

Nuwaysir, E. F., Bittner, M., Trent, J., Barrett, J. C. and Afshari, C. A. (1999). Microarrays and toxicology: The advent of toxicogenomics. *Mol. Carcinogen.* 24,153–159.

Oleksiak, M. F., Churchill, G. A. and Crawford, D. L. (2002). Variation in gene expression within and among natural populations. *Nat. Genet.* 32,261–266.

Oleksiak, M. F., Roach, J. L. and Crawford, D. L. (2005). Natural variation in cardiac metabolism and gene expression in *Fundulus heteroclitus*. *Nat. Genet.* 37,67–72.

Olsson, B., Bradley, B. P., Gilek, M., Reimer, O., Shepard, J. L. and Tedengren, M. (2004). Physiological and proteomic responses in *Mytilus edulis* exposed to PCBs and PAHs extracted from Baltic Sea sediments. *Hydrobiologia* 514,15–27.

Pampanin, D. M., Camus, L., Bjornstad, A., Andersen, O. K. and Gulliksen, B. (2006). A proteomic study using the deep sea amphipod *Eurythenes gryllus*. How do copper, mercury and benzo[a]pyrene affect the protein expression signatures? *Mar. Environ. Res.* 62,S181–S182.

Paules, R. (2003). Phenotypic anchoring: Linking cause and effect. *Environ. Health Perspect.* 111,A338–A339.

Pennie, W., Pettit, S. D. and Lord, P. G. (2004). Toxicogenomics in risk assessment: An overview of an HESI collaborative research program. *Environ. Health Perspect.* 112,417–419.

Perkins, E. J., Furey, J. and Davis, E. (2004). The potential of screening for agents of toxicity using gene expression fingerprinting in *Chironomus tentans*. *Aquat. Ecosyst. Health Manag.* 7,399–405.

Peterson, J. S. K. and Bain, L. J. (2004). Differential gene expression in anthracene-exposed mummichogs (*Fundulus heteroclitus*). *Aquat. Toxicol.* 66,345–355.

Picard, D. J. and Schulte, P. M. (2004). Variation in gene expression in response to stress in two populations of *Fundulus heteroclitus*. *Comp. Biochem. Physiol.* 137,205–216.

Pincetich, C. A., Viant, M. R., Hinton, D. E. and Tjeerdema, R. S. (2005). Metabolic changes in Japanese medaka (*Oryzias latipes*) during embryogenesis and hypoxia as determined by in vivo 31P NMR. *Comp. Biochem. Physiol. C, Comp. Pharmacol.* 140,103–113.

Place, S. P., Zippay, M. L. and Hofmann, G. E. (2004). Constitutive roles for inducible genes: Evidence for the alteration in expression of the inducible hsp70 gene in Antarctic notothenioid fishes. *Am. J. Physiol. Regul. Integr. Physiol.* 287,R429–R436.

Plumb, R. S., Stumpf, C. L., Gorenstein, M. V., Castro-Perez, J. M., Dear, G. J., Anthony, M., Sweatman, B. C. and Haselden, J. N. (2002). Metabonomics: The use of electrospray mass spectrometry coupled to reversed-phase liquid chromatography shows potential for the screening of rat urine in drug development. *Rapid Commun. Mass Spectrom.* 16,1991–1996.

Powell, C. L., Kosyk, O., Ross, P. K., Schoonhoven, R., Boysen, G., Swenberg, J. A., Heinloth, A. N., Boorman, G. A., Cunningham, M. L., Paules, R. S. and Rusyn, I. (2006). Phenotypic anchoring of acetaminophen-induced oxidative stress with gene expression profiles in rat liver. *Toxicol. Sci.* 93,213–222.

Poynton, H. C., Varshavsky, J. R., Chang, B., Cavigiolio, G., Chan, S., Holman, P. S., Loguinov, A. V., Bauer, D. J., Colbourne, J. K., Komachi, K., Theil, E. C., Perkins, E. J., Hughes, O. and Vulpe, C. D. (2007). *Daphnia magna* ecotoxicogenomics provides mechanistic insights into metal toxicity. *Environ. Sci. Technol.* 41,1044–1050.

Prenger, M. D. and Charters, D. W. (1997). *Ecological Risk Assessment Guidance for Superfund: Process for designing and conducting ecological risk assessments*. US Environmental Protection Agency, Washington, DC.

Reichert, K. and Menzel, R. (2005). Expression profiling of five different xenobiotics using a *Caenorhabditis elegans* whole genome microarray. *Chemosphere* 61,229–237.

Reichsman, F. P. and Calabrese, E. J. (1979). Animal extrapolation in environmental health: Its theoretical basis and practical applications. *Rev. Environ. Health* 3,59–78.

Reynders, H., Van Der Ven, K., Moens, L. N., Van Remortel, P., De Coen, W. M. and Blust, R. (2006). Patterns of gene expression in carp liver after exposure to a mixture of waterborne and dietary cadmium using a custom-made microarray. *Aquat. Toxicol.* 80,180–193.

Robertson, D. G. (2005). Metabonomics in toxicology: A review. *Toxicol. Sci.* 85, 809–822.

Rodriguez-Ortega, M. J., Grosvik, B. E., Rodriguez-Ariza, A., Goksoyr, A. and Lopez-Barea, J. (2003). Changes in protein expression profiles in bivalve molluscs (*Chamaelea gallina*) exposed to four model environmental pollutants. *Proteomics* 3,1535–1543.

Roling, J. A., Bain, L. J. and Baldwin, W. S. (2004). Differential gene expression in mummichogs (*Fundulus heteroclitus*) following treatment with pyrene: Comparison to a creosote contaminated site. *Mar. Environ. Res.* 57,377–395.

Samuelsson, L. M., Forlin, L., Karlsson, G., Adolfsson-Eric, M. and Larsson, D. G. J. (2006). Using NMR metabolomics to identify responses of an environmental estrogen in blood plasma of fish. *Aquat. Toxicol.* 78,341–349.

Schena, M., Shalon, D., Davis, R. W. and Brown, P. O. (1995). Quantitative monitoring of gene expression patterns with a complementary DNA microarray. *Science* 270,467–470.

Scott, G. R. and Schulte, P. M. (2005). Intraspecific variation in gene expression after seawater transfer in gills of the euryhaline killifish *Fundulus heteroclitus*. *Comp. Biochem. Physiol.* 141,176–182.

Sheader, D. L., Gensberg, K., Lyons, B. P. and Chipman, K. (2004). Isolation of differentially expressed genes from contaminant exposed European flounder by suppressive, subtractive hybridisation. *Mar. Environ. Res.* 58,553–557.

Sheader, D. L., Williams, T. D., Lyons, B. P. and Chipman, J. K. (2006). Oxidative stress response of European flounder (*Platichthys flesus*) to cadmium determined by a custom cDNA microarray. *Mar. Environ. Res.* 62,33–44.

Shepard, J. L. and Bradley, B. P. (2000). Protein expression signatures and lysosomal stability in *Mytilus edulis* exposed to graded copper concentrations. *Mar. Environ. Res.* 50,457–463.

Shepard, J. L., Olsson, B., Tedengren, M. and Bradley, B. P. (2000). Protein expression signatures identified in *Mytilus edulis* exposed to PCBs, copper and salinity stress. *Mar. Environ. Res.* 50,337–340.

Shrader, E. A., Henry, T. R., Greeley, M. S., Jr. and Bradley, B. P. (2003). Proteomics in zebrafish exposed to endocrine disrupting chemicals. *Ecotoxicology* 12,485–488.

Silvestre, F., Dierick, J. F., Dumont, V., Dieu, M., Raes, M. and Devos, P. (2006). Differential protein expression profiles in anterior gills of *Eriocheir sinensis* during acclimation to cadmium. *Aquat. Toxicol.* 76,46–58.

Snape, J. R., Maund, S. J., Pickford, D. B. and Hutchinson, T. H. (2004). Ecotoxicogenomics: The challenge of integrating genomics into aquatic and terrestrial ecotoxicology. *Aquat. Toxicol.* 67,143–154.

Snell, T. W., Brogdon, S. E. and Morgan, M. B. (2003). Gene expression profiling in ecotoxicology. *Ecotoxicology* 12,475–483.

Soetaert, A., Moens, L. N., van der Ven, K., Van Leemput, K., Naudts, B., Blust, R. and De Coen, W. M. (2006). Molecular impact of propiconazole on *Daphnia magna* using a reproduction-related cDNA array. *Comp. Biochem. Physiol. C, Comp. Pharmacol.* 142,66–76.

Soetaert, A., van der Ven, K., Moens, L. N., Vandenbrouck, T., van Remortel, P. and de Coen, W. M. (2007). *Daphnia magna* and ecotoxicogenomics: Gene expression profiles of the anti-ecdysteroidal fungicide fenarimol using energy-, molting- and life stage-related cDNA libraries. *Chemosphere* 67,60–71.

Steiner, S. and Anderson, N. L. (2000). Expression profiling in toxicology – potentials and limitations. *Toxicol. Lett.* 112–113,467–471.

Stentiford, G. D., Viant, M. R., Ward, D. G., Johnson, P. J., Martin, A., Wenbin, W., Cooper, H. J., Lyons, B. P. and Feist, S. W. (2005). Liver tumors in wild flatfish: A histopathological, proteomic, and metabolomic study. *Omics* 9,281–299.

Sumpter, J. P. and Jobling, S. (1995). Vitellogenesis as a biomarker for estrogenic contamination of the aquatic environment. *Environ. Health Perspect.* 103(Suppl. 7),173–178.

Tennant, R. (2002). The National Center for Toxicogenomics: Using new technologies to inform mechanistic toxicology. *Environ. Health Perspect.* 110,A8–A10.

Thompson, K. L., Afshari, C. A., Amin, R. P., Bertram, T. A., Car, B., Cunningham, M., Kind, C., Kramer, J. A., Lawton, M., Mirsky, M., Naciff, J. M., Oreffo, V., Pine, P. S. and Sistare, F. D. (2004). Identification of platform-independent gene expression markers of cisplatin nephrotoxicity. *Environ. Health Perspect.* 112,488–494.

Tilton, S. C., Givan, S. A., Pereira, C. B., Bailey, G. S. and Williams, D. E. (2006). Toxicogenomic profiling of the hepatic tumor promoters indole-3-carbinol, 17 beta-estradiol and beta-naphthoflavone in rainbow trout. *Toxicol. Sci.* 90,61–72.

Tremblay, A., Bruno, J. B., Osachoff, H. L., Skirrow, R. and van Aggelen, G. C. (2005). Application of molecular tools for linking metal toxicity and xenoestrogenic effects in coho salmon (*Oncorhynchus kisutch*). *Can. Tech. Rep. Fish. Aquat. Sci.* 2617,69–70.

Unlu, M., Morgan, M. E. and Minden, J. S. (1997). Difference gel electrophoresis: A single gel method for detecting changes in protein extracts. *Electrophoresis* 18, 2071–2077.

van der Meer, D. L., van den Thillart, G. E., Witte, F., de Bakker, M. A., Besser, J., Richardson, M. K., Spaink, H. P., Leito, J. T. and Bagowski, C. P. (2005). Gene expression profiling of the long-term adaptive response to hypoxia in the gills of adult zebrafish. *Am. J. Physiol.* 289,R1512–R1519.

van der Ven, K., De Wit, M., Keil, D., Moens, L., Leemput, K. V., Naudts, B. and De Coen, W. (2005). Development and application of a brain-specific cDNA microarray for effect evaluation of neuro-active pharmaceuticals in zebrafish (*Danio rerio*). *Comp. Biochem. Physiol., B* 141,408–417.

van der Ven, K., Keil, D., Moens, L. N., Hummelen, P. V., van Remortel, P., Maras, M. and de Coen, W. (2006). Effects of the antidepressant mianserin in zebrafish: Molecular markers of endocrine disruption. *Chemosphere* 65,1836–1845.

Viant, M. R., Rosenblum, E. S. and Tjeerdema, R. S. (2003a). NMR-based metabolomics: A powerful approach for characterizing the effects of environmental stressors on organism health. *Environ. Sci. Technol.* 37,4982–4989.

Viant, M. R., Werner, I., Rosenblum, E. S., Gantner, A. S., Tjeerdema, R. S. and Johnson, M. L. (2003b). Correlation between heat-shock protein induction and reduced metabolic condition in juvenile steelhead trout (*Oncorhynchus mykiss*) chronically exposed to elevated temperature. *Fish Physiol. Biochem.* 29,159–171.

Viant, M. R., Pincetich, C. A. and Eerderna, R. S. T. (2006a). Metabolic effects of dinoseb, diazinon and esfenvalerate in eyed eggs and alevins of Chinook salmon (*Oncorhynchus tshawytscha*) determined by H-1 NMR metabolomics. *Aquat. Toxicol.* 77,359–371.

72

Viant, M. R., Pincetich, C. A., Hinton, D. E. and Tjeerdema, R. S. (2006b). Toxic actions of dinoseb in medaka (*Oryzias latipes*) embryos as determined by in vivo P-31 NMR, HPLC-UV and H-1 NMR metabolomics. *Aquat. Toxicol.* 76,329–342.

Vijayan, M. M., Wiseman, S., Jorgensen, E. H. and Maule, A. G. (2005). Gene expression patterns in Arctic charr (*Salvelinus alpinus*) population contaminated by PCBs in northern Norway. *Can. Tech. Rep. Fish. Aquat. Sci.* 2617,71–72.

Volz, D. C., Benic, D. C., Hinton, D. E., Law, J. M. and Kullman, S. W. (2005). 2,3,7,8-Tetrachlorodibenzo-*p*-dioxin (TCDD) induces organ-specific differential gene expression in male Japanese medaka (*Oryzias latipes*). *Toxicol. Sci.* 85,572–584.

Volz, D. C., Hinton, D. E., Law, J. M. and Kullman, S. W. (2006). Dynamic gene expression changes precede dioxin-induced liver pathogenesis in medaka fish. *Toxicol. Sci.* 89,524–534.

Vuori, K. A., Koskinen, H., Krasnov, A., Koivumaki, P., Afanasyev, S., Vuorinen, P. J. and Nikinmaa, M. (2006). Developmental disturbances in early life stage mortality (M74) of Baltic salmon fry as studied by changes in gene expression. *BMC Genomics* 7,56.

Walker, C. H., Hopkin, S. P., Sibly, R. M. and Peakall, D. B. (2006). *Principles of Ecotoxicology*. Taylor and Francis Group, Boca raton, FL.

Walter, H., Consolaro, F., Gramatica, P., Scholze, M. and Altenburger, R. (2002). Mixture toxicity of priority pollutants at no observed effect concentrations (NOECs). *Ecotoxicology* 11,299–310.

Waring, J. F., Gum, R., Morfitt, D., Jolly, R. A., Ciurlionis, R., Heindel, M., Gallenberg, L., Buratto, B. and Ulrich, R. G. (2002). Identifying toxic mechanisms using DNA microarrays: Evidence that an experimental inhibitor of cell adhesion molecule expression signals through the aryl hydrocarbon nuclear receptor. *Toxicology* 181–182,537–550.

Warne, M. A., Lenz, E. M., Osborn, D., Weeks, J. M. and Nicholson, J. K. (2000). An NMR-based metabonomic investigation of the toxic effects of 3-trifluoromethyl-aniline on the earthworm *Eisenia veneta*. *Biomarkers* 5,56–72.

Warne, M. A., Lenz, L. E., Osborn, D., Weeks, J. M. and Nicholson, J. K. (2001). Comparative biochemistry and short-term starvation effects on the earthworms *Eisenia veneta* and *Lumbricus terrestris* studied by 1H NMR spectroscopy and pattern recognition. *Soil Biol. Biochem.* 33,1171–1180.

Watanabe, H., Takahashi, E., Nakamura, Y., Oda, S., Tatarazako, N. and Iguchi, T. (2007). Development of *Daphnia magna* DNA microarray for the evaluation of toxicity of environmental chemicals. *Environ. Toxicol. Chem.* 26,669–676.

Waters, M. D. and Fostel, J. M. (2004). Toxicogenomics and systems toxicology: Aims and prospects. *Nat. Rev.* 5,936–948.

Waters, M., Boorman, G., Bushel, P., Cunningham, M., Irwin, R., Merrick, A., Olden, K., Paules, R., Selkirk, J., Stasiewicz, S., Weis, B., Van Houten, B., Walker, N. and Tennant, R. (2003a). Systems toxicology and the chemical effects in biological systems (CEBS) knowledge base. *EHP Toxicogenomics* 111,15–28.

Waters, M. D., Olden, K. and Tennant, R. W. (2003b). Toxicogenomic approach for assessing toxicant-related disease. *Mutat. Res.* 544,415–424.

Wei, Y. D., Tepperman, K., Huang, M. Y., Sartor, M. A. and Puga, A. (2004). Chromium inhibits transcription from polycyclic aromatic hydrocarbon-inducible promoters by blocking the release of histone deacetylase and preventing the binding of p300 to chromatin. *J. Biol. Chem.* 279,4110–4119.

Whitehead, A. and Crawford, D. L. (2005). Variation in tissue-specific gene expression among natural populations. *Genome Biol.* 6,R13.

Williams, T. D., Gensberg, K., Minchin, S. D. and Chipman, J. K. (2003). A DNA expression array to detect toxic stress response in European flounder (*Platichthys flesus*). *Aquat. Toxicol.* 65,141–157.

Williams, T. D., Minchin, S. and Chipman, J. K. (2002). A DNA array to monitor gene expression in European flounder (*Platichthys flesus*). *Mar. Environ. Res.* 54,406–407.

Winneke, G. and Lilienthal, H. (1992). Extrapolation from animals to humans: Scientific and regulatory aspects. *Toxicol. Lett.* 64–65,239–246.

Wintz, H., Yoo, J. L., Loguinov, A., Wu, Y. Y., Steevens, J. A., Holland, R. D., Beger, R. D., Perkins, E. J., Hughes, O. and Vulpe, C. D. (2006). Gene expression profiles in fathead minnow exposed to 2,4-DNT: Correlation with toxicity in mammals. *Toxicol. Sci.* 94,71–82.

Yamamoto, M., Wakatsuki, T., Hada, A. and Ryo, A. (2001). Use of serial analysis of gene expression (SAGE) technology. *J. Immunol. Methods* 250,45–66.

Zhu, J. Y., Huang, H. Q., Bao, X. D., Lin, Q. M. and Cai, Z. W. (2006). Acute toxicity profile of cadmium revealed by proteomics in brain tissue of *Paralichthys olivaceus*: Potential role of transferrin in cadmium toxicity. *Aquat. Toxicol.* 78,127–135.

Fish toxicogenomics

Charles R. Tyler*, Amy L. Filby, Ronny van Aerle, Anke Lange,
Jonathan Ball and Eduarda M. Santos

*Environmental and Molecular Fish Biology, School of Biosciences, The Hatherly
Laboratories, University of Exeter, Exeter, Devon, EX4 4PS, UK*

Abstract. Fish are effective sentinels of pollution in the aquatic environment and are employed widely in biomonitoring and in regulatory testing to assess for health effects of chemical exposure. Molecular biomarkers of chemical exposure in fish have been used effectively for many years. The recent availability of extensive sequence information in fish, however, has facilitated the application of more extensive molecular (including whole-genome) approaches to fish toxicology. Through the application of polymerase chain reaction (PCR), differential display-PCR (DD-PCR), subtractive hybridisation and gene array methodologies (in which the responses of hundreds or even thousands of genes can be measured simultaneously), good insights have been gained into the mode of action (MOA) of a wide range of toxicants in fish and such approaches have illustrated the highly complex nature of some chemical effect pathways. Furthermore, genomic approaches have shown that different classes of toxicants operating through different MOAs can induce unique and diagnostic patterns of gene expression in fish. Studies in fish on transcriptome responses to various chemicals have also indicated the potential of genomics for diagnosing biological effects of chemicals. No transcriptomic studies have, however, been forthcoming to investigate how toxic responses that are consequently deleterious for the individual fish and potentially for the population are distinguished from adaptive responses, which may not affect fish. Rapid advancements have been made, but considerable challenges need to be met before the full potential of toxicogenomics can be realised for studies in fish. These challenges include the need for improved sequence annotation for fish, the application of international standards to arrays for data capture and analysis and appropriate (and more consistent) experimental design to ensure rigour in biological interpretations.

Keywords: toxicology; ecotoxicology; pollutants; endocrine disruption; oestrogens; gene expression; transcriptomics; molecular; genomics; gene array; RT-PCR; real-time RT-PCR; differential display; subtractive hybridisation; biomarker; mode of action; mechanism; phenotypic anchoring; bioinformatics; normalisation.

Corresponding author: Tel.: +44(0)1392-264450. Fax: +44(0)1392-263700.
E-mail: C.R.Tyler@exeter.ac.uk (C.R. Tyler).

ADVANCES IN EXPERIMENTAL BIOLOGY
VOLUME 02   ISSN 1872-2423
DOI: 10.1016/S1872-2423(08)00003-3

Introduction

Fish as sentinels of aquatic pollution

The aquatic environment acts as a sink for a wide range of chemical pollutants derived principally from human activities. Of the more than 200,000 chemicals that are in regular use by humans and released into the aquatic environment, some are discharged inadvertently, for example from landfill seepage and agricultural run-off. Other chemicals, however, are released intentionally from point sources, for example treated wastewater treatment works (WwTWs) effluents and pulp and paper mill effluents. The range of chemical pollutant types discharged includes heavy metals (*e.g.*, copper, zinc, mercury, lead and cadmium), natural and synthetic steroids (*e.g.*, the natural oestrogens oestrone (E_1) and 17β-oestradiol (E_2), the synthetic oestrogen 17α-ethinyloestradiol (EE_2) and the synthetic androgen trenbolone acetate), organophosphates (*e.g.*, the pesticides dimecron and malathion), organochlorines (*e.g.*, the pesticides aldrin, dieldrin and DDT and its metabolites such as DDE and DDD), poly-chlorinated biphenyls (PCBs), polycyclic aromatic hydrocarbons (PAHs), alkylphenol polyethoxylates (APEs) and their breakdown products (*e.g.*, 4-*tert*-nonylphenol (NP) and 4-*tert*-octylphenol (OP)), phthalates, bromi-nated flame retardants (*e.g.*, polybrominated diphenyl ethers (PBDEs)), pharmaceuticals and personal care products (PPCPs; *e.g.*, aspirin, fluoxe-tine, ibuprofen, triclosan and beta blockers) and manufactured nanoparti-cles. All of these chemicals have the capacity to disrupt physiological function in exposed organisms. Furthermore, most are present in the environment as mixtures and this is likely to enhance disruptive effects on biological systems (European Inland Fisheries Advisory Commission, 1987; Scientific Committee on Problems of the Environment, 1987).

Unpolluted water is a precious resource and fish have been used with great success to monitor chemical contamination and potential human health effects associated with chemical exposure. Examples of this con-tamination include the disruption of thyroid function by PCBs and polychlorinated dibenzo-*p*-dioxin (PCDDs) in The Great Lakes – St. Lawrence basin, USA (Leatherland, 1993; reviewed in Leatherland, 1992, 1997; Moccia *et al.*, 1981; Rolland, 2000), blue sac disease induced by exposure to PAHs and dioxins in The Great Lakes (Brown *et al.*, 2004a; Leatherland, 1997; Spitsbergen *et al.*, 1991) disruptions in embryo development induced by exposure to metals and transition metals released by industry, mining and maritime sources (Brown *et al.*, 1994; Jones *et al.*, 2001a; Williams and Holdway, 2000), tumour induction associated with effluent discharges (Grizzle *et al.*, 1984; Kinae *et al.*, 1990) and sexual disruption as a consequence of exposure to natural and pharmaceutical steroidal oestrogens derived from WwTW effluents (Jobling *et al.*, 1998, 2002a; Purdom *et al.*, 1994).

The life history of fish and some features of their physiology make them especially susceptible to pollutant uptake and effects (reviewed in Jobling and Tyler, 2003). Uptake of chemical contaminants into fish can occur via both dermal and gill surfaces for waterborne and sediment-adhesive chemicals, orally through the diet, or maternally, through the transfer of contaminants in the lipid reserves of eggs. Early life stages of fish have been shown to be especially sensitive to the harmful effects of chemical exposure (Liney *et al.*, 2005; Peterson *et al.*, 1993; Pyle *et al.*, 2002; van Aerle *et al.*, 2002) and lipophilic pollutants have been shown to bioconcentrate in fish tissues to levels of 10,000-fold and greater (*e.g.*, organochlorine pesticides, PCBs and APEs; Fox *et al.*, 1994; Hope *et al.*, 1998; Ishibashi *et al.*, 2006), resulting in an enhanced likelihood of adverse effects. In addition, compared with mammals, fish have a limited capacity to metabolise organic chemicals, including PCBs, PCDDs, polychlorinated dibenzofurans (PCDFs) and PAHs, which can make them especially sensitive to the harmful effects of this group of chemicals. This difference between fish and mammals may relate to differences in the cytochrome P450 (CYP) enzymes, which detoxify organic pollutants such as PCBs, PCDDs, PCDFs and PAHs (Hollenberg, 1992). As an example, in mammals there are two genes for the key CYP enzyme CYP1A (*cyp1a1* and *cyp1a2*), which are coordinately regulated by the same aryl hydrocarbon receptor (AhR), but only one *cyp1a* occurs in fish (Buhler and Wang-Buhler, 1998; Nabb *et al.*, 2006; Stegeman, 1995).

The global distribution of fish facilitates geographical comparisons of pollutant effects. Fish are found in almost every aquatic habitat, including fresh, brackish and salt waters and extend from the freezing waters of the Antarctic to hot springs in the tropics, where water temperatures may exceed 45 °C. Some fish species have even adapted to live in highly acidic waters in which the pH is less than 3. Fish occupy many ecological niches and some species are demersal, living on, or even in, the beds of rivers, estuaries or seas where organic chemicals tend to concentrate. Fish also have a significant commercial value for both recreation and food and their exposure to pollutants can affect population sustainability and, via the food chain, have direct implications for human health.

Given their central role in the ecology of aquatic systems, their ability to act as sentinels of aquatic pollution, their commercial value and their capacity to act as model systems for understanding vertebrate biology (*e.g.*, the use of zebrafish (*Danio rerio* Hamilton) embryos as a model for mammalian development; see reviews by Ackermann and Paw, 2003; Dooley and Zon, 2000), it is not surprising that fish have become widely adopted in regulatory testing for hazard identification of chemicals. Furthermore, there are over 26,000 species of fish, representing more than 500 million years of evolution – a radiology that is far greater than for any other class of vertebrates and this situation offers many advantages for studies addressing comparative and evolutionary questions in (eco)toxicology.

Toxicogenomics

Virtually all biological responses to external stressors, including toxicants, involve changes in normal patterns of gene expression (Merrick and Bruno, 2004). Some of these responses are a direct result of the chemical, such as alterations in gene expression caused by the binding of a steroid hormone (or analogue) to a specific steroid hormone receptor, which acts as a transcription factor and subsequently modulates (activates or represses) the transcription of its target genes. Other gene expression responses to toxic chemicals are indirect or compensatory, in that they reflect the response of the organism to molecular damage/cellular dysfunction (Nuwaysir *et al.*, 1999). The time lapsed between the onset of the chemical exposure and the sampling and measuring of the gene expression, therefore, can have a major bearing on the gene responses observed. Importantly, however, different mechanisms of toxicity can generate specific patterns of gene expression that can potentially provide us with molecular biomarkers of disruption of a biological process and be reflective of mechanism or mode of action (MOA; Merrick and Bruno, 2004). The adoption of changes in gene expression in fish toxicology has been facilitated by the availability of genome sequences for a number of key fish species used in both environmental monitoring and regulatory testing. It should be emphasised, however, that, although changes in gene expression are generally rapid and thus potentially provide a capability of a rapid diagnosis of chemical effect, transcription of messenger ribonucleic acid (mRNA) is only an intermediate step in conversion of genetic information into proteins, the biochemical bases of biological function and gene expression and concentration of functional proteins are not necessarily always directly related.

Toxicogenomics is a relatively new terminology, derived in 1999 to describe the application of genomics to toxicology (Nuwaysir *et al.*, 1999). Since that time, there has been an exponential development of this field of research, as shown by a rapid increase in publications, principally derived from studies in mammals. Strictly, the term 'genomics' refers to studies at the whole-genome level: the transcriptome, the proteome and the metabolome (termed transcriptomics, proteomics and metabolomics, respectively). Changes in the expression of smaller suites of transcripts, proteins and metabolites, however, are also often included under the definition of genomics. Changes in transcripts, proteins and metabolites measure responses at different levels of biological organisation and thus provide complementary insights into the biochemical/molecular status of an organism. All three approaches have the potential to define toxicity pathways and are most effective if used in conjunction with one another (MacGregor, 2003; Miracle and Ankley, 2005; Robertson, 2005; Waters and Fostel, 2004). Studies on the proteome and metabolome as applied to toxicology have been initiated in fish (*e.g.*, proteomics: Damodaran *et al.*, 2006; Hogstrand *et al.*, 2002; Shrader *et al.*, 2003; Zhu *et al.*, 2006; metabolomics: Samuelsson *et al.*, 2006; Viant *et al.*, 2006a, 2006b), but are much less developed compared with transcriptomic studies.

Chapter scope

This chapter provides a critical analysis of the state-of-the-art fish toxicogenomics, specifically as applied to transcriptomics. It does not consider either the proteome or the metabolome. The chapter starts by considering the genomic resources available for fish and examines some of the problems associated with their application to toxicological studies. A brief overview of the molecular biomarkers that have been applied in fish toxicology is provided, followed by a detailed analysis of how genomic approaches have been applied to develop our understanding of MOAs of chemicals. In this section, technical aspects relating to the analysis of genomic data sets are considered, including the need to ensure appropriate normalisation of the data generated. The concepts of genetic 'fingerprints' and 'profiles' for predicting the MOA of new or unknown toxicants in fish are described and an emphasis placed on the need to link gene responses induced by chemicals with phenotypic effects (the so-called phenotypic anchoring). The application of genomics to unravel mixture effects of chemicals in fish is also discussed. The final section of this chapter explores the need to understand the basic biology of the organisms used in toxicogenomic studies to ensure appropriate interpretation of gene and genomic responses to toxicological challenges.

In this chapter, there is a strong focus on the use of toxicogenomics as applied to endocrine disruption, because for fish this is where the majority of toxicogenomic data sets have been generated. Nevertheless, many of the principles described are applicable to the molecular responses elicited by chemicals in fish more widely. This chapter is set against a backdrop of the rapidly advancing field of genomics generally, especially with respect to the availability of sequence information in fish, advances in the annotation of the gene sequences (Table 1), technology for array fabrication, bioinformatic approaches and capabilities to analyse and interpret the responses of hundreds to thousands of genes simultaneously to extract biological meaning from the extensive data sets generated with gene microarrays.

Genomic resources for fish

Sequenced genomes

It is now within the realms of any laboratory that is reasonably well equipped with molecular facilities to clone a gene of interest in any species of fish. The process is relatively straightforward and requires a limited amount of comparative genomics to produce nucleotide or amino acid alignments that identify the regions of the gene of interest that are conserved for the available sequences in genomic databases, the design of primers from that alignment to amplify the target gene from RNA template material for the species of interest using reverse transcription polymerase chain reaction (RT-PCR) and, subsequently, gene cloning and sequencing techniques to identify the nucleotide sequence of the amplicon. For some 'model' species, the gene sequence of interest is probably available in databases, which greatly facilitates subsequent expression studies. The ease of gene cloning has led rapidly to the widespread use of specific genes and small suites of genes as biomarkers of chemical exposure and for developing our understanding of MOAs of chemicals in fish (see below). Sequenced genomes, or at least sequences of extensive gene sets, however, are generally required for some of the more advanced molecular approaches in toxicogenomics (*e.g.*, gene arrays – matrices that are used to study the expression responses of hundreds to thousands of genes simultaneously).

Genome sequencing has been completed for three fish species: Japanese pufferfish (*Takifugu rubripes* Temminck and Schlegel) and freshwater pufferfish (*Tetraodon nigroviridis* Marion de Procé) – species chosen originally for sequencing because they have compact genomes – and zebrafish – used widely for chemical testing, as a model for vertebrate development and for studying the genetic basis of some diseases.

Table 1. Genome sequence data available for fish species[a].

GOLDSTAMP	Organism	Type	Size (Kb)	Genome database	Publication	Status
Gc00272	*Danio rerio* Tuebingen	Genome	1,700,000	NCBI; Washington University; University of Oregon; Fishman-MGH; Carnegie Institute of Washington; Stanford University; Sanger Institute	Unpublished (completed in 2005)	Complete
Gi00269	*Gasterosteus aculeatus*	Genome	680,000	NHGRI; Broad Institute; NISC-NIH; NIH	–	Incomplete
Gi00991	*Leucoraja erinacea*	Genome	–	NISC-NIH	–	Incomplete
Gi00419	*Oreochromis niloticus*	Genome	–	Roslin Institute	–	Incomplete
Gi01531	*Oryzias latipes*	Genome	700,000	NCBI, Nagoya University	*Nature* 447, 714–719 (2007)	Quality draft
Gi00529	*Salmo salar*	Genome	–	Genome British Columbia; TIGR	–	Incomplete
Gc00097	*Takifugu rubripes*	Genome	400,000	NCBI; Fugu Genome Consortium; HGMP-RC; Joint Genome Institute; Sanger Institute	*Science* 297, 1301–1310 (2002)	Complete
Gc00229	*Tetraodon nigroviridis*	Genome	385,000	NCBI, Genoscope; Broad Institute, NHGRI	*Nature* 431, 946–957 (2004)	Complete

[a]Data were obtained from the Genomes OnLine Database (GOLD, Release 2.0; as on 21/10/2007; Liolios *et al.*, 2006). HGMP-RC, Human Genome Mapping Project Resource Centre; NHGRI, National Human Genome Research Institute; NIH, National Institutes of Health; NISC, National Intramural Sequencing Center; and TIGR, The Institute for Genomic Research.

In addition, genomes are currently being sequenced for a range of other fish species, including Japanese medaka (*Oryzias latipes* Temminck and Schlegel) – used extensively for chemical testing, three-spined stickleback (*Gasterosteus aculeatus* Linnaeus) – used as a model in the study of adaptive radiation and evolution as well as for studying pollutant effects and the little skate (*Leucoraja erinacea* Mitchill) – used as a model elasmobranch for comparative genomics in helping to understand human health and disease (see Table 1 for full species listing). In addition to these genome projects, extensive gene sequence libraries have become available for many other fish species through expressed sequence tag (EST) projects and they include the channel catfish (*Ictalurus punctatus* Rafinesque), rainbow trout (*Oncorhynchus mykiss* Walbaum), common carp (*Cyprinus carpio* Linnaeus), Nile tilapia (*Oreochromis niloticus*), Atlantic salmon (*Salmo salar* Linnaeus), fathead minnow (*Pimephales promelas* Rafinesque), European flounder (*Platichthys flesus* Linnaeus), largemouth bass (*Micropterus salmoides* Lacepède) and roach (*Rutilus rutilus* Linnaeus). This list comprises species that are used for chemical testing in the laboratory and as sentinels for pollution in the environment and includes representatives of different aquatic environments with different ecological niches. Given the rapid development in sequencing capabilities (for example the new '454 system' (454 Life Sciences, Branford, CA, USA; www.454.com) is 100 times faster than previously available systems and able to sequence 25 million bases, with an accuracy of 99% or more, in a four-hour run (Margulies *et al.*, 2005)), it is very likely that extensive sequence information will become available for many more fish species in the near future.

The availability of sequenced fish genomes in public databases enables *in silico* approaches for assessing likely (and comparative) responses to toxicants. This method is particularly useful for predicting the genome-wide effects of toxicants that have a known MOA, such as oestrogenic and androgenic substances. Such an approach has been applied successfully in mammalian studies to predict the genes responsive to oestrogens in humans (Bourdeau *et al.*, 2004; Kamalakaran *et al.*, 2005; Tang *et al.*, 2004). We have recently started to undertake studies of this nature for oestrogens in fish. In these studies, a comparative genomic approach was adopted to identify conserved oestrogen-responsive elements (EREs) in the regulatory regions of all available genes in the zebrafish. We further evaluated the likely functionality of the EREs identified, by determining their evolutionary conservancy in two other fish species (*Takifugu rubripes* and *Tetraodon nigroviridis*). Genes were categorised as having a functional ERE in their promoter region only when it was present in all three fish species. The bioinformatic predictions were then validated by comparison to an experimentally obtained list of genes differentially expressed in zebrafish ovaries following exposure to the oestrogen EE_2 (Santos et al., 2007b). This example serves to illustrate how, when genome sequences are available, in silico approaches can be used to predict the genes and biological processes directly modulated by a toxicant on genome-wide basis. It should be emphasised, however, that such approaches require not only the availability of genomic sequence but also knowledge of the MOA of the chemicals being considered. Adding further complexity to this approach, even for oestrogens, the modulation of transcription following binding of the toxicant/oestrogen receptor (ER) complex to EREs requires the recruitment of cofactors (co-activators and co-repressors), which results in up- or down-regulation of the transcription of target genes. This process depends not only on the DNA sequence that will allow binding by specific factors, but also on the environment of each specific cell in which specific cofactors may be present (Shibata et al., 1997). The modulation of transcription of each specific gene following oestrogen exposure is, therefore, dependent on each cell type and its physiological status and this finding explains the distinct patterns of modulation encountered following exposure. In addition, the promoters of genes containing functional EREs may be methylated rendering them non-responsive to the toxicant challenge. Mechanisms such as those described here for oestrogenic compounds are likely to exist for many other chemicals with other MOAs and this fact needs to be taken into account when addressing the mechanistic pathways by which chemicals exert their effects. Given the complexities of the toxicological response, hypotheses developed through in silico approaches need to be tested with empirical studies.

Gene arrays

Gene arrays can measure the parallel responses of many thousands of genes and offer considerable power for detecting the MOA of a chemical and for biomarker discovery. In mammalian toxicology research, relatively few study species are used routinely (most notably mouse and rat) and genomic resources (i.e., gene sequences) and experimental tools (i.e., array platforms) for whole-genome investigations are well developed, commercially available and standardised (for examples of assessments of the quality and reproducibility of array data sets, see Guo et al., 2006; Patterson et al., 2006; Shi et al., 2006). In complete contrast, in fish toxicology a wide range of divergent fish species are used for genomic studies and a variety of array platforms have been employed, ranging from custom-made DNA macroarrays, typically containing several hundred genes (*e.g.*, a 200-gene array for the fathead minnow and a 500-gene array for the largemouth bass; www.ecoarray.com, Brown *et al.*, 2004b, 2004c; Klaper *et al.*, 2006; Larkin *et al.*, 2003a, 2003b; Sheader *et al.*, 2006; Williams *et al.*, 2003), to custom-made microarrays (*e.g.*, Handley-Goldstone *et al.*, 2005; Hoyt *et al.*, 2003; Krasnov *et al.*, 2005; Martyniuk *et al.*, 2006; Moens *et al.*, 2006; Reynders *et al.*, 2006; van der Ven *et al.*, 2006a; Williams *et al.*, 2006; Wintz *et al.*, 2006) and commercially sourced oligonucleotide microarrays, containing many thousands of genes (*e.g.*, Hoffmann *et al.*, 2006; Hook *et al.*, 2006a, 2006b; Kishi *et al.*, 2006; Kreiling *et al.*, 2007; Martyniuk *et al.*, 2007; Santos *et al.*, 2007a, 2007b; van der Ven *et al.*, 2006b; Voelker *et al.*, 2007). Furthermore, as yet, there has been little attempt to standardise approaches across these different platforms, making cross-study comparisons difficult.

The development of arrays for non-model fish species, in which there is a comparative paucity of sequence information, has entailed considerable provision of resources and effort from individual laboratories. The construction of such arrays has often included extensive use of PCR, construction of complementary DNA (cDNA) libraries and suppressive subtractive hybridisation between exposed and non-exposed fish, followed by extensive cloning and sequencing, to generate the cDNA probes required (*e.g.*, for the construction of a 160-gene custom microarray for evaluation of the genomic response to pollution in wild European flounder; Williams *et al.*, 2003). An alternative approach, which has now been adopted in several toxicological studies with non-model fish species, is that of 'model hopping', whereby an array available for a model fish species is applied to study the genomic responses in the fish species of interest. As an example, a commercially sourced *Fugu rubripes* gill cDNA array was used to identify genes involved in the response to sub-lethal concentrations of zinc in rainbow trout, a species with a less characterised genome (Hogstrand *et al.*, 2002). A limitation found in this approach, however, was the difficulty in independently verifying the array data generated, as difficulties were often encountered when using cDNA information from the *Fugu* species to design appropriate primers for real-time RT-PCR in the rainbow trout.

Model hopping with gene arrays has been applied with good success to fish species that are more closely allied in terms of their evolution. Rise *et al.* (2004) and von Schalburg *et al.* (2005) investigated the use of a

16K cDNA salmonid (Atlantic salmon/rainbow trout) microarray in four salmonid species (Atlantic salmon, rainbow trout, chinook salmon (*Oncorhynchus tshawytscha* Walbaum) and lake whitefish (*Coregonus clupeaformis* Mitchill)) and found there were no significant differences in the percentage of targets that bound (rainbow trout targets, however, consistently showed a higher overall binding to the microarray; von Schalburg *et al.*, 2005). Recently, we applied a multispecies gene array containing 95 genes with cDNAs derived from three cyprinid fish species (zebrafish, roach and fathead minnow) to investigate the effects of the aromatase inhibitor fenarimol in the fathead minnow and responses measured derived from more than one species representing the same gene clustered together, demonstrating the viability of this approach for closely related fish species (van Aerle *et al.*, 2004). Similar findings were reported for an array containing a mixture of goldfish (*Carassius auratus* Linneaus) and common carp (both cyprinid fish species) cDNA fragments to investigate the effects of EE_2 on the brain gene expression profiles in goldfish (Martyniuk *et al.*, 2006). Thus, the application of gene arrays from one fish species to another is possible, but is perhaps best applied to species that are evolutionary closely allied. Moreover, heterologous arrays further require greater caution in the subsequent validation and interpretation of the data compared with homologous arrays.

Sequence annotation

Even for fish species in which the genome is (or is being) sequenced, there is often difficulty in the interpretation of the biological significance of differentially expressed gene sets following chemical exposure caused by relatively poor annotation of the gene sequences available in public databases. Even for commercial microarrays, annotations of the probes contained in the array are generally incomplete. Frequently, the biological function of the genes represented in the array is unknown or is based on homology to genes in other species and a relatively small proportion of genes are ever firmly identified and functionally described. This problem is particularly evident in fish gene databases, even for the best-characterised genomes like for *Fugu* and zebrafish, in which the annotation of the gene probes contained in a given array is often less than 50%. This annotation can result in data sets in which the biological significance of the gene lists found to be affected by the toxicant being tested is difficult to ascribe.

Molecular biomarkers in fish

Biomarkers are biological responses to chemical exposure that are associated with MOA or toxic effect (Peakall, 1984; van Gestel and van Brummelen, 1996). When induced above a certain threshold, biomarkers can potentially signal for a subsequent effect (normally adverse) at a higher level of biological organisation (Huggett et al., 1992; McCarthy and Shugart, 1990; van der Oost et al., 2005). As such, biomarkers have been used to assess the health status of fish and as early warning signals of environmental risks (Di Giulio et al., 1995; Payne et al., 1987). In our analysis of biomarkers, however, we concluded that, although highly effective as measures of chemical exposure, there are very few examples in which a specific biomarker has been proven to signal for a subsequent adverse and specific biological effect (see section Phenotypic anchoring).

Biomarkers in fish that have been employed for quantifying chemical exposure and in biomonitoring represent a variety of different biochemical/metabolic processes (for a review, see van der Oost et al., 2005). They include phase I biotransformation enzymes (such as CYP enzymes, with CYP1A being particularly well characterised); phase II biotransformation enzymes and related cofactors (glutathione S-transferases (GST) and glutathione (GSH)); the enzymes superoxide dismutase (SOD), catalase (CAT), glutathione peroxidase (GPx) or glutathione reductase (GR) and GSH for oxidative stress; metallothioneins (MTs) for heavy metals; vitellogenin (VTG), vitelline envelope proteins (VEPs), zona radiata proteins (ZRPs) and P450 aromatase (P450arom) for oestrogenic chemicals; cholinesterases for neurotoxic chemicals and measures of DNA damage for genotoxic chemicals. Some of these biomarkers are specific to certain toxicants (or classes of toxicants), for example MTs for metals, cholinesterase activity for organophosphates and carbamates, VTG, VEPs and ZRPs for oestrogenic endocrine active chemicals (EACs) and CYP1A activity (measured as ethoxyresorufin-O-deethylase (EROD) activity) for planar aromatic compounds (Arukwe et al., 1997; Chan, 1995; Hylland et al., 1996; Payne et al., 1996; Sturm et al., 2000; Sumpter and Jobling, 1995; Viarengo et al., 1999; Whyte et al., 2000). Other biomarkers, such as those relating to oxidative stress, are less specific and can be induced by several classes of compounds.

Alteration in the gene expression of some of these specific biomarkers is now commonly used for assessing the toxicity of individual chemicals in the laboratory and for the biomonitoring of exposure in wild fish populations. Methods used to measure changes in the expression of individual genes in fish have included Northern hybridisation (*e.g.*, Jung *et al.*, 2005), ribonuclease (Rnase) protection assays (RPAs) (*e.g.*, Islinger *et al.*, 1999), hybridisation protection assays (HPAs) (*e.g.*, Thomas-Jones *et al.*, 2003a), RT-PCR (*e.g.*, Lee *et al.*, 2002) and real-time RT-PCR (*e.g.*, Celius *et al.*, 2000; Tom *et al.*, 2004), and the latter technique is now the most prevalent. Below, we discuss a selection of the most commonly used molecular biomarkers and their utility in fish toxicology.

vtg and vep mRNAs

Global concern about EACs has driven the search for biomarkers of EAC exposure in fish and in other organisms. Many EACs have been shown to be oestrogenic in nature and are major contaminants in effluents from WwTWs that discharge into rivers globally (*e.g.*, the UK (Purdom *et al.*, 1994), the Netherlands (Vethaak *et al.*, 2005), Norway (Knudsen *et al.*, 1997), the USA (Folmar *et al.*, 1996) and China (Ma *et al.*, 2005)). These oestrogenic compounds include natural and synthetic steroid oestrogens, phyto-oestrogens (plant derived) and myco-oestrogens (fungal derived) and a wide variety of man-made industrial compounds, such as plasticisers, surfactants, bisphenol A (BPA) and various pesticides and herbicides (reviewed in Tyler *et al.*, 1998). Importantly, exposure to oestrogenic WwTW effluents induces the feminisation of wild fish, including the development of an intersex condition (the simultaneous presence of both male and female germ cells in the gonad; Jobling *et al.*, 1998, 2002a) and this has been shown to result in reduced reproductive capacity (Jobling *et al.*, 2002a, 2002b).

Of the oestrogenic biomarkers identified, VTG, a yolk protein precursor, has been the most widely adopted. VTG is normally synthesised in the liver of females only through an ER-mediated pathway, but exposure to exogenous oestrogens also readily activates its synthesis in males (reviewed in Sumpter and Jobling, 1995). Induction of VTG in the plasma and expression of the *vtg* gene in fish liver has been successfully employed to quantify the oestrogenic activity of chemicals both *in vitro* (*e.g.*, in primary hepatocyte cultures; Le Guevel and Pakdel, 2001; reviewed in Navas and Segner, 2006; Okoumassoun *et al.*, 2002; Radice *et al.*, 2002) and *in vivo* in laboratory exposures (Kishida *et al.*, 1992; Pelissero *et al.*, 1991; Sherry *et al.*, 2006; Thorpe *et al.*, 2001), to assess the potency of oestrogenic WwTWs effluents in the field (Diniz *et al.*, 2005; Hemming *et al.*, 2001; Higashitani *et al.*, 2003) and in biomonitoring of oestrogen exposure in wild fish populations (Hashimoto *et al.*, 2000; Hecker *et al.*, 2002). Reductions in plasma VTG levels, as a result of exposure to chemicals that have inhibitory effects on VTG synthesis and/ or secretion (such as the aromatase inhibitor fadrozole), have been found to be linked with adverse effects on fecundity and associated with population-level effects in fish (Miller *et al.*, 2007). Useful features of VTG protein and *vtg* mRNA induction are their specificity for oestrogens, their sensitivity and the magnitude of the response possible (*vtg* mRNA may increase by more than 1,000-fold, a dynamic range not reported for any other molecular biomarkers in use (Filby *et al.*, 2007a; Scholz *et al.*, 2004; Thomas-Jones *et al.*, 2003b). Partial or full-length cDNA sequences for *vtg* have been cloned in numerous different species of fish, including zebrafish (Tong *et al.*, 2004; Wang *et al.*, 2000), blue tilapia (*Oreochromis aureus* Steindachner; Lim *et al.*, 2001), medaka (Tong *et al.*, 2004), mosquitofish (*Gambusia affinis* Baird and Girard; Sawaguchi *et al.*, 2005), rockfish (*Sebastes schlegeli* Hilgendorf; Jung *et al.*, 2005) and conger eel (*Conger myriaster* Brevoort; Mikawa *et al.*, 2006). The temporal dynamics of induction of *vtg* expression has also been described in various fish species, including in the fathead minnow, sheepshead minnow and rainbow trout (Craft *et al.*, 2004; Hemmer *et al.*, 2002; Schmid *et al.*, 2002; Thomas-Jones *et al.*, 2003b) and induction of *vtg* expression (together with VTG protein induction) is being considered by the Organisation for Economic Cooperation and Development (OECD; in Europe) and the Environmental Protection Agency (EPA; in the USA) as an endpoint in the development of standardised tests for EACs in fish.

Expression of the *vep* and *zrp* genes has also been adopted to quantify oestrogen exposure in fish. These transcripts code for glycoproteins that are egg envelope components and form the chorion of the developing egg. Three VEPs have been identified in fish (*vep1*, *vep2* and *vep3*; also known as *vepα*, *vepβ* and *vepγ*), which are normally synthesised in females only during oogenesis but, again, males can (and do) synthesise them in response to exposure to exogenous steroid oestrogens and their mimics (Hyllner *et al.*, 1991). VEPs and ZRPs are synthesised mainly in the liver of most fish (Hamazaki *et al.*, 1985, 1989; Lyons *et al.*, 1993; Shimizu *et al.*, 1998), but, in some teleosts, the oocyte/follicle has been shown to be involved with their synthesis (*e.g.*, pipefish, *Sygnathus fuscus* Linnaeus (Begovac and Wallace, 1989), carp (Chang *et al.*, 1996, 1997) and zebrafish (Wang and Gong, 1999)), as occurs in amphibians and other mammals. VEPs and ZRPs are extremely hydrophobic and are difficult to measure using conventional techniques, thus measurement of their gene expression is a preferred approach to quantify their induction by EACs. Induction of *vep* and *zrp* gene expression has been shown to be equally (or more) responsive to oestrogen stimulation compared with induction of *vtg* gene expression and to respond over a wide dynamic range of up to 500-fold (Arukwe *et al.*, 1997; Celius and Walther, 1998a; Celius *et al.*, 1999; Thomas-Jones *et al.*, 2003b; Westerlund *et al.*, 2001). Partial or full-length *vep* and *zrp* cDNA sequences have also been cloned in a number of species of fish, including rainbow trout (Arukwe *et al.*, 2002; Hyllner *et al.*, 2001), zebrafish (Wang and Gong, 1999) and carp (Chang *et al.*, 1996, 1997, 1999).

spiggin mRNA

The search for biomarkers of (anti)androgens in fish is a relatively recent venture driven by reports that found that androgens, or molecules with androgen-like activities, are present in WwTW effluents, paper and pulp mill effluents and concentrated animal feeding operations (CAFOs) waste waters (Kirk *et al.*, 2002; Orlando *et al.*, 2004; Thomas *et al.*, 2002) and are able to persist in river water downstream of these discharges (Durhan *et al.*, 2002; Jenkins *et al.*, 2001). Furthermore, wild fish living downstream of some WwTWs effluents have been shown to have altered reproductive function consistent with androgenic exposure (Larsson and Forlin, 2002; Orlando *et al.*, 2004; Parks *et al.*, 2001). A reported causative agent in CAFOs was 17β-trenbolone (TRB; a synthetic steroid that is extensively used in the USA as a growth promoter in beef cattle; Ankley *et al.*, 2003). Arguably, the most effective biomarker identified for measuring exposure to (anti)androgens is spiggin, a glue protein used for nest building by male sticklebacks (Jakobsson *et al.*, 1999). Spiggin production is under the control of endogenous androgens and spiggin has also shown to be sensitive for detecting a range of (anti)androgenic xenobiotics (Hahlbeck *et al.*, 2004; Katsiadaki *et al.*, 2002a, 2002b, 2006). Multiple *spiggin* gene transcripts have been cloned (Jones *et al.*, 2001b; Kawahara and Nishida, 2006; Kawasaki *et al.*, 2003) and, so far, have been shown to be highly responsive to androgens (Nagae *et al.*, 2007).

Other genes used in the evaluation of EAC exposure

A wide range of other genes representing central nodes in endocrine networks have been employed to measure exposure to environmental oestrogens, androgens, thyroid disruptors and other EACs in fish. These genes include those coding for ERs (*esr1*, *esr2a* and *esr2b*, formerly known as *ERα*, *ERβ* and *ERγ*; e.g., Andreassen *et al.*, 2005; Filby and Tyler, 2005; Greytak and Callard, 2007; Meucci and Arukwe, 2006;

Sabo-Attwood *et al.*, 2004), androgen receptors (ARs) (*ar*; *e.g.*, Filby *et al.*, 2006, 2007a, 2007b; Seo *et al.*, 2006; Sone *et al.*, 2005), thyroid hormone receptors (TRs) (*thra* and *thrb*; *e.g.*, Filby *et al.*, 2006, 2007a, 2007b), various enzymes involved with sex hormone biosynthesis (including the *cytochrome P450*, family *19* genes (*cyp19a1* and *cyp19a2*)), which code for the two P450 aromatase enzymes that mediate the conversion of testosterone (T) into oestrogen (*e.g.*, Halm *et al.*, 2002; Hinfray *et al.*, 2006; Meucci and Arukwe, 2006; Min *et al.*, 2003) and *cytochrome P450*, family *17* (*cyp17*) (*e.g.*, Filby *et al.*, 2006, 2007a, 2007b; Govoroun *et al.*, 2001), gonadotrophins (*follicle stimulating hormone* (*fshb*) and *luteinising hormone* (*lhb*), which code for hormones synthesised in the pituitary gland that control gonad development and sex steroid synthesis; *e.g.*, Harries *et al.*, 2001) and genes coding for hormones centrally involved in sexual differentiation, including *anti-Müllerian hormone* (*amh*) and *doublesex and mab-3-related transcription factor 1* (*dmrt1*; *e.g.*, Filby *et al.*, 2006, 2007a, 2007b; Schulz *et al.*, 2007). Schulz *et al.* (2007) found that, in zebrafish, alterations in the expression of *amh* and *dmrt1* during sexual differentiation (as a consequence of exposure to EE_2) signalled for subsequent feminisation of the gonadal phenotype. A detailed review on molecular biomarkers applied to studies on endocrine disruption in aquatic organisms generally is provided in Rotchell and Ostrander (2003).

cyp mRNAs

CYP enzymes are widely used as biomarkers of exposure for a wide range of compounds. The most commonly employed are the *cyp1a* genes, which encode the phase I monooxygenase enzyme EROD and are especially responsive to halogenated aromatic hydrocarbons (*e.g.*, 2,3,7,8-tetrachlorodibenzo-*p*-dioxin (TCDD)). Their expression is controlled by the AhR, which is a ligand-activated transcription factor (reviewed in Hahn, 1998). The natural ligand and normal physiological function of this transcription factor however are, as yet, unknown. *cyp1a* expression has been shown to be induced following exposure to 3-methylcholanthrene in rainbow trout and red sea bream (Berndtson and Chen, 1994; Mizukami *et al.*, 1994). Recently, *cyp1a* expression induction was analysed in Atlantic salmon following treatment with β-naphthoflavone and this revealed tissue-specific patterns of gene expression (Rees *et al.*, 2003). In the channel catfish, it has been shown that ligands such as B[*a*]P, TCDD and PCBs 77, 126 and 169 that activate AhR also induce the expression of the *cyp1b* gene (Willett *et al.*, 2006). Hoffmann *et al.* (2006), investigating the effect of exposure to B[*a*]P on the expression of a suite of genes for CYP enzymes, including *cyp1a1*, *cyp11a1*, *cyp17*, *cyp19a1* and *cyp19a2*, in the zebrafish using real-time RT-PCR, however, found altered (increased) expression only for *cyp1a1* and *cyp19a2* (in the head at an exposure concentration of $3.0\,\mu g\,l^{-1}$ B[*a*]P). Many cytochrome P450s are under hormonal control and thus their expression levels can be affected by exposure to EACs. As an example, treatment of rainbow trout with E_2 and T has been show to alter levels of *cyp2k1*, *cyp2m1* and *cyp3a27* (Buhler *et al.*, 2000).

mt mRNA

The *mt* genes are highly responsive to heavy metals and were one of the first molecular biomarkers established in fish (in rainbow trout; Bonham and Gedamu, 1984; Zafarullah *et al.*, 1989). The responsiveness of the *mt* gene to metal exposure is based on the fact that its promoter region contains multiple copies of metal-responsive elements (MREs) that are activated by the binding of metal transcription factors (MTFs), thus activating the *mt* gene. Cheung *et al.* (2004) found that a variety of metal ions, including Cu^{2+}, Cd^{2+}, Hg^{2+}, Ni^{2+}, Pb^{2+} and Zn^{2+}, all induced *mt* mRNA in medaka, but did so at different levels in the different tissues studied (liver, kidney and gill). The most studied fish *mt* promoter regions are those of the rainbow trout genes *mt-a* and *mt-b* (Olsson *et al.*, 1995; Samson and Gedamu, 1995; Zafarullah *et al.*, 1988). In rainbow trout, Zn and Ag have both been demonstrated to activate the *mt-a* promoter (Mayer *et al.*, 2003). In the common carp, Cu^{2+}, Hg^{2+}, Ni^{2+} and Pb^{2+} were shown to similarly induce the *mt-1* gene promoter, but Zn^{2+} was four to five times more potent (Chan *et al.*, 2004). The higher inducibility of *mt* by Zn is related to the function of MT as a storage pool for Zn, which is needed as a cofactor for VTG synthesis (Werner *et al.*, 2003).

Despite being considered as specific biomarkers, MTs can also be induced by hydrogen peroxide (H_2O_2) for instance (Olsson, 1996), which is caused by the presence of further regulatory elements in the *mt* promoter region including the antioxidant response element (ARE) activated by free radicals (Olsson *et al.*, 1995).

Antioxidants

A first attempt to evaluate the possible application of genes encoding antioxidant stress proteins as molecular biomarkers in fish was published recently by Hansen *et al.* (2006). In that study, brown trout (*Salmo trutta*,

Linnaeus) were exposed to different metals in the field and analyses conducted on the expression levels of genes for MT, SOD, catalase, GPx and GR and on MT protein and SOD and catalase enzyme activities. Reasonable correlations were found between levels of the *mt* mRNA and protein, but only poor correlations were found between specific transcripts and enzyme activities and considerably more work is required before transcripts for antioxidant stress proteins can be used with any confidence as molecular biomarkers in fish toxicology.

Multiple molecular biomarker approaches (employing real-time RT-PCR) have been used effectively to identify multiple contaminant exposure in caged fish placed in complex effluents (*e.g.*, measuring *cyp1a*, *mt* and *vtg* expression in caged rainbow trout as biomarkers to signal for exposure to PAHs, heavy metals and oestrogens, respectively; Roberts *et al.*, 2005) and to assess for exposure to mixtures of contaminants in wild populations of fish (*e.g.*, measuring *mt* gene expression, hepatic EROD activity and plasma VTG in European flounders living in UK estuaries to identify exposure to heavy metals, PAHs and oestrogenic compounds, respectively; George *et al.*, 2004). Multiple molecular biomarkers have additionally been applied in exposures conducted for understanding the MOAs of specific chemicals (see section below).

Mechanisms of toxicant action and biomarker discovery using genomics

Traditionally, toxicological assessments with fish have been heavily reliant on analysis of apical (whole fish) endpoints, such as growth, development and reproduction, in determining chemical effects. Undoubtedly, these methods have played an invaluable role in the chemical assessment process but, generally, they have not given insight into the causative mechanism(s) responsible for the effect. Toxicogenomics, in addition to providing useful biomarkers of chemical exposure or effect, can delineate the mechanisms of chemical toxicity in fish and potentially identify new biomarkers of exposure or biological effect. Knowledge of the MOA of a toxicant is especially useful in areas of toxicology for which there is an explicit regulatory focus on chemicals with specific MOAs, for example in assessing chemicals for endocrine disrupting capability (Hutchinson *et al.*, 2006). Moreover, a thorough understanding of MOA is of considerable value for assessments of the comparative responses of different species to chemicals (including their relative sensitivities) and for extrapolating effects across families of chemicals.

Until recently, the majority of studies implementing toxicogenomics to determine the MOAs of chemicals in fish have been conducted on a relatively small scale using the so-called targeted approaches, as for the employment of molecular biomarkers to assess exposure. In such approaches, specific genes have been chosen for study based on their biological interest and known associated functions. Such studies have tended to use RT-PCR/real-time RT-PCR-based methodologies to measure changes in the expression of single genes or small suites of genes. Examples of where this approach has been applied to provide mechanistic insights into the effect of chemicals in fish include for establishing the effects of arsenic on the stress response in killifish (*Fundulus heteroclitus* Linnaeus) (Bears *et al.*, 2006), determining the effect of PCBs on stress-immune system interactions in rainbow trout (Quabius *et al.*, 2005), understanding the inhibitory effects of B[*a*]P on reproduction in zebrafish (Hoffmann and Oris, 2006), assessing the pituitary-mediated mechanisms of environmental oestrogen action in various salmonids (Elango *et al.*, 2006; Harries *et al.*, 2001; Yadetie and Male, 2002), determining the pathways of methylmercury action in zebrafish (Gonzalez *et al.*, 2005) and in the discovery of novel pathways of toxicity of the DDT metabolite, *p,p'*-DDE, in Atlantic salmon (Mortensen and Arukwe, 2006). PCR-based studies with multiple genes have proved especially successful in elucidating specific points in the steroidogenic pathway at which different EACs act to disrupt steroidogenesis in fish (*e.g.*, Arukwe, 2005; Baron *et al.*, 2005a; Buhler *et al.*, 2000; Govoroun *et al.*, 2001; Halm *et al.*, 2002; Kortner and Arukwe, 2007; Lee *et al.*, 2006; Lyssimachou *et al.*, 2006; Meucci and Arukwe, 2006, Yokota *et al.*, 2005).

Recent research has employed expanded PCR-based methodologies with multiple genes to highlight the highly complex nature of some chemical effect pathways in fish. As an example, Garcia-Reyero *et al.* (2006) used multiple (10 gene) real-time RT-PCR to investigate the possible effect mechanisms by which the organochlorine pesticides dieldrin and *p,p'*-DDE alter endogenous hormone levels in largemouth bass. The MOAs they considered were (1) direct interaction with soluble sex steroid hormone receptors (via measurements of *esr1*, *ar* and *vtg* mRNAs); (2) alteration of processes involved in sex steroid biosynthesis (via measurements of *cyp19* and *steroidogenic acute regulatory protein* (*star*) gene expression) and (3) alteration in processes involved in sex steroid metabolism (via measurements of *cyp1a* and *cyp3a* gene expression). They found both pesticides studied affected the expression of genes involved in all three MOAs, suggesting that both chemicals cause their biological effects via multiple mechanisms.

Work in our laboratory has similarly demonstrated multiple mechanisms of effect for environmental oestrogens (E_2 and EE_2) and an

94

antiandrogen (flutamide) in the fathead minnow via multiple-gene real-time RT-PCR methodologies (Filby *et al.*, 2006, 2007a, 2007b). For example, exposure to $35\,\mathrm{ng}\,\mathrm{l}^{-1}$ E_2 for 14 days induced classic oestrogen biomarker responses (hepatic *esr1* mRNA and *vtg* mRNA) and affected the reproductive axis, feminising 'male' steroidogenic enzyme expression profiles and suppressing genes involved in testis differentiation (Filby *et al.*, 2006). E_2 also triggered a cascade of responses for genes that mediate growth, development and thyroid and interrenal function (*growth hormone* (*gh*), *growth hormone receptor* (*ghr*), *insulin-like growth factor 1* (*igf1*), *thyroid hormone receptors alpha* (*thra*) and *beta* (*thrb*) and *glucocorticoid receptor* (*gr*)), across six different tissues, with potential consequences for the normal functioning of many physiological processes, in addition to reproduction. Moreover, the molecular responses to E_2 were further complicated by the fact that most genes showed differential responses between tissues and sexes. A further study in the rainbow trout has shown that several xenobiotics acting via the AhR (α-naphthoflavone and β-naphthoflavone) impair cortisol biosynthesis and, as a result, they also affect the endocrine system by attenuating the expression of *star* and *cyp11a* (which encodes the cytochrome P450 side-chain cleavage enzyme), the rate limiting steps in steroidogenesis (Aluru and Vijayan, 2006; Aluru *et al.*, 2005). These findings demonstrate the importance of developing an integrative understanding of endocrine interactions for unravelling mechanisms of toxicant action in fish and for predicting the health consequences likely to arise as a result of the exposure to these chemicals.

Recent studies investigating reproductive function in fish have expanded the use of real-time RT-PCR further and measured the simultaneous responses of up to 100 genes (Aegerter *et al.*, 2005; Baron *et al.*, 2005b; Bobe *et al.*, 2004) further highlighting the potential for targeted approaches in fish toxicogenomic studies in the future. This scale of analysis of gene expression using the real-time RT-PCR approach is comparable to a macroarray approach in terms of the number of genes analysed, but has greater sensitivity, dynamic range and precision (Chuaqui *et al.*, 2002; Dallas *et al.*, 2005). Given our limited knowledge on chemical effect pathways, targeted PCR approaches, however, have not yet been able to provide a comprehensive evaluation of the molecular pathways mediating the biological effects seen. While multiple PCR studies are likely to remain a valuable approach for the detailed study of specific gene regulatory circuits, the high complexity of chemical effects pathways demonstrated above has highlighted the requirement for larger scale gene analyses to provide a more comprehensive understanding of the mechanisms of chemical toxicity in fish.

95

Such approaches include differential display-PCR (DD-PCR) or subtractive hybridisation methodologies and various gene array methodologies. DD-PCR or subtractive hybridisation methodologies offer good potential for detecting novel genes involved in toxicant effect pathways and identifying new molecular biomarkers, but they have not, as yet, been widely applied in fish toxicogenomics research. Examples in which DD-PCR has been applied to look at alterations in gene expression following chemical exposure include investigations of the effects of natural and anthropogenic environmental oestrogens and pulp and paper mill effluents in the sheepshead minnow (*Cyprinodon variegatus* Lacepède) (Denslow *et al.*, 2001a, 2001b) and largemouth bass (Bowman *et al.*, 2002; Denslow *et al.*, 2004), determining genes differentially expressed in pyrene-exposed (vs. control) killifish and between control and anthracene-exposed fish (Peterson and Bain, 2004), in wild fish from uncontaminated and creosote-contaminated sites (Roling *et al.*, 2004) and between wild fish from uncontaminated and toxic sediment-contaminated sites (Meyer *et al.*, 2005). Subtractive hybridisation approaches have been applied in assessing the MOAs of oestrogens and xenoestrogens in various fish species (Baldwin *et al.*, 2005; Brown *et al.*, 2004b; Pinto *et al.*, 2005), arsenic in juvenile mummichog (Gonzalez *et al.*, 2006), a cocktail of pesticides in European flounder (Marchand *et al.*, 2006) and TCDD in medaka (Volz *et al.*, 2005). Recently, DD-PCR was used to identify two genes involved in the development of intersex in medaka as a consequence of exposure to an environmental oestrogen (T. Iguchi, personal communication).

The application of gene arrays in mechanistic toxicology has far exceeded that for DD-PCR or subtractive hybridisation methodologies and this fact is demonstrated by the exponential growth in publications using this technique since its inception in 1995 (Ju *et al.*, 2007; reviewed in Lettieri, 2006). Array analysis has now been used in fish to elucidate MOAs of various classes of toxicants. Most notable are studies on environmental oestrogens and other EACs (Brown *et al.*, 2004b, 2004c; Gunnarsson *et al.*, 2007; Hoffmann *et al.*, 2006; Hook *et al.*, 2006a, 2006b; Hoyt *et al.*, 2003; Kishi *et al.*, 2006; Knoebl *et al.*, 2006; Kreiling *et al.*, 2007; Larkin *et al.*, 2002, 2003a, 2003b; Martyniuk *et al.*, 2006, 2007; Moens *et al.*, 2006; Santos *et al.*, 2007a, 2007b; Williams *et al.*, 2007). Arrays, however, have also been used to detect gene expression responses to other compounds, including the metals cadmium (Koskinen *et al.*, 2004; Reynders *et al.*, 2006; Sheader *et al.*, 2006; Williams *et al.*, 2006), chromium (Hook *et al.*, 2006a), methylmercury (Klaper *et al.*, 2006) and zinc (Hogstrand *et al.*, 2002), PAHs (Hook *et al.*, 2006a; Krasnov *et al.*, 2005), dioxins (Handley-Goldstone *et al.*, 2005), flame retardants

(Hook *et al.*, 2006a), the hepatotoxin carbon tetrachloride (Koskinen *et al.*, 2004; Krasnov *et al.*, 2004), the human neuropharmaceuticals mianserin and chlorpromazine (van der Ven *et al.*, 2005, 2006a, 2006b), the industrial pollutant 2,4-dinitrotoluene (2,4-DNT) (Wintz *et al.*, 2006), the aquatic herbicide diquat (Hook *et al.*, 2006a) and the hepatic tumour promoters indole-3-carbinol and aflatoxin B1 (Tilton *et al.*, 2005, 2006).

Genetic fingerprints and profiles for chemicals in fish

Genomics has potential for predicting the MOA of new or unknown toxicants. It has been proposed that chemicals that act through distinct MOAs will induce unique and diagnostic gene expression 'fingerprints/ signatures' (in which the genes are not identified) or 'profiles' (in which the genes are identified), even although they may induce the same phenotypic (apical) effects. In contrast, chemicals that act through the same MOA will share extensive commonalities in their gene expression fingerprints/ profiles (Nuwaysir *et al.*, 1999). By comparing the gene expression profile of an unknown toxicant to reference toxicants with known MOAs (in a technique known as comparative transcript profiling), there is, therefore, potential for the ability to identify and classify the unknown chemical according to its MOA.

While the majority of data supporting this hypothesis come from mammalian toxicology (*e.g.*, Burczynski *et al.*, 2000; Hamadeh *et al.*, 2002; Waring *et al.*, 2001), a number of studies have now been published using fish models that further demonstrate this potential. One example comes from our laboratory, where we profiled the expression responses of a suite of 22 genes involved in reproduction, growth and development in the liver and gonad in adult male and female fathead minnow exposed to flutamide (as a model antiandrogen) and EE_2 (as a model oestrogen) (Filby *et al.*, 2007a). Flutamide ($320\,\mu g\,l^{-1}$) and EE_2 ($10\,ng\,l^{-1}$) both produced phenotypic effects indicative of feminisation (induction of plasma VTG, reduced gonadal growth and reduced secondary sex charac- ters in males) but, for the genes studied, flutamide and EE_2 produced distinct expression profiles (Fig. 1), suggesting that they largely operate via distinct molecular mechanisms. As an illustration of this, in liver exposure to EE_2 (but not flutamide) upregulated *esr1* expression in liver of males and females, whereas, at least in males, flutamide increased the expression of *esr2a* and *esr2b*. In the testis, flutamide upregulated the expression of genes coding for enzymes involved in androgen biosynthesis (*cyp17* and *11β-hydroxysteroid dehydrogenase* (*hsd11b*)) implying an inhibitory action on androgen negative feedback pathways. EE_2, in contrast, inhibited the expression of enzymes involved in androgen biosynthesis (*cyp17* and *17β-hydroxysteroid dehydrogenase* (*hsd17b*)). There were also some commonalities in the molecular mechanisms of flutamide and EE_2 action, including the downregulation of gonadal sex steroid receptor expression (gonadal *ar* and ovarian *esr1*), increased expression of genes coding for oestrogen-producing enzymes (*cyp19a* in testis and *cyp19b* in ovary), decreased expression of genes involved in testis differentiation (*amh* and *dmrt1*) and decreased expression of hepatic genes that mediate wider physiological processes such as somatic growth (*ghr*, *igf1*, *igf1r* and, although in females only, *thra* and *thrb*).

Larkin *et al.* (2003a), using a low-density gene array containing 30 oestrogen-responsive genes, demonstrated that EACs that mimic oestrogens exhibit diagnostic genetic fingerprints. In this study, sheepshead minnow were exposed to EE_2 ($109 \, ng \, l^{-1}$) and diethylstilbestrol (DES; $100 \, ng \, l^{-1}$) for 4 days, to E_2 ($65.14 \, ng \, l^{-1}$) and NP ($11.81 \, \mu g \, l^{-1}$) for 5 days and to methoxychlor ($5.59 \, \mu g \, l^{-1}$) and endosulfan ($590.3 \, ng \, l^{-1}$) for 13 days and, in the liver, six genes including *vtg1*, *vtg2* and *esr1* were similarly upregulated (apart from for endosulfan) and three genes downregulated (*transferrin* (*tf*), *beta actin* (*bactin*) and α_1-*microglobulin/bikunin precursor protein* (*ambp*)). Studies by Blum *et al.* (2004) investigating hepatic responses in largemouth bass to two natural androgens, dihydrotestosterone (DHT; $62.5 \, \mu g \, g^{-1}$ body weight) and 11-ketotestosterone (11-KT; $2 \, \mu g \, g^{-1}$) using a cDNA array containing 296 genes (164 of which were known to be androgen-responsive) found that several genes were induced by both treatments (including *vtg2*, *spermine-spermidine-N1-acetyltransferase* (*ssat*), *warm water acclimatisation protein*, *rhamnose binding lectin* and two ESTs named seasonal (SNL) 64 and P4_H07), some genes were downregulated by both androgens (including *ldl receptor*, *rxr interacting protein* and *vtg receptor*, and two ESTs called SNL51 and GP3–11C), but found some genes were differentially affected by the two different androgens. For example, only DHT-enhanced expression of *atpase 6*, *tata box binding protein* (*tbp*) and the ESTs, SNL56, SNL62, P4_C04 and 97–8 and downregulated the expression of *apolipoprotein e*, *glutathione peroxidase I-II*, *haptoglobin*, *igf1* and *hepcidin*, whereas 11-KT stimulated the expression of *solute carrier protein family member 25a5*, *pituitary tumour transforming protein* and the ESTs DHT64, SNLF15, P4_E12 and downregulated *cystatin*, *ribosomal protein Sa* and the ESTs SNL55, SNL11, P4_E09 and P4_F11. The differences in the effects on gene transcription of these different androgens may reflect differences in their endogenous activity (they have specific functions in fish) or may relate to differences in binding affinities of the two fish ARs.

98

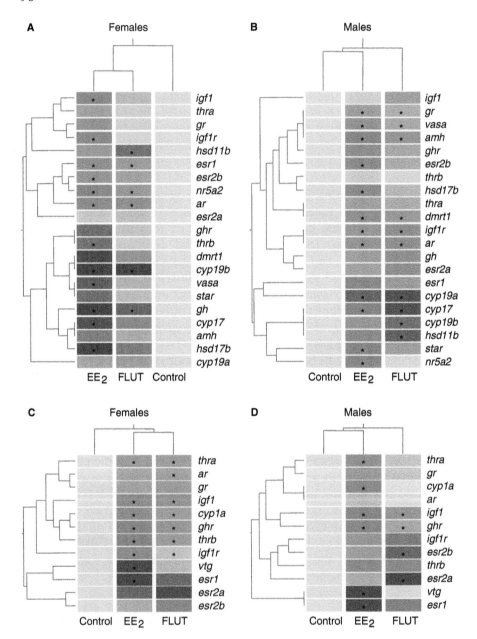

On a more expansive scale, Hook *et al.* (2006a) demonstrated that expression profiles generated using a commercially available 16K cDNA array were sufficiently diverse to distinguish between a suite of model toxicants with varying modes of action in rainbow trout. The fish were exposed to EE_2 ($50\,\mathrm{ng}\,\mathrm{l}^{-1}$), the model androgen TRB ($1\,\mathrm{\mu g}\,\mathrm{l}^{-1}$) and the model PAH B[*a*]P ($1\,\mathrm{\mu g}\,\mathrm{l}^{-1}$), for a period of 7 days, to the thyroid active flame retardant 2,2,4,4'-tetrabromodiphenyl ether (BDE-47), given orally (a dose of approximately $500\,\mathrm{\mu g}\,\mathrm{kg}^{-1}$) and the oxidative stressor diquat and the metal chromium VI, injected intra-peritoneally (at doses of 500 and $25\,\mathrm{\mu g}\,\mathrm{kg}^{-1}$, respectively). The gene expression profile produced by each toxicant was unique, but there was a degree of overlap that correlated with function. For example, the EACs (EE_2 and TRB) had more similar expression patterns to each other compared with the other contaminant classes. Hierarchical clustering of gene expression data showed that unique expression profiles generated for each compound were non-random and that compounds with similar function clustered together and separated from those compounds with disparate function.

Fig. 1. Comparison of the effects of exposure to the oestrogen ethinyloestradiol (EE_2; at $10\,\mathrm{ng}\,\mathrm{l}^{-1}$) or the antiandrogen flutamide (FLUT; at $320\,\mathrm{\mu g}\,\mathrm{l}^{-1}$) for 21 days on the gene expression profiles of the gonad (A and B) and liver (C and D) in adult male and female fathead minnow (*Pimephales promelas*). Gene expression data were generated through real-time RT-PCR. Clustering of both genes and conditions were performed using Pearson's correlation as a similarity measure. The gene trees are displayed horizontally and the condition trees are displayed vertically. Yellow colouration represents a relative expression of 1 (when compared with mean expression of the control group), green colouration represents downregulation and red colouration represents upregulation of gene expression relative to the control group. Each treatment group consisted of eight male and eight female fish and each fish was analysed in triplicate. Statistically significant differences in gene expression between control and EE_2-treated fish and control and flutamide-treated fish, are denoted by an asterix ($P < 0.05$; one-way ANOVA, followed by Dunn's post hoc test). Gene acronyms are as follows: *esr1, oestrogen receptor 1; esr2a, oestrogen receptor 2a; esr2b, oestrogen receptor 2b; ar, androgen receptor; gh, growth hormone; ghr, growth hormone receptor; igf1, insulin-like growth factor 1; igf1r, insulin-like growth factor 1 receptor; thra, thyroid hormone receptor alpha; thrb, thyroid hormone receptor beta; gr, glucocorticoid receptor; cyp17, cytochrome P450 17; cyp19a, cytochrome P450 19a; cyp19b, cytochrome P450 19b; star, steroidogenic acute regulatory protein; hsd11b, hydroxysteroid 11-beta dehydrogenase; hsd17b, hydroxysteroid 17-beta dehydrogenase; vtg, vitellogenin; amh, antiMüllerian hormone; vasa, vasa homologue; dmrt1, doublesex and mab-3-related transcription factor 1; nr5a2, nuclear receptor subfamily 5 group A member 2.* (See color figure 3.1 in color plate section).

Moreover, in another recent study, Moens *et al.* (2006), using a 960 gene custom cDNA microarray consisting of endocrine-related genes, demonstrated unique liver gene expression patterns for 14 OECD-recommended reference EACs, which are known to act via different MOAs (the oestrogens E_2, EE_2, 4-NP and BPA, the antioestrogen tamoxifen, the androgens methyltestosterone and 11-KT, the antiandrogens flutamide and vinclozolin, the cortisol-disrupting chemicals, hydrocortisone and $CuCl_2$, the thyroid-disrupting chemicals, propylthiouracil and tri-iodothyronine/thyroxine and dibutyl phthalate, a compound with an unknown MOA). Each individual compound produced its own unique expression pattern on the array and a small set of 12 genes was identified as being able to discriminate between the different compounds. Although it was noted that some of the compounds preferentially clustered with compounds outside their presumed MOA, some marked expression similarities were observed between groups of analogous compounds.

Chemical mixtures

From an ecological point of view, understanding how mixtures of chemicals disrupt biological homeostasis is of crucial importance in evaluating the risks associated with chemical discharges and in implementing appropriate safety margins and protective measures. Toxicogenomics, with the appropriate validation, has the potential to help us understand mixture effects. Laboratory studies of this nature are starting to emerge, but, caused by the complexity of the experimental designs required for the initial validation of this approach and the bioinformatic challenges posed by the complex data sets, this area of toxicogenomics has been slow to develop. Furthermore, for studies on mixture effects in wild fish inhabiting contaminated environments, the difficulties in interpreting molecular responses and array data are further enhanced by uncertainties in the life history and physiological status of the fish prior to sampling.

Laboratory studies to investigate the mechanistic basis through which complex mixtures effects are mediated in fish have been initiated by Denslow *et al.* (2004) who exposed largemouth bass to a paper mill effluent and used DD-PCR to identify alterations in gene expression. Differentially expressed genes identified by this study included *cyp1a* in males and females (upregulated in effluent-exposed fish) and a generalised downregulation of genes normally expressed during the reproductive season (including *vtg* and *zrp*s) in females, which was consistent with the antioestrogenic nature of the effluent. In our laboratory, we have recently applied real-time RT-PCR with a suite of genes to show that exposure of fathead minnows to UK WwTWs effluents induces patterns of gene expression that are diagnostic for some chemicals contained within those effluents (*e.g.*, environmental oestrogens; Filby *et al.*, 2007b). The responses observed, however, differed between sexes and between tissues.

Although studies using genomics to investigate the effects of real-world chemical mixtures have been initiated, the potential of genomics for understanding the interactive effects of complex mixtures of chemicals has yet to be rigorously tested and it may, in fact, be complex beyond the current capability of bioinformatic analysis (Miracle *et al.*, 2003). As an example, the possible additive, synergistic or antagonistic effects resulting from simultaneous exposure to multiple toxicants may result in the activation of unique gene pathways compared with those observed by exposure to the individual compounds. For example, in a study with rainbow trout primary hepatocytes exposed *in vitro* for 24 hours to the oestrogen EE_2, the AhR agonist TCDD, the oxidative stressor paraquat and the mutagen 4-nitroquinoline-1-oxide, singly or in combination, gene expression profiles generated using a 16K salmonid microarray for the synthetic mixture revealed combination effects that were not predicted by results for the individual chemicals alone and, on average, led to a loss of approximately 60% of the transcriptomic signature found for single chemical exposures (Finne *et al.*, 2007). What is required to progress our understanding of the likelihood of being able to apply genomics in the diagnosis of chemical causation of biological effects for organisms exposed to real-world mixtures and to understand the interactive effects of chemicals on biological function, is a series of carefully conducted experiments that investigate transcriptome responses to a range of chemicals individually and in simple combinations (as in Finne *et al.*, 2007) and to link these responses with measurements of well-characterised phenotypic endpoints. In this way, it may be possible to determine the accuracy of the transcriptomic predictions and to develop a mechanistic understanding of the molecular basis of mixture effects.

Phenotypic anchoring

The concept of phenotypic anchoring first appeared in the literature in the context of toxicological studies in 2003 (Paules, 2003; Waters *et al.*, 2003). Since then, this approach has been used in many toxicological studies in mammals as a technique for understanding chemical action (Amin *et al.*, 2004; Gant *et al.*, 2003; Gottschalg *et al.*, 2006; Hamadeh *et al.*, 2004; Heinloth *et al.*, 2004; Luo *et al.*, 2005; Moggs *et al.*, 2004;

Powell *et al.*, 2006). In these studies, transcriptomic responses were related to conventional and well-established toxicological endpoints, thus anchoring the MOAs of chemicals (as determined by the gene expression profiles) to the subsequent associated biological effect. For example, Moggs *et al.* (2004) studied the temporal associations between the transcriptional programme induced by E_2 and the accompanying histological changes in the uterus of the mouse, to identify groups of genes that drive specific phenotypic events in uterine development (the uterotropic assay in mice and rats is used as a model system for studying the effects of oestrogens in mammals). The establishment of relationships between gene expression signatures or profiles of exposure with phenotypic measures of exposure effect is fundamental for the acceptance of transcriptomics as a technique capable of predicting both MOA of chemicals and in diagnosing specific biological effects.

To date, there is only a handful of examples of studies in fish that have made any concerted effort to link changes in gene expression with effects at higher levels of biological organisation. The reasons why studies employing phenotypic anchoring have been less forthcoming for fish are probably caused by the fact that the phenotypic responses to many classes of toxicants in fish are poorly characterised when compared with mammals and because the resources required for this work are often extensive and are generally less available for studies in fish compared with those for mammalian studies. Examples of studies in which efforts have been made to directly link genome responses with phenotypic effects in fish include the studies of Klaper *et al.* (2006), who used a combination of genomic techniques (macroarray, suppressive subtractive hybridisation and real-time RT-PCR) to identify alterations in gene expression associated with effects of the neurotoxic metal, methylmercury, on steroid levels and reproduction in fathead minnow. Genes identified as differentially expressed by the macroarray included those associated with egg fertilisation and development, sugar metabolism, apoptosis and electron transport and processes directly related to reproduction. van der Ven *et al.* (2006a) recently employed gene array analysis on the brain in parallel with investigations of effects on egg production, fertilisation and hatching and, on a smaller scale, Hoffmann and Oris (2006) related changes in the expression of key steroidogenic enzyme genes (*20β-hydroxysteroid dehydrogenase* (*hsd20b*) and *cyp19a2*) to reduced fecundity and ovarian somatic index in zebrafish exposed to B[*a*]P. In some of our very recent studies, we investigated responses in the gonadal transcriptome together with assessments on gamete quality and reproductive performance in colonies of zebrafish exposed to EE_2 and identified novel molecular mechanisms leading to the observed decrease in egg production and sperm quality (Santos *et al.*, 2007b). Responses to chemical stressors will invoke both toxic and adaptive processes, depending on the chemical, its concentration and the duration of the exposure and this adds a further degree of complexity to the interpretation of the alterations in gene expression profiles following exposure. Comprehensive studies addressing both temporally-related and concentration/dose-related effects of exposure, for a range of chemical types and in which individual variation is taken into account, are much needed to separate the toxic responses (consequently deleterious for the fish) from the adaptive responses (which may not affect the fish). In the current literature, there is no example of study in fish in which the question of toxic vs. adaptive responses has been addressed. This, however, is fundamental for the interpretation of gene expression data sets to generate useful data for future use in the legislative protection of the environment from the impacts of pollutants.

Challenges in the interpretation of toxicogenomic data for fish

To be able to extract biologically relevant information from toxicoge-nomic data, the data first require appropriate normalisation. For gene expression data generated through real-time RT-PCR and small-scale gene arrays (macroarrays), gene responses are typically normalised against genes that are not affected by the toxicant treatment (the so-called internal control or 'housekeeping' genes). This normalisation is performed to adjust for differences in variables such as starting material and hybri-disation efficiency between samples. It is crucial to choose the control genes carefully, ensuring that these genes are not affected by the treatment and this necessity has been frequently overlooked in many of the toxicogenomic studies conducted (and published) in fish. In fact, there is evidence in the literature that many 'housekeeping' genes that have been commonly used as internal controls in toxicology studies are affected by exposure to a wide range of chemicals in fish (*e.g.*, Brown *et al.*, 2004b; Hoffmann *et al.*, 2006; Hook *et al.*, 2006a; Hoyt *et al.*, 2003; Larkin *et al.*, 2003a; Winzer *et al.*, 2002; Yadetie and Male, 2002). A recent study by Filby and Tyler (2007) set out to assess the validity of eight functionally distinct 'housekeeping' genes that have been widely employed as internal controls in fish toxicogenomic research (18S ribosomal RNA (18S rRNA), *ribosomal protein l8* (*rpl8*), *elongation factor 1 alpha* (*ef1a*), *glucose-6-phosphate dehydrogenase* (*g6pd*), *bactin*, *glyceraldehyde-3-phosphate dehydrogenase* (*gapdh*), hypoxanthine *phosphoribosyltransferase 1* (*hprt1*) and (*tbp*)) for use as internal controls in studies on environmental oestrogens in liver and gonad tissues of fathead minnow. Using real-time RT-PCR, this study demonstrated oestrogen-downregulation of *ef1a*, *g6pd*, *bactin* and *gapdh* in the liver and of *bactin* and *gapdh* in the gonad, highlighting the fact that these genes are probably unsuitable for use in normalisation in studies of this nature. In fact, the effects of the environmental oestrogen EE_2 on two biomarker genes, *vtg* and *cyp1a*, in liver were overestimated when *ef1a*, *g6pd*, *bactin* and *gapdh* were used to normalise the expression data (Fig. 2). The authors concluded that, for analyses on liver and gonad tissues, 18S rRNA, *rpl8*, *hprt1* and *tbp* were suitable for use in normalisation and, for analyses on gonad, *ef1a* and *g6pd* were additionally suitable. In a separate study on Atlantic salmon, the expression levels of *bactin*, *beta tubulin*, 18S rRNA and *ef1a* were additionally found to be modulated by exposure to other classes of toxicants, including the thyroid hormone thyroxine and the EDC *p,p'*-DDE, administered both singly and in combination (Arukwe, 2006). These studies clearly highlighted the fact that prevalidation of control genes is critical for accurate assessments of the expression of target genes in exposures of fish to wide ranging classes of toxicants.

Normalisation of gene array data is especially complex and the number of normalisation methods available is steadily increasing because of the importance of this step for robust analysis and subsequent interpretation of gene array data (reviewed in Allison *et al.*, 2006). The nature of the data sets obtained from microarray experiments, in which thousands of measurements are collected for each sample, allows the development of highly robust normalisation methodologies. *Lowess* normalisation methods have been particularly favoured in fish toxicogenomic experiments (Martyniuk *et al.*, 2006; van der Ven *et al.*, 2005; Williams *et al.*, 2006), but other methods have also been employed with good success (*e.g.*, *Limma* normalisation; Santos *et al.*, 2007a, 2007b). It should be noted that the mathematical criteria for application of these methods are based on the assumption that the expression of most genes for sub-lethal exposures to a chemical is expected to remain constant. This situation being the case, caution should be taken when employing these normalisation techniques to data sets based on smaller scale gene arrays and when probes are derived from lists of genes identified based on differential expression following exposure to similar toxicants. The inclusion and measurement of genes that remain unaffected by the chemical exposure, therefore, is important for the data analysis and allows for powerful normalisation of the data set, enhancing the likelihood that the biological responses to certain stressors are identified correctly. It is of fundamental

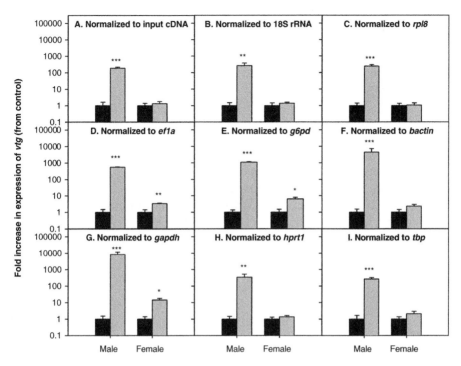

Fig. 2. Effect of normalising real-time RT-PCR expression data for a biomarker gene of oestrogen exposure, *vitellogenin* (*vtg*), with different 'housekeeping' genes. Male and female fathead minnow (*Pimephales promelas*) were exposed to the oestrogen 17α-ethinyloestradiol (EE₂; at $10\,\mathrm{ng\,l^{-1}}$) for 21 days. Hepatic *vtg* expression data are presented: (A) normalised to input cDNA (for details of methods, see Filby and Tyler, 2007), or normalised to the expression of different 'housekeeping' genes: (B) 18S ribosomal RNA (18S rRNA), (C) *ribosomal protein l8* (*rpl8*), (D) *elongation factor 1 alpha* (*ef1a*), (E) *glucose-6-phosphate dehydrogenase* (*g6pd*), (F) *beta actin* (*bactin*), (G) *glyceraldehyde-3-phosphate dehydrogenase* (*gapdh*), (H) *hypoxanthine phosphoribosyltransferase 1* (*hprt1*) and (I) *tata box binding protein* (*tbp*), measured in each sample. The results are represented as mean values ±SEM, expressed as fold increase in expression from the control (which was arbitrarily set as the 1 × expression of the gene) and plotted on a logarithmic scale. Each treatment group (control: shown in black; EE₂: shown in grey) consisted of eight male and eight female fish and each fish was analysed in triplicate. Statistically significant differences in gene expression between control and EE₂-treated fish for each sex are denoted as follows: * $P<0.05$, ** $P<0.01$, *** $P<0.001$ (student's *t*-test).

importance to carefully address the problem of normalisation of gene expression profiles for each individual data set and or each individual technique to generate reliable biological conclusions and avoid technical bias relating to the effects reported.

We compared the robustness of two normalisation scenarios (*Lowess* (GeneSpring, Agilent Technologies, USA) and *Limma* (Linear Models for Microarray Analysis; *R*, *Bioconductor* http://www.bioconductor.org) normalisations) for a data set in which a 17K zebrafish oligonucleotide microarray was employed to determine the differences in gene expression profiles between ovaries and testes in breeding zebrafish (data deposited in the National Center for Biotechnology Information (NCBI) Gene Expression Omnibus (GEO, http://www.ncbi.nlm.nih.gov/geo/) with the series record GSE6063 (Santos *et al.*, 2007a) (Fig. 3). We found that, although both normalisation scenarios produced acceptable results (Fig. 3A), the number of genes identified as differentially expressed between ovaries and testes (using *t*-test ($P < 0.05$) with Multiple Testing Correction (Benjamini and Hochberg False Discovery Rate) was higher when using *Limma* normalisation (2,940 genes)) compared with *Lowess* normalisation (2067 genes; statistical analyses conducted in GeneSpring) (Fig. 3B). In addition, the number of biological processes overrepresented among the differentially expressed genes was greater when the data set was normalised using *Limma* (14 biological processes) compared with Lowess (eight biological processes) normalisation (Fig. 3C). The results obtained clearly illustrate the difference in the identification of differentially expressed genes and loss of biological significance dependent on the choice of normalisation scenario. For this example, however, the overall biological message was remarkably similar for both data sets, as all biological processes found to be overrepresented when using *Lowess* normalisation were also overrepresented when using *Limma*. The major difference was that the *Limma* normalisation approach identified six additional biological processes overrepresented among the genes differentially expressed between ovaries and testes (*i.e.*, had a greater sensitivity).

The biological interpretation of results based on large-scale microarray analysis relies on the genes identified as being differentially expressed between experimental conditions and there are many mathematical models available to biologists to ascribe statistical significance. A number of studies have used fold change to identify differentially expressed genes (*e.g.*, van der Ven *et al.*, 2006b). Care should be taken when using this approach, however, as it does not account for variance in the data set. In addition, the presence of outliers would strongly influence the biological

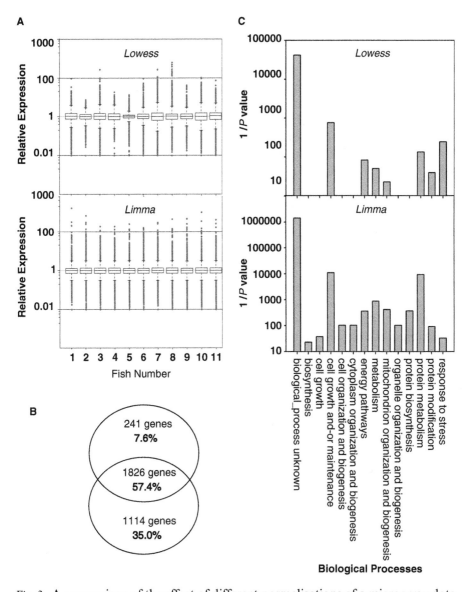

Fig. 3. A comparison of the effect of different normalisations of a microarray data set on the biological information derived. (A) Gene expression profiles of the gonads of five male and six female zebrafish (*Danio rerio*) normalised with *Lowess* (within GeneSpring) and *Limma* (within R). (B) Comparison of the biological processes overrepresented in the list of genes differentially expressed between males and females in the data set normalised using *Lowess* and *Limma*. (C) Numbers of genes differentially expressed between males and females in the data set normalised with *Lowess* (top circle) or *Limma* (bottom circle). Statistical comparisons were performed within GeneSpring using t-test (significant at $P < 0.05$) with multiple testing correction (Benjamini and Hochberg False Discovery Rate). Data were deposited in the National Center for Biotechnology Information (NCBI) Gene Expression Omnibus (GEO, http://www.ncbi.nlm.nih. gov/geo/) with the series record GSE6063.

conclusions obtained. Despite this fact, this method has proven to be effective and consistent in the identification of biological changes for microarray experiments (as demonstrated by a comparative study in rat toxicogenomics; Guo *et al.*, 2006). Statistical analyses in which variance is accounted for have been performed in studies by Hoffmann *et al.* (2006), Martyniuk *et al.* (2006), Santos *et al.* (2007b) and Williams *et al.* (2006) and this approach has been shown to produce results with high reliability in the identification of individual genes differentially regulated by the test chemical. False negatives are likely to occur, however, caused by the low sample number and often high variation between individuals in most fish toxicogenomic studies. In contrast, the likelihood of identifying false positives is reduced significantly when compared with the use of fold change. To increase the number of fish analysed, but keep the associated costs of conducting multiple hybridisations to a minimum, many authors have opted for pooling samples from several individuals. It has been argued that this approach may increase the reliability of the results compared with measurements with a very limited number of individuals. With the occurrence of outliers in most toxicological studies (individuals can react to chemical challenges with very different thresholds), however, this approach is not without its own problems and has the potential to draw misleading conclusions. For this reason, whenever possible, performing hybridisations on individuals, rather than on pooled samples, is likely to result in more reliable biological conclusions (reviewed in Allison *et al.*, 2006).

An issue under active debate concerns the validation of the array results using independent methods. The most commonly used method in fish toxicogenomics is real-time RT-PCR (*e.g.*, in the studies by Gunnarsson *et al.*, 2007; Kreiling *et al.*, 2007; Martyniuk *et al.*, 2006, 2007; Santos *et al.*, 2007a; van der Ven *et al.*, 2006b; Voelker *et al.*, 2007; Williams *et al.*, 2006). Two forms of validation of the microarray results are required. In the first instance, technical validation needs to be performed to assess the accuracy of the gene expression measurements determined (this aspect has been covered in most fish toxicogenomic studies) and to confirm the absence of false positives among the genes deemed to be differentially expressed. The second form of validation derives from the fact that the arrays are often conducted on a small subset of the experimental fish and they may, or may not, represent the average of the population. Independent biological validation of the gene expression measurements is often not considered in fish toxicogenomics, but is advisable, especially when measurements of gene expression are used as the only endpoint. An alternative approach in the validation process is the measurement of phenotypic endpoints in large sets of samples and including those individuals for which the gene expression measurements were conducted, to verify the consistency of the biological responses across a larger number of individuals. This approach can provide an indirect verification of how representative the transcriptomic results are in the context of the whole population.

Ironically, the biggest problem typically encountered in array studies to date concerns how to efficiently and effectively process the sheer volume of data generated into meaningful conclusions. In this respect, the development of Gene Ontology (GO) analysis (Ashburner *et al.*, 2000), that determines if biological processes and systems are altered by the study toxicant, rather than focusing on individual gene annotations, is proving invaluable in mechanistic toxicology studies using fish models. As an example, Hoffmann *et al.* (2006) found that EE_2 significantly affected the expression of 1,622 genes in liver tissue of zebrafish, but the subsequent use of GO analysis was able to delineate that these gene changes represented principally effects on hormone metabolism, vitamin A metabolism, steroid binding, sterol metabolism and cell growth (Hoffmann *et al.*, 2006). In a study in our laboratory on adult breeding zebrafish exposed to EE_2, and in which the expression of hundreds of gonadal genes were altered through the application of GO analysis, we were able to identify that the processes of cell cycle, mitochondrion organisation and biogenesis, and energy pathways were particularly affected by the treatment in females, whereas in males, protein modification, protein metabolism and growth were the biological processes most affected by the treatment. Our findings for the zebrafish demonstrate that toxicants such as EE_2 can not only invoke differences in gene responses between males and females but also sex-specific effects on whole biological processes (Santos *et al.*, 2007b). The approach using GO analyses on gene array data, however, is only valid when it is based on statistical analysis of the overrepresentation of a gene list compared with the proportion of gene ontologies in the list of gene targets contained in the array. This is because the number of genes associated with each GO can be highly variable and therefore categories containing large numbers of genes would appear represented in gene lists derived from chemically induced expression profiles by chance and without representing a true bias of the effects of the exposure towards that biological process. When appropriately applied, however, GO analysis is a powerful methodology to extract relevant information from toxicological data sets and the results obtained using this method have been shown to be comparable when analysing fish chemical exposure between different laboratories and across different microarray platforms (Guo *et al.*, 2006;

van der Ven *et al.*, 2006a, 2006b). GO analysis is likely to be used increasingly in determining the biological significance of the chemical effects observed and with the increased genomic resources available, it is likely to become a tool of choice in standardising the interpretation of microarray data sets across diverse platforms and species.

Pathway Analysis is starting to emerge as a useful tool to extract biological information from chemical exposures. Examples of the application of this in mammalian toxicological studies abound in the literature (*e.g.*, Currie *et al.*, 2005). This approach is restricted in fish, however, because of the limited annotation of the fish genomes and the very sparse information on pathways that has been developed specifically for fish. This situation, in turn, means that fish researchers are largely restricted to using the pathways constructed for other organisms and/or based on homology of the species-specific genes to genes in evolutionary distant species (a number of pathways, however, are available for zebrafish, *Fugu* and tetraodon, KEGG Pathway Maps; http://www.genome.jp/kegg/). The evolutionary distance between fish and the other organisms for which the gene pathways were originally described is likely to mean that the function of some genes and their interactions with neighbouring genes, may be unrelated, especially if a pathway/process is not highly conserved. Despite this finding, such an approach has real potential to help elucidate the gene networks involved with the toxicity of specific chemicals in fish. The generation of pathways specific for fish and their associated networks is crucial to further elucidate the mechanistic basis of chemical exposure effects and to link biomarker responses to predictive responses at the level of the whole organism. The generation of such networks in fish will also allow the identification of genes constituting crucial regulatory nodes in biochemical pathways. These genes, when identified to be differentially expressed following chemical exposure, constitute important candidates for further detailed studies using more robust techniques for the analysis of single genes such as RT-PCR and *in situ* hybridisation.

Good basic biology

A shortfall of many toxicogenomic studies in fish is the lack of information provided on the normal physiology of the test organisms. Most studies have simply included a small set of control samples from which comparisons to the exposed samples have been made. It is difficult to determine if the gene expression profiles measured for the control samples follow the expected patterns or if they are affected in some way by the specific conditions in which the test individuals were kept and sampled. This difficulty is due to the complexity of the gene responses measured, and often the limited number of samples analysed. Many factors may result in bias of data sets including the age, sex, developmental and physiological status, time of day when the samples were collected, photoperiod, temperature, nutritional status and disease status. Some toxicogenomic studies in fish have already demonstrated considerable variation of gene expression responses to toxicants between sexes, tissue types and developmental stages (*e.g.*, Filby *et al.*, 2006, 2007a, 2007b; Hook *et al.*, 2006b; Kishi *et al.*, 2006; Knoebl *et al.*, 2006; Krasnov *et al.*, 2005; Martyniuk *et al.*, 2007; Reynders *et al.*, 2006; van der Ven *et al.*, 2005). Studies on factors (including immune status, sex, genotype and age) that potentially contribute to differences in gene expression among individuals in killifish found that most interindividual variation was accounted for by differences in expression of genes involved with metabolism (reviewed by Whitehead and Crawford, 2005). Almost half (48%) of the 192 metabolic genes studied were differentially expressed among individuals within a population for any one tissue, with fold differences ranging from one- to five-fold and significance values ranging down to 10^{-7}. The need to understand what is normal in gene expression studies is especially relevant when analysing samples collected from wild fish populations, in which the history of the animal is not known.

Individual variation, especially when considering the effects of exposure to environmental chemicals on wild fish, can drive the number of individuals required in the test to numbers incompatible with the cost of genomic technologies. To reduce cost, many studies have opted to pooling samples (see section *Challenges in the interpretation of genomic data for fish*). This approach, however, is not compatible with the evaluation of interindividual variation and may bias the biological interpretation of the data, especially when considering effects of chemical exposure on wild populations of fish.

As a final point, international standards for data capture and analysis have not been applied to studies in fish toxicogenomics. Studies will undoubtedly be enhanced in future years if a stronger emphasis is placed on adhering to the Minimum Information About a Microarray Experiment (MIAME) standards (Brazma *et al.*, 2001), established to maximise the potential use of array data sets beyond the limits of the original experiments (and laboratories generating the data). It is the opinion of these authors that MIAME standards, or an equivalent, should apply in the submission of array data sets into public databases, such as Array Express (http://www.ebi.ac.uk/arrayexpress/) or GEO (http://www.ncbi.nlm.nih.gov/geo/) and be a prerequisite for publication of array data sets in research papers in international scientific journals. MIAME/Tox (http://www.mged.org/) has recently emerged for the capture of data from toxicogenomic experiments and is a system that is likely to enhance not only the utility of toxicogenomic data but also its credibility for use in a legislative context.

Conclusions

Recent developments in the field of fish toxicogenomics hold the promise of a new level of understanding of mechanistic toxicity, in the discovery of new biomarkers of chemical exposure and biological effect and in biomonitoring of exposure in wild fish populations. Genomics also offers the exciting prospect of helping to unravel the interactive effects of chemical toxicants. Significant advances have been made in fish toxi-cogenomics in just a few years, with rapid progression from assessing chemical effects with single genes to the use of microarrays with thousands of genes. Specific needs for the further development of fish toxicogenomics are the enhancement of available genomic resources for key species used in (eco)toxicology and improvements in sequence annotations for fish. Other key issues relate to the need for more standardised approaches for both microarray experiments in fish and the analysis of microarray data. The resources required for comprehensive studies using microarrays are considerable but they have been less forthcoming for studies in fish compared with mammals. This lack of available resources, together with the fact that studies in fish have been less focused (many fish species have been studied and have employed a wider range of arrays and array platforms) than in mammals (in which studies have focused on mouse or rat and commercial and standardised arrays are available) has meant that fish toxicogenomics is not as well advanced as mammalian toxicogeno-mics. Furthermore, it is now well recognised that experiments employing gene microarrays require input from biologists, statisticians and bio-informaticians to extract the most meaningful data and these multi-disciplinary teams have only recently been established for work in fish toxicogenomics. The potential application of genomics in environmental monitoring and in regulatory testing in fish has attracted interest from many stakeholders and it is the opinion of the authors that the scientific community need to ensure that these bodies are fully engaged inter-nationally to ensure that transcriptomics and other genomic approaches in fish (eco)toxicology are supported sufficiently to enable their full potential to be realised.

113

Acknowledgements

A.F. was funded by the Environment Agency, UK to C.R.T. (Project no. SC040078). R.v.A. was funded by the Natural Environment Research Council (NERC) Post Genomics and Proteomics (PG&P) thematic and the Environment Agency, UK (Grant no. NE/C002369/1). A.L. was funded by NERC within the Environmental Genomics programme (Grant no. NER/T/S/2002/00182) to C.R.T. J.B. and E.M.S. were funded by NERC within the PG&P thematic (Grant no. NE/C507661/1) to C.R.T.

References

Ackermann, G. E. and Paw, B. H. (2003). Zebrafish: A genetic model for vertebrate organogenesis and human disorders. *Front Biosci.* 8,d1227–d1253.

Aegerter, S., Jalabert, B. and Bobe, J. (2005). Large scale real-time PCR analysis of mRNA abundance in rainbow trout eggs in relationship with egg quality and after ovulatory ageing. *Mol. Reprod. Dev.* 72,377–385.

Allison, D. B., Cui, X., Page, G. P. and Sabripour, M. (2006). Microarray data analysis: From disarray to consolidation and consensus. *Nat. Rev. Genet.* 7,55–65.

Aluru, N., Renaud, R., Leatherland, J. F. and Vijayan, M. M. (2005). Ah receptor-mediated impairment of interrenal steroidogenesis involved StAR protein and *P450scc* gene attenuation in rainbow trout. *Toxicol. Sci.* 84,260–269.

Aluru, N. and Vijayan, M. M. (2006). Aryl hydrocarbon receptor activation impairs cortisol response to stress in rainbow trout by disrupting the rate-limiting steps in steroidogenesis. *Endocrinology* 147,1895–1903.

Amin, R. P., Vickers, A. E., Sistare, F., Thompson, K. L., Roman, R. J., Lawton, M., Kramer, J., Hamadeh, H. K., Collins, J., Grissom, S., Bennett, L., Tucker, C. J., Wild, S., Kind, C., Oreffo, V., Davis, J. W., Curtiss, S., Naciff, J. M., Cunningham, M., Tennant, R., Stevens, J., Car, B., Bertram, T. A. and Afshari, C. A. (2004). Identification of putative gene based markers of renal toxicity. *Environ. Health Perspect.* 112,465–479.

Andreassen, T. K., Skjoedt, K. and Korsga-ard, B. (2005). Upregulation of estrogen receptor alpha and vitellogenin in eelpout (*Zoarces viviparus*) by waterborne exposure to 4-*tert*-octylphenol and 17β-estradiol. *Comp. Biochem. Physiol. C Toxicol. Pharmacol.* 140,340–346.

Ankley, G. T., Jensen, K. M., Makynen, E. A., Kahl, M. D., Korte, J. J., Hornung, M. W., Henry, T. R., Denny, J. S., Leino, R. L., Wilson, V. S., Cardon, M. C., Hartig, P. C. and Gray, L. E. (2003). Effects of the androgenic growth promoter 17-β-trenbolone on fecundity and reproductive endocrinology of the fathead minnow. *Environ. Toxicol. Chem.* 22,1350–1360.

Arukwe, A. (2005). Modulation of brain steroidogenesis by affecting transcriptional changes of steroidogenic acute regulatory (StAR) protein and cholesterol side chain cleavage (P450scc) in juvenile Atlantic salmon (*Salmo salar*) is a novel aspect of nonylphenol toxicity. *Environ. Sci. Technol.* 39,9791–9798.

114

Arukwe, A. (2006). Toxicological housekeeping genes: Do they really keep the house? *Environ. Sci. Technol.* 40,7944–7949.

Arukwe, A., Knudsen, F. R. and Goksøyr, A. (1997). Fish zona radiata (eggshell) protein: A sensitive biomarker for environmental estrogens. *Environ. Health Perspect.* 105,418–422.

Arukwe, A., Kullman, S. W., Berg, K., Goksoyr, A. and Hinton, D. E. (2002). Molecular cloning of rainbow trout (*Oncorhynchus mykiss*) eggshell zona radiata protein complementary DNA: mRNA expression in 17β-estradiol- and nonylphenol-treated fish. *Comp. Biochem. Physiol. B Biochem. Mol. Biol.* 132,315–326.

Ashburner, M., Ball, C. A., Blake, J. A., Botstein, D., Butler, H., Cherry, J. M., Davis, A. P., Dolinski, K., Dwight, S. S., Eppig, J. T., Harris, M. A., Hill, D. P., Issel-Tarver, L., Kasarskis, A., Lewis, S., Matese, J. C., Richardson, J. E., Ringwald, M., Rubin, G. M. and Sherlock, G. (2000). Gene ontology: Tool for the unification of biology. The Gene Ontology Consortium. *Nat. Genet.* 25,25–29.

Baldwin, W. S., Roling, J. A., Peterson, S. and Chapman, L. M. (2005). Effects of nonylphenol on hepatic testosterone metabolism and the expression of acute phase proteins in winter flounder (*Pleuronectes americanus*): Comparison to the effects of Saint John's Wort. *Comp. Biochem. Physiol. C Toxicol. Pharmacol.* 140,87–96.

Baron, D., Fostier, A., Breton, B. and Guiguen, Y. (2005a). Androgen and estrogen treatments alter steady state messenger RNA (mRNA) levels of testicular steroidogenic enzymes in rainbow trout, *Oncorhynchus mykiss. Mol. Reprod. Dev.* 71,471–479.

Baron, D., Houlgatte, R., Fostier, A. and Guiguen, Y. (2005b). Large-scale temporal gene expression profiling during gonadal differentiation and early gametogenesis in rainbow trout. *Biol. Reprod.* 73,959–966.

Bears, H., Richards, J. G. and Schulte, P. M. (2006). Arsenic exposure alters hepatic arsenic species composition and stress-mediated gene expression in the common killifish (*Fundulus heteroclitus*). *Aquat. Toxicol.* 77,257–266.

Begovac, P. C. and Wallace, R. A. (1989). Major vitelline envelope proteins in pipefish oocytes originate within the follicle and are associated with the Z3 layer. *J. Exp. Zool.* 251,56–73.

Berndtson, A. K. and Chen, T. T. (1994). Two unique cyp1 genes are expressed in response to 3-methylcholanthrene treatment in rainbow trout. *Arch. Biochem. Biophys.* 310,187–195.

Blum, J. L., Knoebl, I., Larkin, P., Kroll, K. J. and Denslow, N. D. (2004). Use of suppressive subtractive hybridization and cDNA arrays to discover patterns of altered gene expression in the liver of dihydrotestosterone and 11-ketotestosterone exposed adult male largemouth bass (*Micropterus salmoides*). *Mar. Environ. Res.* 58,565–569.

Bobe, J., Nguyen, T. and Jalabert, B. (2004). Targeted gene expression profiling in the rainbow trout ovary during maturational competence acquisition and oocyte maturation. *Biol. Reprod.* 71,73–82.

Bonham, K. and Gedamu, L. (1984). Induction of metallothionein and metallothionein messenger RNA in rainbow trout liver following cadmium treatment. *Biosci. Rep.* 4,633–642.

Bourdeau, V., Deschenes, J., Metivier, R., Nagai, Y., Nguyen, D., Bretschneider, N., Gannon, F., White, J. H. and Mader, S. (2004). Genome-wide identification of high-affinity estrogen response elements in human and mouse. *Mol. Endocrinol.* 18,1411–1427.

115

Bowman, C. J., Kroll, K. J., Gross, T. G. and Denslow, N. D. (2002). Estradiol-induced gene expression in largemouth bass (*Micropterus salmoides*). *Mol. Cell. Endocrinol.* 196,67–77.

Brazma, A., Hingamp, P., Quackenbush, J., Sherlock, G., Spellman, P., Stoeckert, C., Aach, J., Ansorge, W., Ball, C. A., Causton, H. C., Gaasterland, T., Glenisson, P., Holstege, F. C., Kim, I. F., Markowitz, V., Matese, J. C., Parkinson, H., Robinson, A., Sarkans, U., Schulze-Kremer, S., Stewart, J., Taylor, R., Vilo, J. and Vingron, M. (2001). Minimum information about a microarray experiment (MIAME) – toward standards for microarray data. *Nat. Genet.* 29,365–371.

Brown, S. B., Evans, R. E., Vandenbyllardt, L., Finnson, K. W., Palace, V. P., Kane, A. S., Yarechewski, A. Y. and Muir, D. C. G. (2004a). Altered thyroid status in lake trout (*Salvelinus namaycush*) exposed to coplanar 3,3′,4,4′,5-pentachlorobiphenyl. *Aquat. Toxicol.* 67,75–85.

Brown, M., Davies, I. M., Moffat, C. F., Robinson, C., Redshaw, J. and Craft, J. A. (2004b). Identification of transcriptional effects of ethynyl oestradiol in male plaice (*Pleuronectes platessa*) by suppression subtractive hybridisation and a nylon macroarray. *Mar. Environ. Res.* 58,559–563.

Brown, M., Robinson, C., Davies, I. M., Moffat, C. F., Redshaw, J. and Craft, J. A. (2004c). Temporal changes in gene expression in the liver of male plaice (*Pleuronectes platessa*) in response to exposure to ethynyl oestradiol analysed by macroarray and real-time PCR. *Mutat. Res.* 552,35–49.

Brown, V., Shurben, D., Miller, W. and Crane, M. (1994). Cadmium toxicity to rainbow trout *Oncorhynchus mykiss* Walbaum and brown trout *Salmo trutta* L. over extended exposure periods. *Ecotoxicol. Environ. Saf.* 29,38–46.

Buhler, D. R. and Wang-Buhler, J. L. (1998). Rainbow trout cytochrome P450s: Purification, molecular aspects, metabolic activity, induction and role in environmental monitoring. *Comp. Biochem. Physiol. C Toxicol. Pharmacol.* 121,107–137.

Buhler, D. R., Miranda, C. L., Henderson, M. C., Yang, Y. H., Lee, S. J. and Wang-Buhler, J. L. (2000). Effects of 17β-estradiol and testosterone on hepatic mRNA/protein levels and catalytic activities of CYP2M1, CYP2K1 and CYP3A27 in rainbow trout (*Oncorhynchus mykiss*). *Toxicol. Appl. Pharmacol.* 168,91–101.

Burczynski, M. E., McMillian, M., Ciervo, J., Li, L., Parker, J. B., Dunn, R. T., Hicken, S., Farr, S. and Johnson, M. D. (2000). Toxicogenomics-based discrimination of toxic mechanism in HepG2 hepatoma cells. *Toxicol. Sci.* 58,399–415.

Celius, T. and Walther, B. T. (1998). Differential sensitivity of zonagenesis and vitellogenesis in Atlantic salmon (*Salmo salar* L) to DDT pesticides. *J. Exp. Zool.* 281,346–353.

Celius, T., Haugen, T. B., Grotmol, T. and Walther, B. T. (1999). A sensitive zonagenetic assay for rapid in vitro assessment of estrogenic potency of xenobiotics and mycotoxins. *Environ. Health Perspect.* 107,63–68.

Celius, T., Matthews, J. B., Giesy, J. P. and Zacharewski, T. (2000). Quantification of rainbow trout (*Oncorhynchus mykiss*) zona radiate and vitellogenin mRNA levels using real-time PCR after *in vivo* treatment with estradiol-17β or α-zearalenol. *J. Steroid Biochem. Mol. Biol.* 75,109–119.

Chan, K. M. (1995). Metallothionein: Potential biomarker for monitoring heavy metal pollution in fish around Hong Kong. *Mar. Pollut. Bull.* 31,411–415.

116

Chan, P. C., Shiu, C. K., Wong, F. W., Wong, J. K., Lam, K. L. and Chan, K. M. (2004). Common carp metallothionein-1 gene: cDNA cloning, gene structure and expression studies. *Biochim. Biophys. Acta* 1676,162–171.

Chang, Y. S., Hsu, C. C., Wang, S. C., Tsao, C. C. and Huang, F. L. (1997). Molecular cloning, structural analysis and expression of carp *ZP2* gene. *Mol. Reprod. Dev.* 46,258–267.

Chang, Y. S., Lu, L. F., Lai, C. Y., Kou, Y. H. and Huang, F. L. (1999). Purification, characterization and molecular cloning of an outer layer protein of carp fertilization envelope. *Mol. Reprod. Dev.* 54,186–193.

Chang, Y. S., Wang, S. C., Tsao, C. C. and Huang, F. L. (1996). Molecular cloning, structural analysis and expression of carp *ZP3* gene. *Mol. Reprod. Dev.* 44,295–304.

Cheung, A. P., Lam, T. H. and Chan, K. M. (2004). Regulation of tilapia metallothionein gene expression by heavy metal ions. *Mar. Environ. Res.* 58,389–394.

Chuaqui, R. F., Bonner, R. F., Best, C. J., Gillespie, J. W., Flaig, M. J., Hewitt, S. M., Phillips, J. L., Krizman, D. B., Tangrea, M. S., Ahram, M., Linehan, W. M., Knezevic, V. and Emmert-Buck, M. R. (2002). After analysis follow-up and validation of microarray experiments. *Nat. Genet.* 32,509–514.

Craft, J. A., Brown, M., Dempsey, K., Francey, J., Kirby, M. F., Scott, A. P., Katsiadaki, I., Robinson, C. D., Davies, I. M., Bradac, P. and Moffat, C. F. (2004). Kinetics of vitellogenin protein and mRNA induction and depuration in fish following laboratory and environmental exposure to oestrogens. *Mar. Environ. Res.* 58,419–423.

Currie, R. A., Orphanides, G. and Moggs, J. G. (2005). Mapping molecular responses to xenoestrogens through gene ontology and pathway analysis of toxicogenomic data. *Reprod. Toxicol.* 20,433–440.

Dallas, P. B., Gottardo, N. G., Firth, M. J., Beesley, A. H., Hoffmann, K., Terry, P. A., Freitas, J. R., Boag, J. M., Cummings, A. J. and Kees, U. R. (2005). Gene expression levels assessed by oligonucleotide microarray analysis and quantitative real-time RT-PCR – how well do they correlate? *BMC Genomics* 6,59.

Damodaran, S., Dlugos, C. A., Wood, T. D. and Rabin, R. A. (2006). Effects of chronic ethanol administration on brain protein levels: A proteomic investigation using 2-D DIGE system. *Eur. J. Pharmacol.* 547,75–82.

Denslow, N. D., Lee, H. S., Bowman, C. J., Hemmer, M. J. and Folmar, L. C. (2001a). Multiple responses in gene expression in fish treated with estrogen. *Comp. Biochem. Physiol. B Biochem. Mol. Biol.* 129,277–282.

Denslow, N. D., Bowman, C. J., Ferguson, R. J., Lee, H. S., Hemmer, M. J. and Folmar, L. C. (2001b). Induction of gene expression in sheepshead minnow (*Cyprinodon variegates*) treated with 17β-estradiol, diethylstilbestrol, or ethinylestradiol: The use of mRNA fingerprints as an indicator of gene regulation. *Gen. Comp. Endocrinol.* 121,250–260.

Denslow, N. D., Kocerha, J., Sepulveda, M. S., Gross, T. and Holm, S. E. (2004). Gene expression fingerprints of largemouth bass (*Micropterus salmoides*) exposed to pulp and paper mill effluents. *Mutat. Res.* 552,19–34.

Di Giulio, R., Benson, B., Sanders, B. M. and van Veld, P. A. (1995). Biochemical mechanisms: Metabolism, adaptation and toxicity. In *Fundamentals of Aquatic Toxicology* (ed. G. Rand), pp. 423–461, Taylor and Francis Publishers, Washington, DC.

117

Diniz, M. S., Peres, I. and Pihan, J. C. (2005). Comparative study of the estrogenic responses of mirror carp (*Cyprinus carpio*) exposed to treated municipal sewage effluent (Lisbon) during two periods in different seasons. *Sci. Total Environ.* 349,129–139.

Dooley, K. and Zon, L. I. (2000). Zebrafish: A model system for the study of human disease. *Curr. Opin. Genet. Dev.* 10,252–256.

Durhan, E. J., Lambright, C., Wilson, V., Butterworth, B. C., Kuehl, O. W., Orlando, E. F., Guillette, L. J., Jr., Gray, L. E. and Ankley, G. T. (2002). Evaluation of androstenedione as an androgenic component of river water downstream of a pulp and paper mill effluent. *Environ. Toxicol. Chem.* 21,1973–1976.

Elango, A., Shepherd, B. and Chen, T. T. (2006). Effects of endocrine disrupters on the expression of growth hormone and prolactin mRNA in the rainbow trout pituitary. *Gen. Comp. Endocrinol.* 145,116–127.

European Inland Fisheries Advisory Commission. (1987). Revised Report on Combined Effects on Freshwater Fish and Other Aquatic Life. EIFAC Technical Paper 37, Rev. 1. Food and Agriculture Organisation (FAO), Rome.

Filby, A. L. and Tyler, C. R. (2005). Molecular characterization of estrogen receptors 1, 2a and 2b and their tissue and ontogenic expression profiles in fathead minnow (*Pimephales promelas*). *Biol. Reprod.* 73,648–662.

Filby, A. L. and Tyler, C. R. (2007). Appropriate 'housekeeping' genes for use in expression profiling the effects of environmental estrogens in fish. *BMC Mol. Biol.* 8,10.

Filby, A. L., Thorpe, K. L. and Tyler, C. R. (2006). Multiple molecular effect pathways of an environmental oestrogen in fish. *J. Mol. Endocrinol.* 37,121–134.

Filby, A. L., Thorpe, K. L., Maack, G. and Tyler, C. R. (2007a). Gene expression profiles revealing the mechanisms of antiandrogen- and estrogen-induced feminization in fish. *Aquat. Toxicol.* 81,219–231.

Filby, A. L., Santos, E. M., Thorpe, K. L., Maack, G. and Tyler, C. R. (2007b). Gene expression profiling for understanding chemical causation of biological effects for complex mixtures: A case study on estrogens. *Environ. Sci. Technol.* 41,8187–8194.

Finne, E. F., Cooper, G. A., Koop, B. F., Hylland, K. and Tollefsen, K. E. (2007). Toxicogenomic responses in rainbow trout (*Oncorhynchus mykiss*) hepatocytes exposed to model chemicals and a synthetic mixture. *Aquat. Toxicol.* 81,293–303.

Folmar, L. C., Denslow, N. D., Rao, V., Chow, M., Crain, D. A., Enblom, J., Marcino, J. and Guillette, L. J., Jr. (1996). Vitellogenin induction and reduced serum testosterone concentrations in feral male carp (*Cyprinus carpio*) captured near a major metropolitan sewage treatment plant. *Environ. Health Perspect.* 104,1096–1101.

Fox, K., Zauke, G. P. and Butte, W. (1994). Kinetics of bioconcentration and clearance of 28 polychlorinated biphenyl congeners in zebrafish (*Brachydanio rerio*). *Ecotoxicol. Environ. Saf.* 28,99–109.

Gant, T. W., Baus, P. R., Clothier, B., Riley, J., Davies, R., Judah, D. J., Edwards, R. E., George, E., Greaves, P. and Smith, A. G. (2003). Gene expression profiles associated with inflammation, fibrosis and cholestasis in mouse liver after griseofulvin. *Environ. Health Perspect.* 111,37–43.

Garcia-Reyero, N., Barber, D. S., Gross, T. S., Johnson, K. G., Sepulveda, M. S., Szabo, N. J. and Denslow, N. D. (2006). Dietary exposure of largemouth bass to OCPs changes expression of genes important for reproduction. *Aquat. Toxicol.* 78, 358–369.

George, S., Gubbins, M., MacIntosh, A., Reynolds, W., Sabine, V., Scott, A. and Thain, J. (2004). A comparison of pollutant biomarker responses with transcriptional responses in European flounders (*Platicthys flesus*) subjected to estuarine pollution. *Mar. Environ. Res.* 58,571–575.

Gonzalez, H. O., Roling, J. A., Baldwin, W. S. and Bain, L. J. (2006). Physiological changes and differential gene expression in mummichogs (*Fundulus heteroclitus*) exposed to arsenic. *Aquat. Toxicol.* 77,43–52.

Gonzalez, P., Dominique, Y., Massabuau, J. C., Boudou, A. and Bourdineaud, J. P. (2005). Comparative effects of dietary methylmercury on gene expression in liver, skeletal muscle and brain of the zebrafish (*Danio rerio*). *Environ. Sci. Technol.* 39,3972–3980.

Gottschalg, E., Moore, N. E., Ryan, A. K., Travis, L. C., Waller, R. C., Pratt, S., Atmaca, M., Kind, C. N. and Fry, J. R. (2006). Phenotypic anchoring of arsenic and cadmium toxicity in three hepatic-related cell systems reveals compound- and cell-specific selective up-regulation of stress protein expression: Implications for fingerprint profiling of cytotoxicity. *Chem. Biol. Interact.* 161,251–261.

Govoroun, M., McMeel, O. M., Mecherouki, H., Smith, T. J. and Guiguen, Y. (2001). 17β-estradiol treatment decreases steroidogenic enzyme messenger ribonucleic acid levels in the rainbow trout testes. *Endocrinology* 142,1841–1848.

Greytak, S. R. and Callard, G. V. (2007). Cloning of three estrogen receptors (ER) from killifish (*Fundulus heteroclitus*): Differences in populations from polluted and reference environments. *Gen. Comp. Endocrinol.* 150,174–188.

Grizzle, J. M., Melius, P. and Strength, D. R. (1984). Papillomas on fish exposed to chlorinated wastewater effluent. *J. Natl. Cancer Inst.* 73,1133–1142.

Gunnarsson, L., Kristiansson, E., Forlin, L., Nerman, O. and Larsson, D. G. (2007). Sensitive and robust gene expression changes in fish exposed to an estrogen – a microarray approach. *BMC Genomics* 8,149.

Guo, L., Lobenhofer, E. K., Wang, C., Shippy, R., Harris, S. C., Zhang, L., Mei, N., Chen, T., Herman, D., Goodsaid, F. M., Hurban, P., Phillips, K. L., Xu, J., Deng, X., Sun, Y. A., Tong, W., Dragan, Y. P. and Shi, L. (2006). Rat toxicogenomic study reveals analytical consistency across microarray platforms. *Nat. Biotechnol.* 24,1162–1169.

Hahlbeck, E., Katsiadaki, I., Mayer, I., Adolfsson-Erici, M., James, J. and Bengtsson, B. E. (2004). The juvenile three-spined stickleback (*Gasterosteus aculeatus* L.) as a model organism for endocrine disruption II – kidney hypertrophy, vitellogenin and spiggin induction. *Aquat. Toxicol.* 70,311–326.

Hahn, M. E. (1998). The aryl hydrocarbon receptor: A comparative perspective. *Comp. Biochem. Physiol. C Toxicol. Pharmacol.* 121,23–53.

Halm, S., Pounds, N., Maddix, S., Rand-Weaver, M., Sumpter, J. P., Hutchinson, T. H. and Tyler, C. R. (2002). Exposure to exogenous 17β-estradiol disrupts P450aromB mRNA expression in the brain and gonad of adult fathead minnows (*Pimephales promelas*). *Aquat. Toxicol.* 60,285–299.

Hamadeh, H. K., Bushel, P. R., Jayadev, S., Martin, K., DiSorbo, O., Sieber, S., Bennett, L., Tennant, R., Stoll, R., Barrett, J. C., Blanchard, K., Paules, R. S. and Afshari, C. A. (2002). Gene expression analysis reveals chemical-specific profiles. *Toxicol. Sci.* 67,219–231.

Hamadeh, H. K., Jayadev, S., Gaillard, E. T., Huang, Q., Stoll, R., Blanchard, K., Chou, J., Tucker, C. J., Collins, J., Maronpot, R., Bushel, P. and Afshari, C. A. (2004).

Integration of clinical and gene expression endpoints to explore furan-mediated hepatotoxicity. *Mutat. Res.* 549,169–183.

Hamazaki, T., Iuchi, I. and Yamagami, K. (1985). A spawning female-specific substance relative to antichorion (egg-envelope) glycoprotein antibody in the teleost, *Oryzias latipes. J. Exp. Zool.* 235,269–279.

Hamazaki, T. S., Nagahama, Y., Iuchi, I. and Yamagami, K. (1989). A glycoprotein from the liver constitutes the inner layer of the egg envelope (zona pellucida interna) of the fish, *Oryzias latipes. Dev. Biol.* 133,101–110.

Handley-Goldstone, H. M., Grow, M. W. and Stegeman, J. J. (2005). Cardiovascular gene expression profiles of dioxin exposure in zebrafish embryos. *Toxicol. Sci.* 85,683–693.

Hansen, B. H., Rømma, S., Garmo, Ø. A., Olsvik, P. A. and Andersen, R. A. (2006). Antioxidative stress proteins and their gene expression in brown trout (*Salmo trutta*) from three rivers with different heavy metal levels. *Comp. Biochem. Physiol. C Toxicol. Pharmacol.* 143,263–274.

Harries, C. A., Santos, E. M., Janbakhsh, A., Pottinger, T. G., Tyler, C. R. and Sumpter, J. P. (2001). Nonylphenol affects gonadotropin levels in the pituitary gland and plasma of female rainbow trout. *Environ. Sci. Technol.* 35,2909–2916.

Hashimoto, S., Bessho, H., Hara, A., Nakamura, M., Iguchi, T. and Fujita, K. (2000). Elevated serum vitellogenin levels and gonadal abnormalities in wild male flounder (*Pleuronectes yokohamae*) from Tokyo Bay, Japan. *Mar. Environ. Res.* 49,37–53.

Hecker, M., Tyler, C. R., Hoffmann, M., Maddix, S. and Karbe, L. (2002). Plasma biomarkers in fish provide evidence for endocrine modulation in the Elbe River, Germany. *Environ. Sci. Technol.* 36,2311–2321.

Heinloth, A. N., Irwin, R. D., Boorman, G. A., Nettesheim, P., Fannin, R. D., Sieber, S. O., Snell, M. L., Tucker, C. J., Li, L., Travlos, G. S., Vansant, G., Blackshear, P. E., Tennant, R. V., Cunningham, M. L. and Paules, R. S. (2004). Gene expression profiling of rat livers reveals indicators of potential adverse effects. *Toxicol. Sci.* 80,193–202.

Hemmer, M. J., Bowman, C. J., Hemmer, B. L., Friedman, S. D., Marcovich, D., Kroll, K. J. and Denslow, N. D. (2002). Vitellogenin mRNA regulation and plasma clearance in male sheepshead minnows, (*Cyprinodon variegatus*) after cessation of exposure to 17β-estradiol and *p*-nonylphenol. *Aquat. Toxicol.* 58,99–112.

Hemming, J. M., Waller, W. T., Chow, M. C., Denslow, N. D. and Venables, B. (2001). Assessment of the estrogenicity and toxicity of a domestic wastewater effluent flowing through a constructed wetland system using biomarkers in male fathead minnows (*Pimephales promelas* Rafinesque, 1820). *Environ. Toxicol. Chem.* 20,2268–2275.

Higashitani, T., Tamamoto, H., Takahashi, A. and Tanaka, H. (2003). Study of estrogenic effects on carp (*Cyprinus carpio*) exposed to sewage treatment plant effluents. *Water Sci. Technol.* 47,93–100.

Hinfray, N., Palluel, O., Turies, C., Cousin, C., Porcher, J. M. and Brion, F. (2006). Brain and gonadal aromatase as potential targets of endocrine disrupting chemicals in a model species, the zebrafish (*Danio rerio*). *Environ. Toxicol.* 21,332–337.

Hoffmann, J. L. and Oris, J. T. (2006). Altered gene expression: A mechanism for reproductive toxicity in zebrafish exposed to benzo[*a*]pyrene. *Aquat. Toxicol.* 78,332–340.

Hoffmann, J. L., Torontali, S. P., Thomason, R. G., Lee, D. M., Brill, J. L., Price, B. B., Carr, G. J. and Versteeg, D. J. (2006). Hepatic gene expression profiling using genechips in zebrafish exposed to 17α-ethynylestradiol. *Aquat. Toxicol.* 79, 233–246.

Hogstrand, C., Balesaria, S. and Glover, C. N. (2002). Application of genomics and proteomics for study of the integrated response to zinc exposure in a nonmodel fish species, the rainbow trout. *Comp. Biochem. Physiol. B Biochem. Mol. Biol.* 133,523–535.

Hollenberg, P. F. (1992). Mechanisms of cytochrome P450 and peroxidase-catalyzed xenobiotic metabolism. *FASEB J.* 6,686–694.

Hook, S. E., Skillman, A. D., Small, J. A. and Schultz, I. R. (2006a). Gene expression patterns in rainbow trout, *Oncorhynchus mykiss*, exposed to a suite of model toxicants. *Aquat. Toxicol.* 77,372–385.

Hook, S. E., Skillman, A. D., Small, J. A. and Schultz, I. R. (2006b). Dose–response relationships in gene expression profiles in rainbow trout, *Oncorhynchus mykiss*, exposed to ethynylestradiol. *Mar. Environ. Res.* 62,S151–S155.

Hope, B., Scatolini, S. and Titus, E. (1998). Bioconcentration of chlorinated biphenyls in biota from the North Pacific Ocean. *Chemosphere* 36,1247–1261.

Hoyt, P. R., Doktycz, M. J., Beattie, K. L. and Greeley, M. S. (2003). DNA microarrays detect 4-nonylphenol-induced alterations in gene expression during zebrafish early development. *Ecotoxicology* 12,469–474.

Huggett, R. J., Kimerle, R. A., Mehrle, P. M., Jr. and Bergman, H. L. (1992). *Biomarkers – Biochemical, Physiological and Histological Markers of Anthropogenic Stress*. Lewis Publishers, Boca Raton, FL.

Hutchinson, T. H., Ankley, G. T., Segner, H. and Tyler, C. R. (2006). Screening and testing for endocrine disruption in fish – biomarkers as 'signposts,' not 'traffic lights,' in risk assessment. *Environ. Health Perspect.* 114(Suppl. 1),106–114.

Hylland, K., Sandvik, M., Skare, J. U., Beyer, J., Ega-as, E. and Goksøyr, A. (1996). Biomarkers in flounder (*Platichthys flesus*): An evaluation of their use in pollution monitoring. *Mar. Environ. Res.* 42,223–227.

Hyllner, S. J., Oppen Berntsen, D. O., Helvik, J. V., Walther, B. T. and Haux, C. (1991). Oestradiol-17-β induces the major vitelline envelope proteins in both sexes in teleosts. *J. Endocrinol.* 131,229–236.

Hyllner, S. J., Westerlund, L., Olsson, P. E. and Schopen, A. (2001). Cloning of rainbow trout egg envelope proteins: Members of a unique group of structural proteins. *Biol. Reprod.* 64,805–811.

Ishibashi, H., Hirano, M., Matsumura, N., Watanabe, N., Takao, Y. and Arizono, K. (2006). Reproductive effects and bioconcentration of 4-nonylphenol in medaka fish (*Oryzias latipes*). *Chemosphere* 65,1019–1026.

Islinger, M., Pawlowski, S., Hollert, H., Volkl, A. and Braunbeck, T. (1999). Measurement of vitellogenin-mRNA expression in primary cultures of rainbow trout hepatocytes in a nonradioactive dot blot/RNase protection-assay. *Sci. Total Environ.* 233,109–122.

Jakobsson, S., Borg, B., Haux, C. and Hyllner, S. J. (1999). An 11-ketotestosterone induced kidney-secreted protein: The nest building glue from male three-spined stickleback, *Gasterosteus aculeatus*. *Fish Physiol. Biochem.* 20,79–85.

Jenkins, R., Angus, R. A., McNatt, H., Howell, W. M., Kemppainen, J. A., Kirk, M. and Wilson, E. M. (2001). Identification of androstenedione in a river containing paper mill effluent. *Environ. Toxicol. Chem.* 20,1325–1331.

Jobling, S. and Tyler, C. R. (2003). Endocrine disruption in wild freshwater fish. *Pure Appl. Chem.* 75,2219–2234.

Jobling, S., Beresford, N., Nolan, M., Rodgers-Gray, T., Brighty, G. C., Sumpter, J. P. and Tyler, C. R. (2002a). Altered sexual maturation and gamete production in wild roach (*Rutilus rutilus*) living in rivers that receive treated sewage effluents. *Biol. Reprod.* 66,272–281.

Jobling, S., Coey, S., Whitmore, J. G., Kime, D. E., Van Look, K. J. W., McAllister, B. G., Beresford, N., Henshaw, A. C., Brighty, G., Tyler, C. R. and Sumpter, J. P. (2002b). Wild intersex roach (*Rutilus rutilus*) have reduced fertility. *Biol. Reprod.* 67,515–524.

Jobling, S., Nolan, M., Tyler, C. R., Brighty, G. and Sumpter, J. P. (1998). Widespread sexual disruption in wild fish. *Environ. Sci. Technol.* 32,2498–2506.

Jones, I., Kille, P. and Sweeney, G. (2001a). Cadmium delays growth hormone expression during rainbow trout development. *J. Fish Biol.* 59,1015–1022.

Jones, I., Lindberg, C., Jakobsson, S., Hellqvist, A., Hellman, U., Borg, B. and Olsson, P. E. (2001b). Molecular cloning and characterization of spiggin. An androgen-regulated extraorganismal adhesive with structural similarities to von Willebrand factor-related proteins. *J. Biol. Chem.* 276,17857–17863.

Ju, Z., Wells, M. C. and Walter, R. B. (2007). DNA microarray technology in toxicogenomics of aquatic models: Methods and applications. *Comp. Biochem. Physiol. C Toxicol. Pharmacol.* 145,5–14.

Jung, J. H., Jeon, J. K., Shim, W. J., Oh, J. R., Lee, J. Y., Kim, B. K. and Han, C. H. (2005). Molecular cloning of vitellogenin cDNA in rockfish (*Sebastes schlegeli*) and effects of 2,2′,4,4′5,5′-hexachlorobiphenyl (PCB 153) on its gene expression. *Mar. Pollut. Bull.* 51,794–800.

Kamalakaran, S., Radhakrishnan, S. K. and Beck, W. T. (2005). Identification of estrogen-responsive genes using a genome-wide analysis of promoter elements for transcription factor binding sites. *J. Biol. Chem.* 280,21491–21497.

Katsiadaki, I., Morris, S., Squires, C., Hurst, M. R., James, J. D. and Scott, A. P. (2006). Use of the three-spined stickleback (*Gasterosteus aculeatus*) as a sensitive in vivo test for detection of environmental antiandrogens. *Environ. Health Perspect.* 114(Suppl. 1),115–121.

Katsiadaki, I., Scott, A. P., Hurst, M. R., Matthiessen, P. and Mayer, I. (2002a). Detection of environmental androgens: A novel method based on enzyme-linked immunosorbent assay of spiggin, the stickleback (*Gasterosteus aculeatus*) glue protein. *Environ. Toxicol. Chem.* 21,1946–1954.

Katsiadaki, I., Scott, A. P. and Mayer, I. (2002b). The potential of the three-spined stickleback (*Gasterosteus aculeatus* L.) as a combined biomarker for oestrogens and androgens in European waters. *Mar. Environ. Res.* 54,725–728.

Kawahara, R. and Nishida, M. (2006). Multiple occurrences of spiggin genes in sticklebacks. *Gene* 373,58–66.

Kawasaki, F., Katsiadaki, I., Scott, A. P., Matsubara, T., Osatomi, K., Soyano, K., Hara, A., Arizono, K. and Nagae, M. (2003). Molecular cloning of two types of spiggin cDNA in the three-spined stickleback, *Gasterosteus aculeatus*. *Fish Physiol. Biochem.* 28,425.

Kinae, N., Yamashita, M., Tomita, I., Kimura, I., Ishida, H., Kumai, H. and Nakamura, G. (1990). A possible correlation between environmental chemicals and pigment cell neoplasia in fish. *Sci. Total Environ.* 94,143–153.

122

Kirk, L. A., Tyler, C. R., Lye, C. M. and Sumpter, J. P. (2002). Changes in estrogenic and androgenic activities at different stages of treatment in wastewater treatment works. *Environ. Toxicol. Chem.* 21,972–979.

Kishi, K., Kitagawa, E., Onikura, N., Nakamura, A. and Iwahashi, H. (2006). Expression analysis of sex-specific and 17β-estradiol-responsive genes in the Japanese medaka, *Oryzias latipes*, using oligonucleotide microarrays. *Genomics* 88,241–251.

Kishida, M., Anderson, T. R. and Specker, J. L. (1992). Induction by β-estradiol of vitellogenin in striped bass (*Morone saxatilis*): Characterization and quantification in plasma and mucus. *Gen. Comp. Endocrinol.* 88,29–39.

Klaper, R., Rees, C. B., Drevnick, P., Weber, D., Sandheinrich, M. and Carvan, M. J. (2006). Gene expression changes related to endocrine function and decline in reproduction in fathead minnow (*Pimephales promelas*) after dietary methylmercury exposure. *Environ. Health Perspect.* 114,1337–1343.

Knoebl, I., Blum, J. L., Hemmer, M. J. and Denslow, N. D. (2006). Temporal gene induction patterns in sheepshead minnows exposed to 17β-estradiol. *J. Exp. Zool. A Ecol. Genet. Physiol.* 305,707–719.

Knudsen, F. R., Schou, A. E., Wiborg, M. L., Mona, E., Tollefsen, K. E., Stenersen, J. and Sumpter, J. P. (1997). Increase of plasma vitellogenin concentration in rainbow trout (*Oncorhynchus mykiss*) exposed to effluents from oil refinery treatment works and municipal sewage. *Bull. Environ. Contam. Toxicol.* 59,802–806.

Kortner, T. M. and Arukwe, A. (2007). The xenoestrogen, 4-nonylphenol, impaired steroidogenesis in previtellogenic oocyte culture of Atlantic cod (*Gadus morhua*) by targeting the StAR protein and P450scc expressions. *Gen. Comp. Endocrinol.* 150,419–429.

Koskinen, H., Pehkonen, P., Vehniainen, E., Krasnov, A., Rexroad, C., Afanasyev, S., Molsa, H. and Oikari, A. (2004). Response of rainbow trout transcriptome to model chemical contaminants. *Biochem. Biophys. Res. Commun.* 320,745–753.

Krasnov, A., Koskinen, H., Rexroad, C., Afanasyev, S., Molsa, H. and Oikari, A. (2005). Transcriptome responses to carbon tetrachloride and pyrene in the kidney and liver of juvenile rainbow trout (*Oncorhynchus mykiss*). *Aquat. Toxicol.* 74,70–81.

Kreiling, J. A., Creton, R. and Reinisch, C. (2007). Early embryonic exposure to polychlorinated biphenyls disrupts heat-shock protein 70 cognate expression in zebrafish. *J. Toxicol. Environ. Health* A70,1005–1013.

Larkin, P., Folmar, L. C., Hemmer, M. J., Poston, A. J., Lee, H. S. and Denslow, N. D. (2002). Array technology as a tool to monitor exposure of fish to xenoestrogens. *Mar. Environ. Res.* 54,395–399.

Larkin, P., Folmar, L. C., Hemmer, M. J., Poston, A. J. and Denslow, N. D. (2003a). Expression profiling of estrogenic compounds using a sheepshead minnow cDNA macroarray. *EHP Toxicogenomics* 111,29–36.

Larkin, P., Sabo-Attwood, T., Kelso, J. and Denslow, N. D. (2003b). Analysis of gene expression profiles in largemouth bass exposed to 17-β-estradiol and to anthropogenic contaminants that behave as estrogens. *Ecotoxicology* 12,463–468.

Larsson, D. G. J. and Forlin, L. (2002). Male-biased sex ratios of fish embryos near a pulp mill: Temporary recovery after a short-term shutdown. *Environ. Health Perspect.* 110,739–742.

Leatherland, J. F. (1992). Endocrine and reproductive function in Great Lakes salmon. In *Chemically Induced Alternations in Sexual and Functional Development: The*

123

Wildlife/Human Connection (eds T. Colborn and C. Clement), pp. 129–145, Princeton Scientific Publishing Company, Princeton, NJ.

Leatherland, J. F. (1993). Field observations on reproductive and developmental dysfunction in introduced and native salmonids from the Great Lakes. *J. Great Lakes Res.* 19,737–751.

Leatherland, J. F. (1997). Endocrine and reproductive function in Great Lakes salmon. *J. Clean Technol. Environ. Toxicol. Occup. Med.* 6,381–395.

Lee, C., Na, J. G., Lee, K. C. and Park, K. (2002). Choriogonin mRNA induction in male medaka, *Oryzias latipes* as a biomarker of endocrine disruption. *Aquat. Toxicol.* 61,233–241.

Lee, Y. M., Seo, J. S., Kim, I. C., Yoon, Y. D. and Lee, J. S. (2006). Endocrine disrupting chemicals (bisphenol A, 4-nonylphenol, 4-*tert*-octylphenol) modulate expression of two distinct cytochrome P450 aromatase genes differently in sex types of the hermaphroditic fish *Rivulus marmoratus*. *Biochem. Biophys. Res. Commun.* 345,894–903.

Le Guevel, R. and Pakdel, F. (2001). Assessment of oestrogenic potency of chemicals used as growth promoter by in-vitro methods. *Hum. Reprod.* 16,1030–1036.

Lettieri, T. (2006). Recent applications of DNA microarray technology to toxicology and ecotoxicology. *Environ. Health Perspect.* 114,4–9.

Lim, E. H., Teo, B. Y., Lam, T. J. and Ding, J. L. (2001). Sequence analysis of a fish vitellogenin cDNA with a large phosvitin domain. *Gene* 277,175–186.

Liney, K. E., Jobling, S., Shears, J. A., Simpson, P. and Tyler, C. R. (2005). Assessing the sensitivity of different life stages for sexual disruption in roach (*Rutilus rutilus*) exposed to effluents from wastewater treatment works. *Environ. Health Perspect.* 113,1299–1307.

Liolios, K., Tavernarakis, N., Hugenholtz, P. and Kyrpides, N. C. (2006). The genomes on line database (GOLD) v.2: A monitor of genome projects worldwide. *Nucleic Acids Res.* 34,D332–D334.

Luo, W., Fan, W., Xie, H., Jing, L., Ricicki, E., Vouros, P., Zhao, L. P. and Zarbl, H. (2005). Phenotypic anchoring of global gene expression profiles induced by *N*-hydroxy-4-acetylaminobiphenyl and benzo[*a*]pyrene diol epoxide reveals correlations between expression profiles and mechanism of toxicity. *Chem. Res. Toxicol.* 18,619–629.

Lyons, C. E., Payette, K. L., Price, J. L. and Huang, R. C. (1993). Expression and structural analysis of a teleost homolog of a mammalian zona pellucida gene. *J. Biol. Chem.* 268,21351–21358.

Lyssimachou, A., Jenssen, B. M. and Arukwe, A. (2006). Brain cytochrome P450 aromatase gene isoforms and activity levels in Atlantic salmon after waterborne exposure to nominal environmental concentrations of the pharmaceutical ethynylestradiol and antifoulant tributyltin. *Toxicol. Sci.* 91,82–92.

Ma, T. W., Wan, X. Q., Huang, Q. H., Wang, Z. J. and Liu, J. K. (2005). Biomarker responses and reproductive toxicity of the effluent from a Chinese large sewage treatment plant in Japanese medaka (*Oryzias latipes*). *Chemosphere* 59,281–288.

MacGregor, J. T. (2003). The future of regulatory toxicology: Impact of the biotechnology revolution. *Toxicol. Sci.* 75,236–248.

Marchand, J., Tanguy, A., Charrier, G., Quiniou, L., Plee-Gauthier, E. and Laroche, J. (2006). Molecular identification and expression of differentially regulated genes of European flounder, *Platichthys flesus*, submitted to pesticide exposure. *Mar. Biotechnol.* 8,275–294.

Margulies, M., Eghold, M., Altman, W. E., Attiya, S., Bader, J. S., Bemben, L. A., Berka, J., Braverman, M. S., Chen, Y. J., Chen, Z., Dewell, S. B., Du, L., Fierro, J. M., Gomes, X. V., Godwin, B. C., He, W., Helgesen, S., Ho, C. H., Irzyk, G. P., Jando, S. C., Alenquer, M. L., Jarvie, T. P., Jirage, K. B., Kim, J. B., Knight, J. R., Lanza, J. R., Leamon, J. H., Lefkowitz, S. M., Lei, M., Li, J., Lohman, K. L., Lu, H., Makhijani, V. B., McDade, K. E., McKenna, M. P., Myers, E. W., Nickerson, E., Nobile, J. R., Plant, R., Puc, B. P., Ronan, M. T., Roth, G. T., Sarkis, G. J., Simons, J. F., Simpson, J. W., Srinivasan, M., Tartaro, K. R., Tomasz, A., Vogt, K. A., Volkmer, G. A., Wang, S. H., Wang, Y., Weiner, M. P., Yu, P., Begley, R. F. and Rothberg, J. M. (2005). Genome sequencing in microfabricated high-density picolitre reactors. *Nature* 437,376–380.

Martyniuk, C. J., Xiong, H., Crump, K., Chiu, S., Sardana, R., Nadler, A., Gerrie, E. R., Xia, X. and Trudeau, V. L. (2006). Gene expression profiling in the neuroendocrine brain of male goldfish (*Carassius auritus*) exposed to 17α-ethinylestradiol. *Physiol. Genomics* 27,328–336.

Martyniuk, C. J., Gerrie, E. R., Popesku, J. T., Ekker, M. and Trudeau, V. L. (2007). Microarray analysis in the zebrafish (*Danio rerio*) liver and telencephalon after exposure to low concentration of 17alpha ethinylestradiol. *Aquat. Toxicol.* 84,38–49.

Mayer, G. D., Leach, A., Kling, P., Olsson, P. E. and Hogstrand, C. (2003). Activation of the rainbow trout metallothionein-A promoter by silver and zinc. *Comp. Biochem. Physiol. B Biochem. Mol. Biol.* 134,181–188.

McCarthy, J. F. and Shugart, L. R. (1990). Biomarkers of environmental contamination. In *Biomarkers of Environmental Contamination* (eds J. F. McCarthy and L. R. Shugart), pp. 3–16, Lewis Publishers, Chelsea, MI.

Merrick, B. A. and Bruno, M. E. (2004). Genomic and proteomic profiling for biomarkers and signature profiles of toxicity. *Curr. Opin. Mol. Ther.* 6,600–607.

Meucci, V. and Arukwe, A. (2006). Transcriptional modulation of brain and hepatic estrogen receptor and P450arom isotypes in juvenile Atlantic salmon (*Salmo salar*) after waterborne exposure to the xenoestrogen, 4-nonylphenol. *Aquat. Toxicol.* 77,167–177.

Meyer, J. N., Volz, D. C., Freedman, J. H. and Di Giulio, R. T. (2005). Differential display of hepatic mRNA from killifish (*Fundulus heteroclitus*) inhabiting a superfund estuary. *Aquat. Toxicol.* 73,327–341.

Mikawa, N., Utoh, T., Horie, N., Okamura, A., Yamada, Y., Akazawa, A., Tanaka, S., Tsukamoto, K., Hirono, I. and Aoki, T. (2006). Cloning and characterization of vitellogenin cDNA from the common Japanese conger (*Conger myriaster*) and vitellogenin gene expression during ovarian development. *Comp. Biochem. Physiol. B Biochem. Mol. Biol.* 143,404–414.

Miller, D. H., Jensen, K. M., Villeneuve, D. L., Kahl, M. D., Makynen, E. A., Durhan, E. J. and Ankley, G. T. (2007). Linkage of biochemical responses to population-level effects: A case study with vitellogenin in the fathead minnow (*Pimephales promelas*). *Environ. Toxicol. Chem.* 26,521–527.

Min, J., Lee, S. K. and Gu, M. B. (2003). Effects of endocrine disrupting chemicals on distinct expression patterns of estrogen receptor, cytochrome P450 aromatase and p53 genes in *Oryzias latipes* liver. *J. Biochem. Mol. Toxicol.* 17,272–277.

Miracle, A. L. and Ankley, G. T. (2005). Ecotoxicogenomics: Linkages between exposure and effect in linking risks of aquatic contaminants to fish. *Reprod. Toxicol.* 19,321–326.

Miracle, A. L., Toth, G. P. and Lattier, D. L. (2003). The path from molecular indicators of exposure to describing dynamic biological systems in an aquatic organism: Microarrays and the fathead minnow. *Ecotoxicology* 12,457–462.

Mizukami, Y., Okauchi, M., Arizono, K., Ariyoshi, T. and Kito, H. (1994). Isolation and sequence of cDNA encoding a 3-methylcholanthrene-inducible cytochrome-P450 from wild Red Sea bream, *Pagrus major*. *Mar. Biol.* 120,343–349.

Moccia, R. D., Leatherland, J. F. and Sonstegard, R. A. (1981). Quantitative interlake comparison of thyroid disease in Great Lakes coho (*Oncorhynchus kisutch*) and chinook (*Oncorhynchus tschawytscha*) salmon. *Cancer Res.* 41,2200–2210.

Moens, L. N., van der Ven, K., Van Remortel, P., Del-Favero, J. and De Coen, W. M. (2006). Expression profiling of endocrine-disrupting compounds using a customized *Cyprinus carpio* cDNA microarray. *Toxicol. Sci.* 93,298–310.

Moggs, J. G., Tinwell, H., Spurway, T., Chang, H. S., Pate, I., Lim, F. L., Moore, D. J., Soames, A., Stuckey, R., Currie, R., Zhu, T., Kimber, I., Ashby, J. and Orphanides, G. (2004). Phenotypic anchoring of gene expression changes during estrogen-induced uterine growth. *Environ. Health Perspect.* 112,1589–1606.

Mortensen, A. S. and Arukwe, A. (2006). The persistent DDT metabolite, 1,1-dichloro-2,2-bis(*p*-chlorophenyl)ethylene, alters thyroid hormone-dependent genes, hepatic cytochrome P4503A and pregnane X receptor gene expressions in Atlantic salmon (*Salmo salar*) parr. *Environ. Toxicol. Chem.* 25,1607–1615.

Nabb, D. L., Mingoia, R. T., Yang, C. H. and Han, X. (2006). Comparison of basal level metabolic enzyme activities of freshly isolated hepatocytes from rainbow trout (*Oncorhynchus mykiss*) and rat. *Aquat. Toxicol.* 80,52–59.

Nagae, M., Kawasaki, F., Tanaka, Y., Ohkubo, N., Matsubara, T., Soyono, K., Hara, A., Arizono, K., Scott, A. P. and Katsiadaki, I. (2007). Detection and assessment of androgenic potency of endocrine-disrupting chemicals using three-spined stickleback, *Gasterosteus aculeatus*. *Environ. Sci.* 14,255–261.

Navas, J. M. and Segner, H. (2006). Vitellogenin synthesis in primary cultures of fish liver cells as endpoint for in vitro screening of the (anti)estrogenic activity of chemical substances. *Aquat. Toxicol.* 80,1–22.

Nuwaysir, E. F., Bittner, M., Trent, J., Barrett, J. C. and Afshari, C. A. (1999). Microarrays and toxicology: The advent of toxicogenomics. *Mol. Carcinog.* 24,153–159.

Okoumassoun, L. E., Averill-Bates, D., Gagne, F., Marion, M. and Denizeau, F. (2002). Assessing the estrogenic potential of organochlorine pesticides in primary cultures of male rainbow trout (*Oncorhynchus mykiss*) hepatocytes using vitellogenin as a biomarker. *Toxicology* 178,193–207.

Olsson, P. E. (1996). Metallothioneins in fish: Induction and use in environmental monitoring. In *Toxicology of Aquatic Pollution: Physiological, Cellular and Molecular Approaches* (ed. E. W. Taylor), pp. 187–203, Cambridge University Press, Cambridge.

Olsson, P. E., Kling, P., Erkell, L. J. and Kille, P. (1995). Structural and functional analysis of the rainbow trout (*Oncorhynchus mykiss*) metallothionein-A gene. *Eur. J. Biochem.* 230,344–349.

Orlando, E. F., Kolok, A. S., Binzcik, G. A., Gates, J. L., Horton, M. K., Lambright, C. S., Gray, L. E., Soto, A. M. and Guillette, L. J. (2004). Endocrine-disrupting effects of cattle feedlot effluent on an aquatic sentinel species, the fathead minnow. *Environ. Health Perspect.* 112,353–358.

126

Patterson, T. A., Lobenhofer, E. K., Fulmer-Smentek, S. B., Collins, P. J., Chu, T. M., Bao, W., Fang, H., Kawasaki, E. S., Hager, J., Tikhonova, I. R., Walker, S. J., Zhang, L., Hurban, P., de Longueville, F., Fuscoe, J. C., Tong, W., Shi, L. and Wolfinger, R. D. (2006). Performance comparison of one-color and two-color platforms within the MicroArray Quality Control (MAQC) project. *Nat. Biotechnol.* 24,1140–1150.

Parks, L. G., Lambright, C. S., Orlando, E. F., Guillette, L. J., Ankley, G. T. and Gray, L. E. (2001). Masculinization of female mosquitofish in kraft mill effluent-contaminated fenholloway river water is associated with androgen receptor agonist activity. *Toxicol. Sci.* 62,257–267.

Paules, R. (2003). Phenotypic anchoring: Linking cause and effect. *Environ. Health Perspect.* 111,A338–A339.

Payne, J. F., Fancey, L. L., Rahimtula, A. D. and Porter, E. L. (1987). Review and perspective on the use of mixed-function oxygenase enzymes in biological monitoring. *Comp. Biochem. Physiol. C Toxicol. Pharmacol.* 86,233–245.

Payne, J. F., Mathieu, A., Melvin, W. and Fancey, L. L. (1996). Acetylcholinesterase, an old biomarker with a new future? Field trials in association with two urban rivers and a paper mill in Newfoundland. *Mar. Pollut. Bull.* 32,225–231.

Peakall, D. W. (1984). Biomarkers: The way forward in environmental assessment. *Toxicol. Ecotoxicol. News* 1,55–60.

Pelissero, C., Bennetau, B., Babin, P., Le Menn, F. and Dunogues, J. (1991). The estrogenic activity of certain phytoestrogens in the Siberian sturgeon *Acipenser baeri*. *J. Steroid Biochem. Mol. Biol.* 38,293–299.

Peterson, J. S. K. and Bain, L. J. (2004). Differential gene expression in anthracene-exposed mummichogs (*Fundulus heteroclitus*). *Aquat. Toxicol.* 66,345–355.

Peterson, R. E., Theobald, H. M. and Kimmel, G. L. (1993). Developmental and reproductive toxicity of dioxins and related compounds – cross-species comparisons. *Crit. Rev. Toxicol.* 23,283–335.

Pinto, P. I. S., Teodosio, H. R., Galay-Burgos, M., Power, D. M., Sweeney, G. E. and Canario, A. V. M. (2005). Identification of estrogen-responsive genes in the testis of sea bream (*Sparus auritus*) using suppression subtractive hybridization. *Mol. Reprod. Dev.* 73,318–329.

Powell, C. L., Kosyk, O., Ross, P. K., Schoonhoven, R., Boysen, G., Swenberg, J. A., Heinloth, A. N., Boorman, G. A., Cunningham, M. L., Paules, R. S. and Rusyn, I. (2006). Phenotypic anchoring of acetaminophen-induced oxidative stress with gene expression profiles in rat liver. *Toxicol. Sci.* 93,213–222.

Purdom, C. E., Hardiman, P. A., Bye, V. J., Eno, N. C., Tyler, C. R. and Sumpter, J. P. (1994). Oestrogenic effects of effluents from sewage treatment works. *Chem. Ecol.* 8,275–285.

Pyle, G. G., Swanson, S. M. and Lehmkuhl, D. M. (2002). Toxicity of uranium mine receiving waters to early life stage fathead minnows (*Pimephales promelas*) in the laboratory. *Environ. Pollut.* 116,243–255.

Quabius, E. S., Krupp, G. and Sevombes, C. J. (2005). Polychlorinated biphenyl 126 affects expression of genes involved in stress-immune interaction in primary cultures of rainbow trout anterior kidney cells. *Environ. Toxicol. Chem.* 24,3053–3060.

Radice, S., Ferraris, M., Marabini, L. and Chiesara, E. (2002). Estrogenic activity of procymidone in primary cultured rainbow trout hepatocytes (*Oncorhynchus mykiss*). *Toxicol. In Vitro* 16,475–480.

Rees, C. B., McCormick, S. D., Vanden Heuvel, J. P. and Li, W. M. (2003). Quantitative PCR analysis of CYP1A induction in Atlantic salmon (*Salmo salar*). *Aquat. Toxicol.* 62,67–78.

Reynders, H., van der Ven, K., Moens, L. N., van Remortel, P., De Coen, W. M. and Blust, R. (2006). Patterns of gene expression in carp liver after exposure to a mixture of waterborne and dietary cadmium using a custom-made microarray. *Aquat. Toxicol.* 80,180–193.

Rise, M. L., von Schalburg, K. R., Brown, G. D., Mawer, M. A., Devlin, R. H., Kuipers, N., Bushby, M., Beetz-Sargent, M., Alberto, R., Gibbs, A. R., Hunt, P., Shukin, R., Zeznick, J. A., Nelson, C., Jones, S. R., Smailus, D. E., Jones, S. J., Schein, J. E., Marra, M. A., Butterfield, Y. S., Stott, J. M., Ng, S. H., Davidson, W. S. and Koop, B. F. (2004). Development and application of a salmonid EST database and cDNA microarray: Data mining and interspecific hybridization characteristics. *Genome Res.* 14,478–490.

Roberts, A. P., Oris, J. T., Burton, G. A. and Clements, W. H. (2005). Gene expression in caged fish as a first-tier indicator of contaminant exposure in streams. *Environ. Toxicol. Chem.* 24,3092–3098.

Robertson, D. G. (2005). Metabonomics in toxicology: A review. *Toxicol. Sci.* 85, 809–822.

Roling, J. A., Bain, L. J. and Baldwin, W. S. (2004). Differential gene expression in mummichogs (*Fundulus heteroclitus*) following treatment with pyrene: Comparison to a creosote contaminated site. *Mar. Environ. Res.* 57,377–395.

Rolland, R. M. (2000). A review of chemically induced alterations in thyroid and vitamin A status from field studies of wildlife and fish. *J. Wildl. Dis.* 36,615–635.

Rotchell, J. M. and Ostrander, G. K. (2003). Molecular markers of endocrine disruption in aquatic organisms. *J. Toxicol. Environ. Health B* 6,453–496.

Sabo-Attwood, T., Kroll, K. J. and Denslow, N. D. (2004). Differential expression of largemouth bass (*Micropterus salmoides*) estrogen receptor isotypes alpha, beta and gamma by estradiol. *Mol. Cell. Endocrinol.* 218,107–118.

Samson, S. L. A. and Gedamu, L. (1995). Metal-responsive elements of the rainbow trout metallothionein-B gene-function for basal and metal-induced activity. *J. Biol. Chem.* 270,6864–6871.

Samuelsson, L. M., Forlin, L., Karlsson, G., Adolfsson-Erici, M. and Larsson, D. G. (2006). Using NMR metabolomics to identify responses of an environmental estrogen in blood plasma of fish. *Aquat. Toxicol.* 78,341–349.

Santos, E. M., Workman, V. L., Paull, G. C., Filby, A. L., Van Look, K. J., Kille, P. and Tyler, C. R. (2007a). Molecular basis of sex and reproductive status in breeding zebrafish. *Physiol. Genomics* 30,111–122.

Santos, E. M., Paull, G. C., Van Look, K. J., Workman, V. L., Holt, W. V., van Aerle, R., Kille, P. and Tyler, C. R. (2007b). Gonadal transcriptome responses and physiological consequences of exposure to oestrogen in breeding zebrafish (*Danio rerio*). *Aquat. Toxicol.* 83,134–142.

Sawaguchi, S., Koya, Y., Yoshizaki, N., Ohkubo, N., andoh, T., Hiramatsu, N., Sullivan, C. V., Hara, A. and Matsubara, T. (2005). Multiple vitellogenins (Vgs) in mosquitofish (*Gambusia affinis*): Identification and characterization of three functional Vg genes and their circulating and yolk protein products. *Biol. Reprod.* 72,1045–1060.

128

28

Schmid, T., Gonzalez-Valero, J., Rufli, H. and Dietrich, D. R. (2002). Determination of vitellogenin kinetics in male fathead minnows (*Pimephales promelas*). *Toxicol. Lett.* 131,65–74.

Scholz, S., Kordes, C., Hamann, J. and Gutzeit, H. O. (2004). Induction of vitellogenin in vivo and in vitro in the model teleost medaka (*Oryzias latipes*): Comparison of gene expression and protein levels. *Mar. Environ. Res.* 57,235–244.

Schulz, R. W., Bogerd, J., Male, R., Ball, J., Fenske, M., Olsen, L. C. and Tyler, C. R. (2007). Estrogen-induced alterations in *amh* and *dmrt1* expression signal for disruption in male sexual development in the zebrafish. *Environ. Sci. Technol.* 41,6305–6310.

Scientific Committee on Problems of the Environment. (1987). *Methods for Assessing the Effects of Mixtures of Chemicals*. Wiley, Chichester.

Seo, J. S., Lee, Y. M., Jung, S. O., Kim, I. C., Yoon, Y. D. and Lee, J. S. (2006). Nonylphenol modulates expression of androgen receptor and estrogen receptor genes differently in sex types of the hermaphroditic fish *Rivulus marmoratus*. *Biochem. Biophys. Res. Commun.* 346,213–223.

Sheader, D. L., Williams, T. D., Lyons, B. P. and Chipman, J. K. (2006). Oxidative stress response of European flounder (*Platichthys flesus*) to cadmium determined by a custom cDNA microarray. *Mar. Environ. Res.* 62,33–44.

Sherry, J. P., Whyte, J. J., Karrow, N. A., Gamble, A., Boerman, H. J., Bol, N. C., Dixon, D. G. and Solomon, K. R. (2006). The effect of creosote on vitellogenin production in rainbow trout (*Oncorhynchus mykiss*). *Arch. Environ. Contam. Toxicol.* 50,65–68.

Shi, L., Reid, L. H., Jones, W. D., Shippy, R., Warrington, J. A., Baker, S. C., Collins, P. J., de Longueville, F., Kawasaki, E. S., Lee, K. Y., Luo, Y., Sun, Y. A., Willey, J. M., Setterquist, R. A., Fischer, G. M., Tong, W., Dragan, Y. P., Dix, D. J., Frueh, F. W., Goodsaid, F. M., Herman, D., Jensen, R. V., Johnson, C. D., Lobenhofer, E. K., Puri, R. K., Schrf, U., Thierry-Mieg, J., Wang, C., Wilson, M., Wolber, P. K., Zhang, L., Slikker, W., Shi, L. and Reid, L. H.MAQC Consortium. (2006). The MicroArray Quality Control (MAQC) project shows inter- and intraplatform reproducibility of gene expression measurements. *Nat. Biotechnol.* 24,1151–1161.

Shibata, H., Spencer, T. E., Onate, S. A., Jenster, G., Tsai, S. Y. and O'Malley, B. W. (1997). Role of coactivators and corepressors in the mechanism of steroid/thyroid receptor action. *Recent Prog. Horm. Res.* 52,141–164.

Shimizu, M., Fujita, T., Hiramatsu, N. and Hara, A. (1998). Immunochemical detection, estrogen induction and occurrence in serum of vitelline envelope-related proteins of sakhalin taimen *Hucho perryi*. *Fisheries Sci.* 64,600–605.

Shrader, E. A., Henry, T. R., Greeley, M. S. and Bradley, B. P. (2003). Proteomics in zebrafish exposed to endocrine disrupting chemicals. *Ecotoxicology* 12,485–488.

Sone, K., Hinago, M., Itamoto, M., Katsu, Y., Watanabe, H., Urushitani, H., Tooi, O., Guillette, L. J., Jr. and Iguchi, T. (2005). Effects of an androgenic growth promoter 17beta-trenbolone on masculinization of mosquitofish (*Gambusia affinis affinis*). *Gen. Comp. Endocrinol.* 143,151–160.

Spitsbergen, J. M., Walker, M. K., Olson, J. R. and Peterson, R. E. (1991). Pathological alterations in early life stages of lake trout, *Salvelinus namaycush*, exposed to 2,3,7,8-tetrachlorodibenzo-para-dioxin as fertilized eggs. *Aquat. Toxicol.* 19,41–71.

Stegeman, J. J. (1995). Diversity and regulation of cytochrome P450 in aquatic species. In *Molecular Aspects of Oxidative Drug Metabolizing Enzymes: Their Significance in Environmental Toxicology, Chemical Carcinogenesis and Health* (eds E. Ariniç, J. B. Schenkman and E. Hodgson), pp. 273–276, Springer, Heidelberg.

Sturm, A., Wogram, J., Segner, H. and Liess, M. (2000). Different sensitivity to organophosphates of acetylcholinesterase and butyrylcholinesterase from three-spined stickleback (*Gasterosteus aculeatus*): Application in biomonitoring. *Environ. Toxicol. Chem.* 19,1607–1615.

Sumpter, J. P. and Jobling, S. (1995). Vitellogenesis as a biomarker for estrogenic contamination of the aquatic environment. *Environ. Health Perspect.* 103,173–178.

Tang, S., Tan, S. L., Ramadoss, S. K., Kumar, A. P., Tang, M. H. and Najic, V. B. (2004). Computational method for discovery of estrogen responsive genes. *Nucleic Acids Res.* 32,6212–6217.

Thomas, K. V., Hurst, M. R., Matthiessen, P., McHugh, M., Smith, A. and Waldock, M. J. (2002). An assessment of in vitro androgenic activity and the identification of environmental androgens in United Kingdom estuaries. *Environ. Toxicol. Chem.* 21,1456–1461.

Thomas-Jones, E., Walkley, N., Morris, C., Kille, P., Cryer, J., Weeks, I. and Woodhead, J. S. (2003a). Quantitative measurement of fathead minnow vitellogenin mRNA using hybridization protection assays. *Environ. Toxicol. Chem.* 22,992–995.

Thomas-Jones, E., Thorpe, K., Harrison, N., Thomas, G., Morris, C., Hutchinson, T., Woodhead, S. and Tyler, C. (2003b). Dynamics of estrogen biomarker responses in rainbow trout exposed to 17β-estradiol and 17α-ethinylestradiol. *Environ. Toxicol. Chem.* 22,3001–3008.

Thorpe, K. L., Hutchinson, T. H., Hetheridge, M. J., Scholze, M., Sumpter, J. P. and Tyler, C. R. (2001). Assessing the biological potency of binary mixtures of environmental estrogens using vitellogenin induction in juvenile rainbow trout (*Oncorhynchus mykiss*). *Environ. Sci. Technol.* 35,2476–2481.

Tilton, S. C., Gerwick, L. G., Hendricks, J. D., Rosato, C. S., Corley-Smith, G., Givan, S. A., Bailey, G. S., Bayne, C. J. and Williams, D. E. (2005). Use of a rainbow trout oligonucleotide microarray to determine transcriptional patterns in aflatoxin B1-induced hepatocellular carcinoma compared to adjacent liver. *Toxicol. Sci.* 88,319–330.

Tilton, S. C., Givan, S. A., Pereira, C. B., Bailey, G. S. and Williams, D. E. (2006). Toxicogenomic profiling of the hepatic tumor promoters indole-3-carbinol, 17β-estradiol and beta-naphthoflavone in rainbow trout. *Toxicol. Sci.* 90,61–72.

Tom, M., Chen, N., Segev, M., Herut, B. and Rinkevich, B. (2004). Quantifying fish metallothionein transcript by real time PCR for its utilization as an environmental biomarker. *Mar. Pollut. Bull.* 48,705–710.

Tong, Y., Shan, T., Poh, Y. K., Yan, T., Wang, H., Lam, S. H. and Gong, Z. (2004). Molecular cloning of zebrafish and medaka vitellogenin genes and comparison of their expression in response to 17β-estradiol. *Gene* 328,25–36.

Tyler, C. R., Jobling, S. and Sumpter, J. P. (1998). Endocrine disruption in wildlife: A critical review of the evidence. *Crit. Rev. Toxicol.* 28,319–361.

van Aerle, R., Ball, J., Katsu, Y., Iguchi, T., Santos, E. M., Paull, G. C., Filby, A. L. and Tyler, C. R. (2004). Establishing a multi-fish species array to identify the effects of endocrine disrupting chemicals (EDCs) on sexual development in fish.

In *Proceedings of the 5th International symposium on Fish Endocrinology*, September 5–9, Castellon, Spain.

van Aerle, R., Pounds, N., Hutchinson, T. H., Maddix, S. and Tyler, C. R. (2002). Window of sensitivity for the estrogenic effects of ethinylestradiol in early life-stages of fathead minnow, *Pimephales promelas*. *Ecotoxicology* 11,423–434.

van der Oost, R., Porte-Visa, C. and van den Brink, N. W. (2005). Biomarkers in environmental assessment. In *Ecotoxicological Testing of Marine and Freshwater Ecosystems* (eds M. Munawar and P. J. Den Besten), pp. 87–152, CRC Press, Boca Raton.

van der Ven, K., De Wit, M., Keil, D., Moens, L., Van Leemput, K., Naudts, B. and De Coen, W. (2005). Development and application of a brain-specific cDNA microarray for effect evaluation of neuro-active pharmaceuticals in zebrafish (*Danio rerio*). *Comp. Biochem. Physiol. B Biochem. Mol. Biol.* 141,408–417.

van der Ven, K., Keil, D., Moens, L. N., Van Leemput, K., van Remortel, P. and De Coen, W. M. (2006a). Neuropharmaceuticals in the environment: Mianserin-induced neuroendocrine disruption in zebrafish (*Danio rerio*) using cDNA microarrays. *Environ. Toxicol. Chem.* 25,2645–2652.

van der Ven, K., Keil, D., Moens, L. N., Van Hummelen, P., van Remortel, P., Maras, M. and De Coen, W. (2006b). Effects of the antidepressant mianserin in zebrafish: Molecular markers of endocrine disruption. *Chemosphere* 65,1836–1845.

van Gestel, C. A. M. and van Brummelen, T. C. (1996). Incorporation of the biomarker concept in ecotoxicology calls for a redefinition of terms. *Ecotoxicology* 5,217–225.

Vethaak, A. D., Lahr, J., Schrap, S. M., Belfroid, A. C., Rijs, G. B. J., Gerritsen, A., de Boer, J., Bulder, A. S., Grinwis, G. C. M., Kuiper, R. V., Legler, J., Murk, T. A. J., Peijnenburg, W., Verha-ar, H. J. M. and deVoogt, P. (2005). An integrated assessment of estrogenic contamination and biological effects in the aquatic environment of The Netherlands. *Chemosphere* 59,511–524.

Viant, M. R., Pincetich, C. A. and Tjeerdema, R. S. (2006a). Metabolic effects of dinoseb, diazinon and esfenvalerate in eyed eggs and alevins of Chinook salmon (*Oncorhynchus tshawytscha*) determined by 1H NMR metabolomics. *Aquat. Toxicol.* 77,359–371.

Viant, M. R., Pincetich, C. A., Hinton, D. E. and Tjeerdema, R. S. (2006b). Toxic actions of dinoseb in medaka (*Oryzias latipes*) embryos as determined by in vivo P-31 NMR, HPLC-UV and H-1 NMR metabolomics. *Aquat. Toxicol.* 76,329–342.

Viarengo, A., Burlando, B., Dondero, F., Marro, A. and Fabbri, R. (1999). Metallo-thionein as a tool in biomonitoring programmes. *Biomarkers* 4,455–466.

Voelker, D., Vess, C., Tillmann, M., Nagel, R., Otto, G. W., Geisler, R., Schirmer, K. and Scholz, S. (2007). Differential gene expression as a toxicant-sensitive endpoint in zebrafish embryos and larvae. *Aquat. Toxicol.* 81,355–364.

Volz, D. C., Bencic, D. C., Hinton, D. E., Law, J. M. and Kullman, S. W. (2005). 2,3,7,8-tetrachlorodibenzo-*p*-dioxin (TCDD) induces organ-specific differential gene expression in male Japanese medaka (*Oryzias latipes*). *Toxicol. Sci.* 85,572–584.

von Schalburg, K. R., Rise, M. L., Cooper, G. A., Brown, G. D., Gibbs, A. R., Nelson, C. C., Davidson, W. S. and Koop, B. F. (2005). Fish and chips: Various methodologies demonstrate utility of a 16,006 salmonid microarray. *BMC Genomics* 6,126.

Wang, H. and Gong, Z. (1999). Characterization of two zebrafish cDNA clones encoding egg envelope proteins ZP2 and ZP3. *Biochim. Biophys. Acta* 1446,156–160.

Wang, H., Yan, T., Tan, J. T. and Gong, Z. (2000). A zebrafish vitellogenin gene (vg3) encodes a novel vitellogenin without a phosvitin domain and may represent a primitive vertebrate vitellogenin gene. *Gene* 256,303–310.

Waring, J. F., Ciurlionis, R., Jolly, R. A., Heindel, M. and Ulrich, R. G. (2001). Microarray analysis of hepatotoxins in vitro reveals a correlation between gene expression profiles and mechanisms of toxicity. *Toxicol. Lett.* 120,359–368.

Waters, M., Boorman, G., Bushel, P., Cunningham, M., Irwin, R., Merrick, A., Olden, K., Paules, R., Selkirk, J., Stasiewicz, S., Weis, B., Van Houten, B., Walker, N. and Tennant, R. (2003). Systems toxicology and the Chemical Effects in Biological Systems (CEBS) knowledge base. *Environ. Health Perspect.* 111,15–28.

Waters, M. D. and Fostel, J. M. (2004). Toxicogenomics and systems toxicology: Aims and prospects. *Nat. Rev. Genet.* 5,936–948.

Werner, J., Wautier, K., Evans, R. E., Baron, C. L., Kidd, K. and Palace, V. (2003). Waterborne ethynylestradiol induces vitellogenin and alters metallothionein expression in lake trout (*Salvelinus namaycush*). *Aquat. Toxicol.* 62,321–328.

Westerlund, L., Hyllner, S. J., Schopen, A. and Olsson, P. E. (2001). Expression of three vitelline envelope protein genes in arctic char. *Gen. Comp. Endocrinol.* 122, 78–87.

Whitehead, A. and Crawford, D. L. (2005). Variation in tissue-specific gene expression among natural populations. *Genome Biol.* 6,R13.

Whyte, J. J., Jung, R. E., Schmitt, C. J. and Tillitt, D. E. (2000). Ethoxyresorufin-*O*-deethylase (EROD) activity in fish as a biomarker of chemical exposure. *Crit. Rev. Toxicol.* 30,347–570.

Willett, K. L., Ganesan, S., Patel, M., Metzger, C., Quiniou, S., Waldbieser, G. and Scheffler, B. (2006). In vivo and in vitro CYP1B mRNA expression in channel catfish. *Mar. Environ. Res.* 62,S332–S336.

Williams, N. D. and Holdway, D. A. (2000). The effects of pulse-exposed cadmium and zinc on embryo hatchability, larval development and survival of Australian crimson spotted rainbow fish (*Melanotaenia fluviatilis*). *Environ. Toxicol.* 15,165–173.

Williams, T. D., Gensberg, K., Minchin, S. D. and Chipman, J. K. (2003). A DNA expression array to detect toxic stress response in European flounder (*Platichthys flesus*). *Aquat. Toxicol.* 65,141–157.

Williams, T. D., Diab, A. M., George, S. G., Sabine, V., Conesa, A., Minchin, S. D., Watts, P. C. and Chipman, J. K. (2006). Development of the GENIPOL European flounder (*Platichthys flesus*) microarray and determination of temporal transcriptional responses to cadmium at low dose. *Environ. Sci. Technol.* 40,6479–6488.

Williams, T. D., Diab, A. M., George, S. G., Sabine, V. and Chipman, J. K. (2007). Gene expression responses of European flounder (*Platichthys flesus*) to 17β estradiol. *Toxicol. Lett.* 168,236–248.

Wintz, H., Yoo, L. J., Loquinov, A., Wu, Y. Y., Steevens, J. A., Holland, R. D., Beger, R. D., Perkins, E. J., Hughes, O. and Vulpe, C. D. (2006). Gene expression profiles in fathead minnow exposed to 2,4-DNT: Correlation with toxicity in mammals. *Toxicol. Sci.* 94,71–82.

132

Winzer, K., Van Noorden, C. J. F. and Koher, A. (2002). Glucose-6-phosphate dehydrogenase: The key to sex-related xenobiotics toxicity in hepatocytes of European flounder (*Platichthys flesus* L.). *Aquat. Toxicol.* 56,275–288.

Yadetie, F. and Male, R. (2002). Effects of 4-nonylphenol on gene expression of pituitary hormones in juvenile Atlantic salmon (*Salmo salar*). *Aquat. Toxicol.* 58,113–129.

Yokota, H., Abe, T., Nakai, M., Murakami, H., Eto, C. and Yakabe, Y. (2005). Effects of 4-*tert*-pentylphenol on the gene expression of P450 11β-hydroxylase in the gonad of medaka (*Oryzias latipes*). *Aquat. Toxicol.* 71,121–132.

Zafarullah, M., Bonham, K. and Gedamu, L. (1988). Structure of the rainbow trout metallothionein B-gene and characterization of its metal-responsive region. *Mol. Cell. Biol.* 8,4469–4476.

Zafarullah, M., Olsson, P. E. and Gedamu, L. (1989). Endogenous and heavy metal ion-induced metallothionein gene expression in salmonid tissues and cell lines. *Gene* 83,85–93.

Zhu, J. Y., Huang, H. Q., Bao, X. D., Lin, Q. M. and Cai, Z. (2006). Acute toxicity profile of cadmium revealed by proteomics in brain tissue of *Paralichthys olivaceus*: Potential role of transferrin in cadmium toxicity. *Aquat. Toxicol.* 78,127–135.

Current research in soil invertebrate ecotoxicogenomics

David J. Spurgeon[1,*], A. John Morgan[2] and Peter Kille[2]

[1]Centre for Ecology and Hydrology, Monks Wood, Abbots Ripton, Huntingdon, Cambridgeshire, PE28 2LS, UK
[2]Cardiff School of Biosciences, BIOSI 1, University of Cardiff, P.O. Box 915, Cardiff, CF10 3TL, UK

Abstract. Soil species, such as earthworms, potworms, springtails and to a lesser extent, carabids, molluscs and oribatid and predator mites are widely used to assess the toxicity of soil pollutants in academic and regulatory contexts. A recent extension of this work has been to provide a mechanistic component to these studies. Initially, terrestrial ecotoxicogenomic studies used the nematode *Caenorhabditis elegans* (Maupas). This was simply because it was the first soil-dwelling species for which the required tools for detailed study (principally sequence information) were available. Latterly, sufficient sequencing information to allow routine assessment of the expression of individual genes/proteins and the production of custom cDNA microarrays has become available for species such as the earthworms *Lumbricus rubellus* (Hoffmeister), *Eisenia fetida* (Savigny), the potworm *Enchytraeus albidus* (Henle), and springtails *Orchesella cincta* (Linnaeus) and *Folsomia candida* (Willem). Initial transcriptomic studies with such species have confirmed the sensitivity of gene expression as an endpoint for chemical exposure and its value for identifying the mechanism of toxic effects. Combined with the development of methods for proteomics and metabolomics this means that it is now feasible to use soil invertebrates in studies that can advance fundamental knowledge of important aspects of ecotoxicology, such as the biochemical basis of species sensitivity, the prevalence of multiple (and unexpected) modes of action, the basis and consequences of chemical-induced change at the population and community level, and deriving better understanding of the combined effects of pollutants.

Keywords: transcriptomics; proteomics; metabolomics; expressed sequence tags; metallothionein; heat shock protein; *Caenorhabditis elegans*; *Lumbricus rubellus*; *Folsomia candida*; *Eisenia fetida*; earthworm; Collembola; mollusc; potworm; microarray; nuclear magnetic resonance spectroscopy; mixture toxicity; metals; pesticides; oribatid mites; messenger RNA; vitellogenin; phytochelatin synthase; ferritin; cytochrome p450; glutathione S-transferase; biomarker; surface enhanced laser desorption ionisation; species sensitivity; mode of action.

Introduction

In identifying and assessing pollutant impacts on organisms in natural ecosystems, the classic approach is to use toxicity testing (Van Leeuwen

---

Corresponding author: Tel: 44(0)-1487-772561. Fax: 44(0)-1487-773467.
E-mail: dasp@ceh.ac.uk (D.J. Spurgeon).

ADVANCES IN EXPERIMENTAL BIOLOGY
VOLUME 02   ISSN 1872-2423
DOI: 10.1016/S1872-2423(08)00004-5

and Hermens, 1995). While suitable for describing acute effects on gross ecological endpoints, the focus on short-term acute tests fails to identify biochemical effects underpinning subtle changes in the life cycle. The need to understand the processes governing complex biological responses resulting from exposure to industrial chemicals and specifically acting compounds (*e.g.*, pesticides, biocides, pharmaceuticals, and endocrine disruptors) has resulted in the transfer of methods and approaches from mechanistic toxicology into ecotoxicology. The promise of this transfer for ecotoxicology was that it would bring a better understanding of the functional cascade linking exposure (potentially to a complex mixture of chemicals) to biological effect(s) (Ankley *et al.*, 2006; Eggen *et al.*, 2004).

To date, the focus in ecotoxicogenomics has been on method development and validation. The principal driver has been that the first wave of sequenced organisms (*e.g.*, *Escherichia coli* (Migula)), *Arabidopsis thaliana* (Heynh), *Saccharomyces cerevisiae* (Gasp), *Caenorhabditis elegans* (Maupas), *Drosophila melanogaster* (Meigen), *Danio rerio* (Hamilton-Buchanan) and *Homo sapiens* (Linnaeus) did not map onto the species most commonly used in ecotoxicological testing (*e.g.*, *Vibrio fischeri* (Beijerinck), various freshwater and saltwater algae, *Daphnia* spp., *Apis mellifera* (Linnaeus) (honey bees), *Eisenia fetida* (Savigny) (earthworms), *Folsomia* spp. (springtails), *Pimephales promelas* (Rafinesque) (fathead minnow) and *Coturnix* spp. (quail)). While this has been recently addressed to some extent through genome sequencing projects for *V. fischeri* (http://ergo.integratedgenomics.com/Genomes/VFI/index.html), *Daphnia pulex* (De Geer) (http://cgb.indiana.edu/genomics/projects/7) and *A. mellifera* (http://www.hgsc.bcm.tmc.edu/projects/honeybee/), there remain other ecotoxicological species for which complete (or draft) genome sequences have yet to be published. For ecotoxicogenomic studies with soil organisms, the issue concerning the availability of sequence information remains particularly acute. This is because no concerted genome sequencing effort has been initiated for those species for which standardised methods for toxicity testing exist, namely the earthworms *E. fetida* and *Eisenia andrei* (Bouche), the potworm *Enchytraeid albidus* (Henle) and the springtail *Folsomia candida* (Willem).

Rather than move the research focus in terrestrial ecotoxicogenomics to laboratory model species (*e.g.*, *Drosophila* and *C. elegans*), researchers have so far mainly chosen to focus on bridging the sequence information gap for the traditional terrestrial ecotoxicology species (for a summary of current developments for soil invertebrate ecotoxicology species used most commonly in soil ecotoxicology see Table 1). This has been done through the execution of expressed sequencing tag (EST) programmes (see further). This trend clearly shows that the terrestrial ecotoxicology field has developed mainly through the movement of traditional ecotoxicology into toxicogenomics, rather than through the application of existing toxicogenomic methods to questions in soil ecotoxicology. Perhaps the principal reason for this is the complex nature of the soil medium itself and the influence this has on exposure. To interpret toxicogenomic data correctly requires at least a partial understanding of the interaction that occurs between a species and its soil environment, and how the outcomes of the interactions determine the dynamics of exposure and uptake. This is the value of the traditional ecotoxicological test species where issues relating to exposure assessment have been comparatively well defined and addressed (Jager *et al.*, 2003; Steenbergen *et al.*, 2005; Van Gestel and Koolhaas, 2004). Thus, it may be easier to transfer observed toxicogenomic methods to traditional ecotoxicological species than it is to relate toxicogenomic effects on laboratory models exposed to field soil conditions.

This review summarises the current state of the field of terrestrial toxicogenomics, spanning the full spectrum of studies from the development of single molecular biomarkers to the application of transcriptomic, proteomic and metabolomic methods in both laboratory model species (especially *C. elegans*) and traditional soil ecotoxicology species (especially annelids and Collembola). It details the resources and methods currently available to the soil invertebrate community, the range of situations in which these have, to date, been applied and critically reviews the results so far obtained. It ends with a brief discussion on the future direction of the field.

Molecular genetic assays in soil invertebrates by single gene transcript quantification

A central paradigm in (eco)toxicology is that biological systems display a response escalade, from gene to physiological and higher organisational levels, after exposure to chemicals (Spurgeon *et al.*, 2005). Thus, the initial responses to stress are manifest at the level of gene transcription and protein synthesis or post-translational modification. At higher exposure concentrations, these molecular effects conflate to yield cell-level disruptions, followed by tissue dysfunction. Tissue damage leads ultimately to changes in the organism's life cycle, potentially leading to

Table 1. Summary of current status of sequencing and toxicogenomic studies in 15 of the soil invertebrate species most commonly used in ecotoxicology.

		Genome sequence	Available sequences in web repositories or Genbank	Toxicogenomic data available (transcriptomics)	Toxicogenomic data available (proteomics)	Toxicogenomic data available (metabolomics)
Nematode	*Caenorhabditis elegans* (Maupas)	Final	Full genome	Yes	Yes	No
Earthworm	*Eisenia fetida* (Savigny)	No	50	Yes (not fully published)	Yes	No
	Eisenia andrei (Bouche)	No	1,108	No	Yes	No
	Eisenia veneta (Rosa)	No	0	No	No	Yes
	Lumbricus rubellus (Hoffmeister)	No	17,225	Yes (not fully published)	Yes (not fully published)	Yes
	Lumbricus terrestris (Linnaeus)	No	26	No	No	Yes
Collembola	*Folsomia candida* (Willem)	No	> 7,000	No	No	No
	Orchesella cincta (Linnaeus)	No	500	No	No	No

Enchytraeid	*Enchytraeus albidus* (Henle)	No	3	Yes (not fully published)	No	No
	Cognetia sphagnetorum (Vejd)	No	0	No	No	No
Mollusc	*Helix aspersa* (Müller)	No	298	No	No	No
	Helix pomatia (Linnaeus)	No	20	No	No	No
Isopod	*Porcellio scaber* (Latreille)	No	27	No	No	No
	Oniscus asellus (Linnaeus)	No	4	No	No	No
Orabatid mite	*Platynothrus peltifer* (Koch)	No	49	No	No	No

altered population dynamics and community structure (Newman and Unger, 2003).

Recognising that changes in the expression of genes and proteins are the first indications of toxicity, the measurement of gene expression changes has been commonly used for assessing the responses of a range of aquatic and terrestrial species to chemical exposures in the laboratory and field (Kammenga et al., 2000; Peakall, 1992; Snape et al., 2004). Measuring the transcription of individual genes or their expressed (protein) products is a relatively recent trend in ecotoxicology that brings to the field of environmental diagnostics the biomarker concept that has been an accepted core 'tool' in clinical diagnostics. The measurement of biochemical response as a means of assessing exposure and/or effects of ecologically relevant animal species to pollutants has had both its proponents and detractors (Forbes et al., 2006; Triebskorn et al., 2002; Weeks et al., 2004). Indeed an active debate still remains as to whether such measurements can ever move from the research-centred laboratory into the regulatory toolbox (Weeks, 1995).

Biomarker assays can in principle be developed for a vast range of different target transcripts or proteins using a variety of different protocols for detection and quantification. A suitable method for assessing changes in the expression of any individual gene (i.e., quantification of the amount of specific mRNA present in cells) is reverse transcriptase PCR. Given reliable sequence information for the gene product in question, this method can provide a quantitative measurement of the concentrations of transcript present in tissue (Bustin, 2000). Protein abundance or activity can be quantified using a range of biochemical protocols that have been transferred, in many cases, to environmentally relevant species from mammalian or other laboratory model species.

Three generalised groups of gene- or protein-based biomarkers are commonly found to be differentially expressed in some organisms following toxic chemical exposure. These are biomarkers of exposure, of physiological compensation and of effect. Examples of studies, with soil invertebrate species where available, that have measured gene- or protein-response patterns in each of these categories are given below.

Biomarkers of exposure

As this category includes genes encoding protein known to be involved in pollutant handling and detoxification, a number of established biomarkers are included. Well-known metal-responsive examples are the genes encoding metallothioneins, phytochelatin synthase, and ferritin.

Metallothionein-like proteins have been found to be upregulated by metal-exposed earthworms (Galay-Burgos *et al.*, 2003; Sturzenbaum *et al.*, 2004), enchytraeids (Willuhn *et al.*, 1994), molluscs (Dallinger *et al.*, 2004) and springtails (Hensbergen *et al.*, 1999). Evidence for the presence of high metallothionein expression phenotypes has been found in springtail populations collected from metal-contaminated sites, indicating that these sulphydryl-rich proteins play a role in the development of metal tolerance (Sterenborg and Roelofs, 2003). The presence of a gene encoding phytochelatin synthase was first demonstrated in *C. elegans* and has since been found in other animal taxa. Initial work has demonstrated that phytochelatin is involved in the development of tolerance to Cd in *C. elegans* (Vatamaniuk *et al.*, 2005).

Primary detoxification responses to organic compounds in many species are mediated through phase 1 and phase 2 metabolic enzymes. In phase 1 metabolism, cytochrome P450s play key roles in the initial catabolism of organic compounds to polar products. In *C. elegans* it has been demonstrated that a spectrum of different cytochrome P450s may be upregulated following exposure to different organic compounds (Menzel *et al.*, 2001). In some soil mesofauna, such as Collembola, the measurement of metabolites has indicated the presence of an active cytochrome P450 system. This is demonstrated by the relative sensitivity of these species to organophosphate pesticides such as chlorpyrifos that require metabolic activation. Other taxa, such as the isopods, show intermediate cytochrome P450 activity, as indicated by measurement of polycyclic aromatic hydrocarbon (PAH) metabolites (Stroomberg *et al.*, 2003). Yet other taxa, such as earthworms may, according to preliminary evidence, show a limited upregulation of first-phase metabolism (PAHs exposure) (Achazi *et al.*, 1998; Eason *et al.*, 1998).

The main enzymes involved in phase 2 metabolism are the glutathione S-transferases (GSTs), whose role is the conjugation of phase I metabolites. The response of soil invertebrate GSTs appears, like that of the cytochrome P450s, to be species specific. For example, some species such as lacewing larvae show high upregulation (Rumpf *et al.*, 1997), while species such as *C. elegans* (Reichert and Menzel, 2005) and earthworms (SaintDenis *et al.*, 1998) show lower upregulation. Interesting patterns of GST induction have been found within the earthworms: Borgeraas *et al.*, (1996) observed that exposure to *trans*-stilbene oxide, 3-methylcholanthrene and phenobarbital did not change GST levels in the closely related species *E. andrei* and *Eisenia veneta* (Rosa); while Hans *et al.*, (1993) observed that exposure to aldrin, endosulphan and lindane did induce GST in *Pheretima posthuma* (Vaillant). That this induction was transient may explain the apparent differences between the species, although an alternative interpretation may be that the multiple GST isoenzymes known to be present in *Eisenia* may each be substrate specific (Borgeraas *et al.*, 1996).

Biomarkers of physiological compensation

This category includes a range of genes involved in adaptive biochemical pathways. Best known and most established as biomarker candidates are the heat shock family of protein chaperones (HSPs). Expression of HSPs has been found to be altered following exposure to a range of pollutants under both laboratory and field conditions in species such as earthworms (Marino *et al.*, 1999; Nadeau *et al.*, 2001), oribatid mites (Kohler *et al.*, 2005), woodlice (Arts *et al.*, 2004; Knigge and Kohler, 2000) and nematodes (Arts *et al.*, 2004). Other genes involved in compensatory responses to chemical exposure are certain mitochondrial genes (Galay-Burgos *et al.*, 2003), whose upregulation is a consequence of increased energy demand or direct toxic effects on the organelle. Another network of compensatory genes are linked to lysosomal function (Kille *et al.*, 1999; Liao *et al.*, 2002). Upregulation of these represents an increase in lysosomal activity in pollutant-exposed cells that is associated with the oxyradical-mediated toxic effects of the chemicals on the structural and functional integrity of the lysosomal membrane (Svendsen *et al.*, 2004). Metal-containing enzymes can also be upregulated in response to pollutant exposure, notably when the pollutant competes for binding in the active site. Finally, genes encoding proteins in the antioxidant system, such as superoxide dismutase, catalase and peroxidase, are also commonly found to be responsive to chemical and other stress-inducing agents in a diverse range of species. Indeed, quantitative changes in the levels of antioxidant enzymes, as well as in the non-enzyme components of the system, are amongst the most widely observed physiological responses to chemical-evoked stress in plants and animals (De Coen and Janssen, 2003).

Biomarkers of effect

Biomarkers reflecting toxicosis are by definition involved in biochemical and physiological events. A well-known example of a biomarker linked to perturbations of reproductive output is the egg yolk protein vitellogenin. Vitellogenin assays provide well-established biomarkers of exposure to environmental oestrogens in aquatic invertebrate and vertebrate species (Brion *et al.*, 2004; Celius *et al.*, 2000; Porte *et al.*, 2006; Wheeler *et al.*, 2005). Despite tentative early observations suggestive of the presence of vitellogenin-like proteins in terrestrial invertebrate species (Okuno *et al.*, 2000; Rouabahsadaoui and Marcel, 1995), there are to date no definitive studies reporting quantitative changes in these proteins following chemical exposure. Other genes associated with reproduction that have been quantified as indictors of toxicant effect are the zona pellucida proteins (Arukwe *et al.*, 2001a, 2001b; Denslow *et al.*, 2001). Although identified in the freshwater bivalve *Unio elongatulus* (Pfeiffer) (Focarelli *et al.*, 2001), these protein have yet to be well characterised in any terrestrial invertebrate species. Annetocin, an oxytocin-related peptide isolated from the earthworm *E. fetida*, is another gene functionally linked to reproduction. Annetocin is known to induce a series of egg-laying-related behaviours in this oligochaete species (Inomata *et al.*, 2000; Oumi *et al.*, 1994). Ricketts *et al.* (2004) compared annetocin mRNA levels in two groups of reference earthworms exposed to an unpolluted soil and a metalliferous mine soil. Both reproductive output (number of egg capsules produced per unit time) and the level of annetocin expression were significantly lowered by exposure to the mine soil.

As mentioned above, measuring changes in the abundance of a single gene product or transcribed protein as potential biomarkers of exposure and/or effect remains a controversial concept in ecotoxicology (Forbes *et al.*, 2006). While it is incontrovertible that measurement of molecular-level responses can provide fundamental insights into the mechanistic basis of toxicosis and homoeostatic regulation, such measurements have to date rarely been used in formal risk assessment and in post-release monitoring. In aquatic ecotoxicology, measurement of plasma vitello-genin has become an established part of the assessment for environmen-tally mediated endocrine disruption in fish and has been subject to validation as a potential screening method. In soil invertebrates, however, there is no single biomarker that is yet established as a cornerstone of any prospective or retrospective risk assessment, although measurement of earthworm metallothionein induction is included in the suite of tests that can be used to support the assessment of the ecological effects of contaminated land in the UK (Environment Agency, 2003).

A perceived hurdle for the application of biomarkers in terrestrial risk assessment is that a number of the most commonly measured biomarkers are also responsive to environmental variables such as soil pH, heat and cold shock, disease, starvation and pH (Korsloot *et al.*, 2004). This absence of specificity presents a confounding factor in the interpretation of biomarker performance in wild populations. In an attempt to counter these limitations, reviews of the application of biomarkers in soil ecotoxicology have recommended the use of suites of markers rather than single measurements (Kammenga, 2000; Wharfe et al., 2004). The 'toolbox' of techniques available in contemporary toxicogenomics (transcriptomics, proteomics and metabolomics) each have the capacity to provide extensive, multivariate, datasets and, unsurprisingly, environmental toxicologists have begun to exploit these analytical resources for assessing chemical effects.

The power of the 'omic' technologies has quickly established them as core new tools in toxicology research. DNA microarrays, for example, have been used in toxicology for clustering chemicals according to their modes of action (Parker et al., 2003; Steiner et al., 2004) and in the categorisation of chemicals into those causing hepatic damage and those compromising renal function (Fielden et al., 2005; Natsoulis et al., 2005). Natsoulis et al. (2005) established a database of thousands of separate microarray analyses that were combined with histological observations and a supervised multivariate data analysis method (support vector machines) to identify a minimal transcript signature of early-stage liver and renal tissue damage. These signatures were then used as the effects-driven basis on which to categorise unknown compounds according to the risks they pose to the two target organs. This powerful and integrative approach provides a stimulating precedent for environmental biology and pollution monitoring, and could present a new paradigm for ecotoxicological effect assessment.

Soil invertebrate transcriptomics

Microarray technology is increasingly finding application in fundamental and applied ecotoxicological research. The early availability of a full-genome sequence for C. elegans (www.wormbase.org) has meant that this species was among the first for which microarrays were commercially available. Consequently, the soil-dwelling nematode, though most familiar as a laboratory model organism, was also the first to be used in terrestrial ecotoxicogenomics. Because it is more often seen living on agar in the laboratory, it can be easily overlooked that C. elegans is a 'real' free-living species, inhabiting a wide range of soil types and climatic regions (Sivasundar and Hey, 2003). That said, it is chiefly the advantages that made C. elegans well suited for studies in functional genetics (a fully sequenced genome, short life cycle, finite number of cells, easy handling, etc.) that have also made the species the subject of theoretical ecotoxicological work. Research areas where *C. elegans* has been used include assessment of the effects of chemical challenge on strains with different life-history characteristics (Alvarez *et al.*, 2005; Kammenga and Riksen, 1996) and modelling the effects of chemical mixtures (Jonker *et al.*, 2004a). While such theoretical work is most usually undertaken using the same agar culturing medium in most functional genetics work, a method for assessing chemical effects in soil has also been developed and standardised (Jonker *et al.*, 2004b; Peredney and Williams, 2000).

Given the volume of ecotoxicological and 'omic' resources that are presently available, it is slightly surprising that, although first to be used, *C. elegans* has so far been somewhat underemployed in ecotoxicogenomics. Custodia *et al.*, (2001) have assessed the responses of *C. elegans* to the hormones testosterone, oestrogen and progesterone that are frequent contaminants of sewage effluent. An extension of this work also considered the nematode's responses to Cd, the principal hallmark of which was the upregulation of stress response pathways and metallothioneins (Novillo *et al.*, 2005). Reichert and Menzel (2005) used a full-genome microarray to study the effects of single doses of five different xenobiotics on gene expression in *C. elegans*. They found that 203 genes belonging to different families (like the cytochrome P450s, UDP-glucoronosyltransferases, GSTs, carboxylesterases, collagens, C-type lectins) were significantly induced, and 153 genes significantly repressed. In addition to microarray, other methods, such as subtractive suppression hybridisation or SAGE, for measuring differential expression have been used to identify genes involved in metal and organic chemical handling (Liao and Freedman, 1998; Menzel *et al.*, 2005). *C. elegans* is, of course, amenable to genetic manipulation. This provides a myriad of functional approaches that may be brought to bear to examine the role of these candidate genes in pollutant handling and detoxification (Swain *et al.*, 2004).

Although the use of *C. elegans* in terrestrial ecotoxicogenomics is increasing, the focus of the field still remains firmly on traditionally studied taxa such as earthworms, potworms and springtails and to a lesser extent carabids, molluscs and oribatid and predator mites. For these species, the fact that no microarray studies are currently published is a clear indication that terrestrial toxicogenomics lags behind the aquatic field, in which recent papers have demonstrated the application of partial genome microarrays to investigate the basis of stress and fitness (reproductive) responses. These include gene expression change in

Daphnia magna (Müller) after exposure to propiconazole (Soetaert *et al.*, 2006) and in European flounder (*Platichthys flesus* (Linnaeus)) after exposures to Cd and complexly polluted river waters (Sheader *et al.*, 2006; Williams *et al.*, 2003). While there remains (as at end 2006) no published transcriptomic study in any ISI cited journal on the response of any invertebrate species other than *C. elegans* to chemical exposure in soil, it is clear from analysis of internet sites and conference proceedings that much of the groundwork required for the generation of microarrays suitable for toxicogenomic studies in a number of different soil taxa is well under way.

The website www.earthworms.org reports the sequencing and annotation of genes from four annelid species, two species of earthworm (*Lumbricus rubellus* (Hoffmeister), *E. andrei*), a marine polycheate (*Nereis virens* (Sars)) and a freshwater leech (*Haementeria depressa* (Ringuelet)). For *E. andrei*, one of the two species (with *E. fetida*) recommended by the OECD for use in standardised laboratory toxicity tests (OECD, 2004), 1,108 ESTs clustered into 690 gene objects are included in the database. These ESTs have all been sequenced from a library constructed using tissue taken from the mid-hindgut region. Since this area is away from the anterior region of the earthworm, where many of the major organs such as the calciferous glands, gonads and main nerve ganglia are located, these ESTs are likely to represent a somewhat limited population of the range of genes in earthworms. As such, it is a matter of priority for the earthworm ecotoxicogenomic community to extend the size and tissue diversity of the sequence resource for this species to allow transcriptomic and proteomic methods to be developed for this high-profile 'standard' test species.

For the other test species, *E. fetida*, there is presently a dearth of sequence data lodged in public repositories. However, even though there are no peer-reviewed publications describing transcriptomic profiling in *E. fetida*, a search of conference proceeding indicates that efforts to utilise the species for toxicogenomics are under way (see http://abstracts.co. allenpress.com/pweb/setac2005/document/?ID = 57039). As far as we are able to ascertain, these studies have to date focused on the development of small-scale microarray-containing clones enriched for genes affected by selected metal and energetic compounds. Initial application of the array to investigate the transcription effects of trinitrotoluene concluded that short-term exposure to it inhibits oxygen transport systems, while longer-term exposure results in significant damage to cellular proteins and macromolecules. This focus on functionality within this short report suggests that annotation of the microarray has been (at least partially) conducted.

145

The genomic information available for the epigeic earthworm *L. rubellus* is far more comprehensive than for other members of the taxon or indeed any other species of soil-dwelling invertebrate except *C. elegans*. The LumbriBASE resources available for this species at www.earthworms.org currently contains details of over 17,225 single-pass EST sequences (summarised in Fig. 1). These sequences have been generated from nine cDNA libraries constructed from a range of tissues, developmental stages, and from chemical (Cd, Cu, fluoranthene, atrazine)-exposed worms. The current database builds on an initial EST project (Sturzenbaum *et al.*, 2003), and has since been extended such that EST clustering, predicted translations, partial protein mass fingerprints, functional annotation and additional tools to facilitate searching and data visualisation are all available. Informatic analysis estimates that the complement of ESTs so far obtained for *L. rubellus* derives from 8,129 distinct gene objects, of which 41% have been assigned to a gene ontology term.

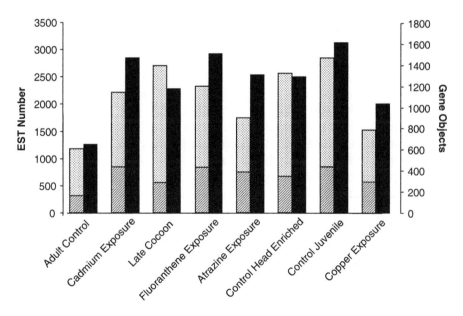

Fig. 1. Summary of expressed sequence tag (EST) information for the earthworm *L. rubellus* held in the LumbriBase web repository (www.earthworms.org). The lower shaded area of the left-hand bars indicates the number of ESTs which do not lie within a cluster (singletons) while the top lightly shaded area indicates the number of ESTs grouped into clusters. The total height of the left-hand bar therefore indicates the total number of ESTs sequenced from each library. The right-hand solid bars indicate the number of gene objects represented in each library.

The fact that fully annotated sequence information and clone sets are available for *L. rubellus* has allowed fabrication of a custom glass slide microarray for this species. Summary data for exposures conducted with two metals (Cd and Cu), the non-polar PAH fluoranthene and the herbicide atrazine for the complete set of gene objects are all accessible through www.earthworms.org. To illustrate the value of the dataset, we analysed the results from dual-labelled microarray experiments in which labelled RNA samples from groups of worms exposed in soil to a control (eight replicates of three pooled worms) and four Cd concentrations of 0, 13, 43, 148, 500 $\mu g\,g^{-1}$ (five replicates of three pooled worms for each concentration) were hybridised against a universal reference. Following mean polishing per chip and per gene normalisation and filter for discretionary profiles, analysis by principal components analysis (PCA) indicated a wide spread of the replicates. Nonetheless the biological replicates exposed to Cd concentrations above $148\,\mu g\,g^{-1}$ were separated from control and lowest exposure concentration ($13\,\mu g^{-1}$ soil) along PCA component 1 (Fig. 2). Analysis of the 27 genes showing a >10-fold change in worms exposed to Cd at $500\,\mu g\,g^{-1}$ soil indicated that these included three different isoforms of metallothioneins. This finding supports the previous quantitative PCR observation of a large metallothionein upregulation in *L. rubellus* following Cd exposure (Spurgeon *et al.*, 2005; Sturzenbaum *et al.*, 2004), thus providing independent verification of the suitability of the microarray for ecotoxicological applications. Since the same consortium (see: www.earthworms.org) has also conducted exposures with the same compounds in *C. elegans*, it may be expected that insight into the evolutionary cross-taxa conservation of response mechanisms can emerge.

For the other group of annelids commonly used in terrestrial ecotoxicology, the potworms or enchytraeids (Römbke and Moser, 2002), the development of resources for toxicogenomics current lags behind that for earthworms. Nonetheless at the Society of Environmental Toxicology and Chemistry 16th Annual European meeting in The Hague, a poster was presented that featured results on the initial cloning of *E. albidus* mRNAs, as well as the fabrication of a glass slide microarray and initial results of hybridisations for control and Cu-exposed worms (M. Amorim, personal communication). While no details are available in the open literature on the success or otherwise of this project, the fact that the work was conducted as a collaboration between ecotoxicologists and molecular biologists indicates that the application of genomic tools for assessing toxicosis in potworms can be anticipated in the near future.

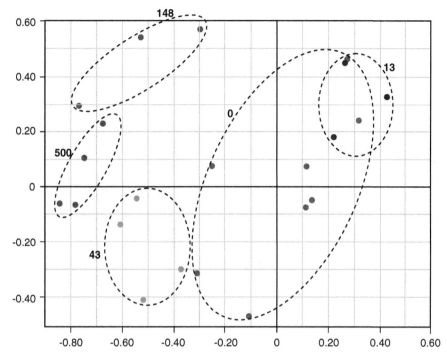

x-axis: PCA component 1 (39.17% variance)
y-axis: PCA component 2 (14.36% variance)

Fig. 2. Principal components analysis (PCA) of mean polishing per chip and per gene normalised cDNA microarray data for earthworms exposed to Cd at the shown concentrations of 13, 43, 148 and 500 $\mu g\,g^{-1}$. Each biological replicate consists of a cDNA sample generated by pooling tissues from three adult worms from a single box exposed for 28 days at 12°C and under a 16 h:8 h light:dark photoperiod. (See color figure 4.2 in color plate section).

Collembola are, like the annelids, important standard test species for soil ecotoxicology (ISO, 1999). Studies have confirmed the presence of heritable Cd-resistance in *Orchesella cincta* (Linnaeus) populations with multigenerational exposure histories (Posthuma *et al.*, 1993). Furthermore, recent evidence suggests that Cd-resistance in these ecotypes may be mediated through polymorphisms present in the promoter region of the metallothionein gene (Roelofs *et al.*, 2006). Given both the availability of a published toxicity testing methods for *O. cincta* (Nottrot *et al.*, 1987) and evidence describing in structural detail the role of the genome in determining Cd-resistance it is not a surprise that this species has been the focus of recent work to develop transcriptomic tools for Collembola toxicogenomics. According to the website www.collembase.org, as of October 2006 >7,000 *F. candida* ESTs had been sequenced, including almost 800 from springtails exposed to Cd and phenanthrene. In addition, over 500 *O. cincta* EST sequences had also been generated. The inclusion of ESTs in the database clearly indicates the intention of the researchers leading this project to use the generated sequences as probes for mechanistic toxicological work in pollutant-exposed springtails.

Among the remaining groups of soil invertebrates, the paucity of clone sets and annotated sequence information remains a hurdle to transcriptomic studies. For carabid beetles, soil-dwelling mites, woodlice, slugs, snails, centipedes and millipedes sequence resources remain extremely limited. In each of these cases, individual species have been the subject of either very limited EST sequencing (Davison and Blaxter, 2005; Hughes *et al.*, 2006) or, worse, no EST sequencing programme at all. Consequently, annotated microarrays remain unavailable for the majority of ecologically relevant soil-dwelling taxa. The ascent of these taxa into the pantheon of toxicogenomic organisms remains a priority task for the future.

Soil invertebrate proteomics

Proteomics has been made possible by the advent of analytical platforms for the separation of proteins based on physical characteristics (such as size, or charge, or both) and identification using time-of-flight mass spectrometry (for review see Ferguson and Smith, 2003). The most rapid identification of excised proteins is possible when information is available on the complete sequenced genome of a particular species. Unfortunately, with the exception of *C. elegans*, this ideal is presently unattainable for soil invertebrates. For unsequenced or partially sequenced species, however, it may be possible to use data from ESTs for protein identification (Wasmuth and Blaxter, 2004). Alternatively, more time-consuming tandem mass spectrometry methods can be used for identification.

The two-dimensional separation (2-DE) and mass spectrometry approaches to proteomics have been applied in human health research (Downes *et al.*, 2006; Fu and Van Eyk, 2006; Solassol *et al.*, 2006). In aquatic systems, there has been a number of recent papers that have used 2-DE to describe mechanisms of Cd toxicity in *Chironomus riparius* (Meigen) and *Eriocheir sinensis* (Milne-Edwards) (Lee *et al.*, 2006;

Silvestre *et al.*, 2006). These studies indicated that the expression levels of only a relatively small number of proteins (<25) were significantly altered (upregulated or downregulated) following exposures. The small number of downregulated proteins compared with the relatively large number of downregulated transcripts following other soil species (see aforementioned) suggests that transcriptional changes may not be directly related to protein expression. However, it is noteworthy that the 2-DE method has a low relative sensitivity, and is consequently unable to detect changes in the expression of low-abundance proteins.

So far, for terrestrial invertebrates there is only a single published study reporting the use of 2-DE proteomics in ecotoxicology. In this work, Kuperman *et al.* (2004) used the approach to identify differentially expressed proteins in earthworms (*E. fetida*) exposed to chemical warfare agents. The study failed to find a clear concentration–response relationship between exposure concentrations and the numbers of proteins induced or suppressed. The authors suggest this may be because very low concentration ranges were used for the exposure. They go on to further suggest that extending the concentration range to encompass concentrations that produce significant impacts on earthworm survival and/or reproduction may produce measurable changes in protein expression, though the value of identifying the proteomic changes associated with such gross effects is mainly of academic interest rather than of practical use for risk assessment or monitoring. More comprehensive studies of global- and sub-proteome (GST superfamily) have also been undertaken in *L. rubellus* exposed to Cd, fluoranthene, and atrazine. Both tissue- and exposure-specific biomarkers have been highlighted, and the differentially expressed protein identities established by comparing peptide mass fingerprints against the available EST database (P. M. Brophy, personal communication).

Surface enhanced laser desorption ionisation (SELDI) technology offers an alternative approach to identify diagnostic or prognostic patterns of protein expression associated with chemical exposure or disease. The SELDI approach has several advantages over traditional 2-DE proteomics, including: (i) higher-throughput, (ii) versatility, (iii) ease of use, and (iv) comparative low cost. Using a SELDI analysis, Pampanin (2004) analysed haemolymph (blood) samples from a number of marine species. The aim of the study was to establish if the technology could be used to differentiate protein profiles from exposure and control samples. Results confirmed this potential. For example, it was possible to differentiate the protein profiles of spider crabs (*Hyas araneus* (Linnaeus)) exposed to diallyl phthalate, bisphenol A and polybrominated diphenyl ether 47 from controls, and of shore crabs (*Carcinus maenas* (Linnaeus)) exposed to crude oil and crude oil spiked with alkylphenols and 4-nonylphenol from matched controls. No SELDI-derived data appear to have been published in the context of terrestrial ecotoxicology, even for *C. elegans*. In fact soil-dwelling invertebrates have received but scant proteomic attention.

Soil invertebrate metabolomics

The comprehensive measurement of the non-polymeric substrates and molecular products of the biochemical activities in cells and tissues is the raison d'être of metabolomics. By comparing metabolite profiles between exposed and unexposed individuals quantitative/qualitative informa- tion about stressor-induced shifts in biochemical pathways can be gained, and hypotheses generated for later testing through targeted work. In common with proteomics, there are several chromatography, mass spectrometry and spectroscopy platforms suitable for metabolomics. Of the available methods, however, two approaches have so far predomi- nated in soil invertebrate studies. These are nuclear magnetic resonance (NMR) spectroscopy on the one hand and gas or liquid chromatography separation followed by mass spectrometry (GC-MS, LC-MS) on the other.

NMR spectroscopy is probably the most widely used metabolomics platform. The technique is based on the fact that atomic nuclei orientated by a strong magnetic field absorb radiation at characteristic frequencies (Robertson *et al.*, 2005). In different atomic environments, nuclei of the same element give rise to distinct spectral lines. This makes it possible to observe and measure signals from individual atoms in complex macro- molecules and, from these, to interpret molecular structure. There are only a limited number of NMR-active nuclei: ^{1}H is the most sensitive and stable nucleus; ^{13}C may also be used, but has low sensitivity and abundance; ^{31}P and ^{19}F can be used for specific applications (Robertson *et al.*, 2005).

The use of NMR spectroscopy in ecotoxicology and environmental sciences is undeveloped compared with its application in toxicology and drug discovery (Lindon *et al.*, 2003; Nicholson and Wilson, 2003; Nicholson *et al.*, 2002). The method has also been used in environmental epidemiology to classify diseased shellfish, and to identify the mechan- isms and putative early biomarkers of the disease phenotype (Viant *et al.*, 2003a, 2003b). In soil invertebrate ecotoxicology, ^{1}H-NMR has been used in earthworms to detect elevated free histidine levels in tissue extracts of *L. rubellus* (Bundy *et al.*, 2001; Gibb *et al.*, 1997), and to define metabolite profiles for *E. veneta* exposed to different substituted anilines

(Bundy *et al.*, 2001; Warne *et al.*, 2000). NMR was also used by Bundy *et al.* (2002) to classify the metabolite commonalities and distinctions in response to different fluorinated organic compounds. NMR-based metabolomics has been used to describe site-specific phenotype differences in earthworms (*L. rubellus*) at sampling stations along a pollution gradient (Bundy *et al.*, 2004). Interrogation of the acquired metabolomic profiles indicated that changes in sugar and histidine tissue contents were the main variables 'driving' the observed statistical separations (Bundy *et al.*, 2004).

Although mass spectrometery-based metabolomics has been used in a variety of toxicological applications (Griffin and Bollard, 2004; Lenz *et al.*, 2005), there are only a limited number of examples of the use of GC-MS or LC-MS in ecotoxicology. A notable exception is that GC-MS has recently been used in combination with NMR to investigate the physiological basis of pyrene toxicity in the earthworm *L. rubellus* (O. Jones, personal communication). Alterations in metabolite concentrations were sufficient to allow statistical separation of worms exposed to a soil pyrene concentration of $40\,mg\,kg^{-1}$ from controls. This applied whether metabolite measurements were made using either GC-MS or NMR. Detailed analysis of the data indicated that pyrene decreased free fatty acid and increased free amino acid contents, traits indicative of a switch in metabolism toward the consumption of lipids and enhanced proteolysis. These changes were more pronounced in worms exposed to higher concentration of pyrene in soil. The combined use of GC-MS and NMR proved valuable for identifying the increased number of metabolite species present at different concentrations in the exposed and control worms.

Future perspectives in soil invertebrate ecotoxicogenomics

Research activity in terrestrial invertebrate ecotoxicogenomics is roughly on a par, currently, with that for aquatic invertebrate species. While the initial focus was understandably on the exploitation of *C. elegans* as a 'representative' soil species, resources for application in other soil taxa, such as earthworm and springtails, are burgeoning. Nevertheless, investment and commitment in the fields of terrestrial and aquatic ecotoxicology lag far behind those in the toxicology field, where the commercial gains to be made in drug discovery and drug safety evaluation are a motivating force (Fielden *et al.*, 2005; Lindon *et al.*, 2005; Natsoulis *et al.*, 2005).

Deriving some guidance from the evolution of ecotoxicogenomics in the field of biomedicine, a number of areas where ecotoxicogenomics can be anticipated to exert a major impact can be predicted. Some of these areas are highlighted below.

Species sensitivity

That species differing in their sensitivity to toxicants is one of the bedrock concepts in ecotoxicology, and provides the foundation of a swathe of risk assessment policies that incorperate the species sensitivity paradigm (Van Straalen and Denneman, 1989). Ecotoxicogenomics has potential to provide a sounder knowledge of the basis of species sensitivity in support of such risk assessments. As an example, toxicogenomics may be used to establish the activity of pathways involved in first- and second-phase metabolism in different taxa, and these data may be then used to infer toxicological sensitivity for metabolically inactivated and activated compounds (*e.g.*, pesticides). A major obstacle in achieving this goal is the lack of sequence data among species belonging to some of the most important taxonomic groups represented in the soil fauna (*e.g.*, molluscs, centipedes, millipedes, and oribatid mites). The universality of metabolites means that metabolomics is at present probably the best technique for comparing responses between closely- and distantly-related species. That said, the development of very high-throughput sequencing technologies (Margulies *et al.*, 2005), and the resulting deflationary effect on costs, means it is likely that the number of species for which large numbers of annotated ESTs become available for custom microarray fabrication will increase rapidly in the foreseeable future. A concerted effort will be needed within the research community to share these resources to maximise the value of the advances and to allow truly comparative ecotoxicogenomic studies to take place.

Unexpected modes of action

The ecotoxicological literature is dominated by examples of compounds that cause effects in non-mammalian species that were not foreseen by the findings of laboratory tests undertaken for safety assessment purposes (*e.g.*, tri-butyl tin in marine molluscs, organochlorines in birds of prey, diclofenac in vultures). Toxicogenomics certainly has a role to play in identifying the mechanisms responsible for such unexpected responses through pathway analyses of transcriptomic/proteomic/ metabolomic profiles. Evidence of significant changes in unexpected key functional pathways (*e.g.*, calcium metabolism, an endocrine axis) could act as stimuli and signposts for further detailed, directed assessments. To realise this potential, much work needs to be done to sequence and annotate ESTs in receptor species judged to be most vulnerable to environmental contaminant exposures, or whose essential ecological services are most likely to be perturbed. The clone sets for these sentinel or keystone species should then be made available as fully annotated microarrays for laboratory assessments or field deployment purposes.

Consequences of exposure for populations and communities

Ultimately, the aim of regulatory ecotoxicology is to protect populations and communities. To ensure that current risk assessment procedures are effectively protecting soil communities and processes, field studies of the individual, population, and community consequences of exposure are needed. Can ecotoxicogenomics contribute to such assessments? Currently, ecotoxicogenomic field assessments are based on response profiling in sentinel species. Whilst providing a useful species-specific picture of the exposure–response relationship, this approach is only a surrogate for proper assessment of the ecological status (richness, evenness, diversity and rank abundance characteristics) of the exposed community. The emerging field of metagenomics makes it possible to undertake holistic assessments to yield insights into the diversity and functional status of ecological communities (Venter *et al.*, 2004). Elevating such methods to quantitative status could provide novel, high-throughput, means of objectively describing site-specific faunistic composition. This would be a timely as well as a powerful biomonitoring advance in a research climate where the number of specialist taxonomists available to undertake the traditional, morphology-based, studies needed to do this in all soil surface and sub-surface taxa (*e.g.*, spiders, beetles, earthworms, springtails, mites, and nematodes) is in serious decline.

Applications in mixture toxicity assessment

Two concepts, concentration addition and independent action, are frequently used to model the joint effects of multiple chemicals present in pollutant mixtures (Jonker *et al.*, 2005). The choice of which model to use is driven by the availability of prior information concerning the mode of action of the compounds in the mixture. Compounds with the same mode of action are modelled using the concentration addition model; those with different modes of action using the independent action model.

154

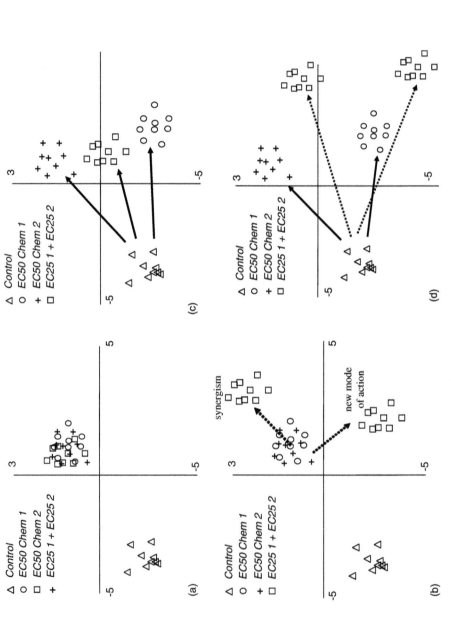

Fig. 3. A theoretical framework for toxicogenomic responses in mixture tests for similarly and dissimilarly acting compounds with and without synergistic interaction (see text for full description).

155

Ecotoxicogenomics can provide insight into the circumstances under which these alternative models succeed and, more importantly, fail to describe joint effects (*e.g.*, due to synergism or antagonism between chemicals).

For concentration addition, exposure of a given species to the respective EC_{50}s of two compounds with similar modes of action would be expected to result in a similar shift in physiological response profile (*e.g.*, its transcriptome) away from the control condition to a similar status point (*e.g.*, in a PCA plot) (Fig. 3a). When exposed to a mixture of half the EC_{50}s of both compounds, the concentration addition model states that these exposures will sum to give a 50% effect. When an interaction causes toxicity to deviate from model predictions (*e.g.*, synergism), this would be expected to yield a response profile differing more from control. If this interaction occurs through the main mode of action the consequential deviation may lead to greater perturbation of the pathways already affected (Fig. 3b). When the interaction causes a new mode of toxicity to occur then the interaction may move the organism's physiology to a different state (Fig. 3b).

In situations conforming to the independent action model, exposure of the species to the respective EC_{50}s of differently acting compounds would result in physiological changes that separate individuals from the control and each other (Fig. 3c). Simultaneous exposure to half the EC_{50} of two such compounds produces an intermediate phenotype. If the interaction is, for example, synergistic, then the phenotype would remain intermediate, but the response of individuals would differ more than anticipated from those of the controls. This prediction holds unless the interaction causes toxicity through a new mode of action, in which case the response patterns may mean that individuals cluster in another part of the response space (Fig. 3d). Using ecotoxicogenomic protocols may provide a robust and efficient means of determining the nature of mixture effects (concentration addition, independent action). Ecotoxicogenomics may also provide effective tools for determining how specific interactions cause deviations from predicted joint effects.

Acknowledgements

This work is an extension of a review article originally commissioned by the Environment Agency of England and Wales under a project entitled 'Review of biomonitoring techniques, supported by fieldwork, to investigate use of environmental outcomes in regulating air pollution'. Experimental data of earthworm transcriptomics were collected during a research grant from the UK Natural Environmental Research Council (NER/T/S/2001/00021) with additional support from the AstraZeneca Brixham Environmental Laboratory.

References

Achazi, R. K., Flenner, C., Livingstone, D. R., Peters, L. D., Schaub, K. and Scheiwe, E. (1998). Cytochrome P450 and dependent activities in unexposed and PAH-exposed terrestrial annelids. *Comp. Biochem. Physiol. B* 121,339–350.

Alvarez, O. A., Jager, T., Kooijman, S. A. L. M. and Kammenga, J. E. (2005). Responses to stress of *Caenorhabditis elegans* populations with different reproductive strategies. *Funct. Ecol.* 19,656–664.

Ankley, G. T., Daston, G. P., Degitz, S. J., Denslow, N. D., Hoke, R. A., Kennedy, S. W., Miracle, A. L., Perkins, E. J., Snape, J., Tillitt, D. E., Tyler, C. R. and Versteeg, D. (2006). Toxicogenomics in regulatory ecotoxicology. *Environ. Sci. Technol.* 40,4055–4065.

Arts, M.-J., Schill, R. O., Knigge, T., Eckwert, H., Kammenga, J. E. and Köhler, H.-R. (2004). Stress proteins (hsp70, hsp60) induced in isopods and nematodes by field exposure to metals in a gradient near Avonmouth, UK. *Ecotoxicology* 13,739–755.

Arukwe, A., Kullman, S. W. and Hinton, D. E. (2001a). Differential biomarker gene and protein expressions in nonylphenol and estradiol-17 beta treated juvenile rainbow trout (*Oncorhynchus mykiss*). *Comp. Biochem. Physiol. C Pharmacol. Toxicol.* 129,1–10.

Arukwe, A., Yadetie, F., Male, R. and Goksoyr, A. (2001b). In vivo modulation of nonylphenol-induced zonagenesis and vitellogenesis by the antiestrogen, 3,3′ 4,4′-tetrachlorobiphenyl (PCB-77) in juvenile fish. *Environ. Toxicol. Pharmacol.* 10,5–15.

Borgeraas, J., Nilsen, K. and Stenersen, J. (1996). Methods for purification of glutathione transferases in the earthworm genus *Eisenia*, and their characterization. *Comp. Biochem. Physiol C Pharmacol. Toxicol.* 114,129–140.

Brion, F., Tyler, C. R., Palazzi, X., Laillet, B., Porcher, J. M., Garric, J. and Flammarion, P. (2004). Impacts of 17 beta-estradiol, including environmentally relevant concentrations, on reproduction after exposure during embryo-larval-, juvenile- and adult-life stages in zebrafish (*Danio rerio*). *Aquat. Toxicol.* 68,193–217.

Bundy, J. G., Lenz, E. M., Bailey, N. J., Gavaghan, C. L., Svendsen, C., Spurgeon, D., Hankard, P. K., Osborn, D., Weeks, J. M., Trauger, S. A., Speir, P., Sanders, I., Lindon, J. C., Nicolson, J. K. and Tang, H. (2002). Metabonomic assessment of toxicity of 4-fluoraniline, 3,5-difluoraniline and 2-fluoro-4-methylaniline to the earthworm *Eisenia veneta* (Rosa): Identification of new endogenous biomarkers. *Environ. Toxicol. Chem.* 21,1966–1972.

Bundy, J. G., Osborn, D., Week, J. M., Lindon, J. C. and Nicholson, J. K. (2001). NMR-based metabonomic approach to the investigation of coelomic fluid biochemistry in earthworms under toxic stress. *FEBS Lett.* 500,31–35.

Bundy, J. G., Spurgeon, D. J., Svendsen, C., Hankard, P. K., Weeks, J. M., Osborn, D., Lindon, J. C. and Nicholson, J. K. (2004). Environmental metabonomics: Applying combination biomarker analysis in earthworms at a metal contaminated site. *Ecotoxicology* 13,797–806.

Bustin, S. A. (2000). Absolute quantification of mRNA using real-time reverse transcription polymerase chain reaction assays. *J. Mol. Endocrinol.* 25,169–193.

Celius, T., Matthews, J. B., Giesy, J. P. and Zacharewski, T. R. (2000). Quantification of rainbow trout (*Oncorhynchus mykiss*) zona radiata and vitellogenin mRNA levels using real-time PCR after in vivo treatment with estradiol-17 beta or alpha-zearalenol. *J. Steroid Biochem. Mol. Biol.* 75,109–119.

Custodia, N., Won, S. J., Novillo, A., Wieland, M., Li, C. and Callard, I. P. (2001). *Caenorhabditis elegans* as an environmental monitor using DNA microarray analysis. Environmental hormones: The scientific basis of endocrine disruption. *Ann. N. Y. Acad. Sci.* 948,32–42.

Dallinger, R., Lagg, B., Egg, M., Schipflinger, R. and Chabicovsky, M. (2004). Cd accumulation and Cd-metallothionein as a biomarker in *Cepaea hortensis* (Helicidae, Pulmonata) from laboratory exposure and metal-polluted habitats. *Ecotoxicology* 13,757–772.

Davison, A. and Blaxter, M. L. (2005). An expressed sequence tag survey of gene expression in the pond snail *Lymnaea stagnalis*, an intermediate vector of *Fasciola hepatica*. *Parasitology* 130,539–552.

De Coen, W. M. and Janssen, C. R. (2003). A multivariate biomarker-based model predicting population-level responses of *Daphnia magna*. *Environ. Toxicol. Chem.* 22,2195–2201.

Denslow, N. D., Bowman, CJ., Ferguson, R. J., Lee, H. S., Hemmer, M. J. and Folmar, L. C. (2001). Induction of gene expression in sheepshead minnows (*Cyprinodon variegatus*) treated with 17 beta-estradiol, diethylstilbestrol, or ethinylestradiol: The use of mRNA fingerprints as an indicator of gene regulation. *Gen. Comp. Endocrinol.* 121,250–260.

Downes, M. R., Byrne, J. C., Dunn, M. J., Fitzpatrick, J. M., Watson, R. W. G. and Pennington, S. R. (2006). Application of proteomic strategies to the identification of urinary biomarkers for prostate cancer: A review. *Biomarkers* 11,406–416.

Eason, C. T., Booth, L. H., Brennan, S. and Ataria, J. (1998). Cytochrome P450 activity in 3 earthworm species. In *Advances in Earthworm Ecotoxicology* (eds S. Sheppard, J. Bembridge, M. Holmstrup and L. Posthuma), pp. 191–198, SETAC Press, Pensacola, FL.

Eggen, R. I. L., Behra, R., Burkhardt-Holm, P., Escher, B. I. and Schwegert, N. (2004). Challenges in ecotoxicology. *Environ. Sci. Technol.* 38,59A–64A.

Environment Agency. (2003). *A public consultation on a framework and methods for assessing harm to ecosystems from contaminants in soil*. Environment Agency for England and Wales, Bristol, UK.

Ferguson, P. L. and Smith, R. D. (2003). Proteome analysis by mass spectrometry. *Annu. Rev. Biophys. Biomol. Struct.* 32,399–424.

Fielden, M. R., Pearson, C., Brennan, R. and Kolaja, K. L. (2005). Preclinical drug safety analysis by chemogenomic profiling in the liver. *Am. J. Pharmacogenomics* 5,161–171.

Focarelli, R., La Sala, G. B., Balasini, M. and Rosati, F. (2001). Carbohydrate-mediated sperm–egg interaction and species specificity: A clue from the *Unio elongatulus* model. *Cells Tissues Organs* 168,76–81.

Forbes, V. E., Palmqvist, A. and Bach, L. (2006). The use and misuse of biomarkers in ecotoxicology. *Environ. Toxicol. Chem.* 25,272–280.

Fu, Q. and Van Eyk, J. E. (2006). Proteomics and heart disease: Identifying biormarkers of clinical utility. *Expert Rev. Proteomics* 3,237–249.

Galay-Burgos, M., Spurgeon, D. J., Weeks, J. M., Stürzenbaum, S. R., Morgan, A. J. and Kille, P. (2003). Developing a new method for soil pollution monitoring using molecular genetic biomarkers. *Biomarkers* 8,229–239.

Gibb, J. O. T., Holmes, E., Nicholson, J. K. and Weeks, J. M. (1997). Proton NMR spectroscopic studies on tissue extracts of invertebrate species with pollution indicator potential. *Comp. Biochem. Physiol. C* 118,587–598.

Griffin, J. L. and Bollard, M. E. (2004). Metabonomics: Its potential as a tool in toxicology for safety assessment and data integration. *Curr. Drug Metab.* 5,389–398.

Hans, R. K., Khan, M. A., Farooq, M. and Beg, M. U. (1993). Glutathione S transferase activity in an earthworm (*Pheretima posthuma*) exposed to 3 insecticides. *Soil Biol. Biochem.* 25,509–511.

Hensbergen, P. J., Donker, M. H., Van Velzen, M. J. M., Roelofs, D., Van DerSchors, R. C., Hunziker, P. E. and Van Straalen, N. M. (1999). Primary structure of a cadmium-induced metallothionein from the insect *Orchesella cincta* (Collembola). *Eur. J. Biochem.* 259,197–203.

Hughes, J., Longhorn, S. J., Papadopoulou, A., Theodorides, K., de Riva, A., Mejia-Chang, M., Foster, P. G. and Vogler, A. P. (2006). Dense taxonomic EST sampling and its applications for molecular systematics of the Coleoptera (beetles). *Mol. Biol. Evol.* 23,268–278.

Inomata, K., Kobari, F., Yoshida-Noro, C., Myohara, M. and Tochinai, S. (2000). Possible neural control of asexually reproductive fragmentation in *Enchytraeus japonensis* (Oligochaeta, Enchytraeidae). *Invertebr. Reprod. Dev.* 37,35–42.

ISO. (1999). Soil quality – Inhibition of reproduction of collembola (*Folsomia candida*) by soil pollutants. Rep. No. ISO11267. International Organization for Standardization, Geneva.

Jager, T., Fleuren, R., Hogendoorn, E. A. and De Korte, G. (2003). Elucidating the routes of exposure for organic chemicals in the earthworm, *Eisenia andrei* (Oligochaeta). *Environ. Sci. Technol.* 37,3399–3404.

Jonker, M. J., Piskiewicz, A. M., Castella, N. I. I. and Kammenga, J. E. (2004a). Toxicity of binary mixtures of cadmium-copper and carbendazim-copper to the nematode *Caenorhabditis elegans*. *Environ. Toxicol. Chem.* 23,1529–1537.

Jonker, M. J., Sweijen, R. and Kammenga, J. E. (2004b). Toxicity of simple mixtures to the nematode *Caenorhabditis elegans* in relation to soil sorption. *Environ. Toxicol. Chem.* 23,480–488.

Jonker, M. J., Svendsen, C., Bedaux, J. J. M., Bongers, M. and Kammenga, J. E. (2005). Significance testing of synergistic/antagonistic, dose level-dependent, or dose ratio-dependent effects in mixture dose-response analysis. *Environ. Toxicol. Chem.* 24, 2701–2713.

Kammenga, J. (2000). Potential and limitations of soil invertebrate biomarkers: final report of the EU BIOPRINT field project. In *SETAC Europe 11th Annual Meeting*, p. 47, Madrid.

Kammenga, J. E. and Riksen, J. A. G. (1996). Comparing differences in species sensitivity to toxicants: Phenotypic plasticity versus concentration–response relationships. *Environ. Toxicol. Chem.* 15,1649–1653.

Kammenga, J. E., Dallinger, R., Donker, M. H., Kohler, H. R., Simonsen, V. R. T. and Weeks, J. M. (2000). Biomarkers in terrestrial invertebrates: Potential and limitations for ecotoxicological soil risk assessment. *Rev. Environ. Contam. Toxicol.* 164,93–147.

Kille, P., Sturzenbaum, S. R., Galay, M., Winters, C. and Morgan, A. J. (1999). Molecular diagnosis of pollution impact in earthworms – towards integrated biomonitoring. *Pedobiologia* 43,602–607.

Knigge, T. and Kohler, H. R. (2000). Lead impact on nutrition, energy reserves, respiration and stress protein (hsp 70) level in *Porcellio scaber* (Isopoda) populations differently preconditioned in their habitats. *Environ. Pollut.* 108,209–217.

Kohler, H. R., Alberti, G., Seniczak, S. and Seniczak, A. (2005). Lead-induced hsp70 and hsp60 pattern transformation and leg malformation during postembryonic development in the oribatid mite, *Archegozetes longisetosus* Aoki. *Comp. Biochem. Physiol. C Toxicol. Pharmacol.* 141,398–405.

Korsloot, A., Van Gestel, C. A. M. and Van Straalen, N. M. (2004). *Environmental Stress and Cellular Response in Arthropods.* CRC Press, London, UK.

Kuperman, R. G., Checkai, R. T., Ruth, L. M., Henry, T., Simini, M., Kimmel, D. G., Phillips, C. T. and Bradley, B. P. (2004). A proteome based assessment of the earthworm *Eisenia fetida* response to chemical warfare agents (CWA) in a sandy loam soil. *Pedobiologia* 47,617–621.

Lee, S. E., Yoo, D. H., Son, J. and Cho, K. (2006). Proteomic evaluation of cadmium toxicity on the midge *Chironomus riparius* Meigen larvae. *Proteomics* 6,945–957.

Lenz, E. M., Bright, J., Knight, R., Westwood, F. R., Davies, D., Major, H. and Wilson, I. D. (2005). Metabonomics with H-1-NMR spectroscopy and liquid chromatography-mass spectrometry applied to the investigation of metabolic changes caused by gentamicin-induced nephrotoxicity in the rat. *Biomarkers* 10,173–187.

Liao, V. H. C. and Freedman, J. H. (1998). Cadmium-regulated genes from the nematode *Caenorhabditis elegans* – identification and cloning of new cadmium-responsive genes by differential display. *J. Biol. Chem.* 273,31962–31970.

Liao, V. H. C., Dong, J. and Freedman, J. H. (2002). Molecular characterization of a novel, cadmium-inducible gene from the nematode *Caenorhabditis elegans* – a new gene that contributes to the resistance to cadmium toxicity. *J. Biol. Chem.* 277,42049–42059.

Lindon, J. C., Keun, H. C., Ebbels, T. M. D., Pearce, J. M. T., Holmes, E. and Nicholson, J. K. (2005). The Consortium for Metabonomic Toxicology (COMET): Aims, activities and achievements. *Pharmacogenomics* 6,691–699.

Lindon, J. C., Nicholson, J. K., Holmes, E., Antti, H., Bollard, M. E., Keun, H., Beckonert, O., Ebbels, T. M., Reilly, M. D., Robertson, D., Stevens, G. J., Luke, P., Breau, A. P., Cantor, G. H., Bible, R. H., Niederhauser, U., Senn, H., Schlotterbeck, G., Sidelmann, U. G., Laursen, S. M., Tymiak, A., Car, B. D., Lehman-Mckeeman, L., Colet, J. M., Loukaci, A. and Thomas, C. (2003). Contemporary issues in toxicology – the role of metabonomics in toxicology and its evaluation by the COMET project. *Toxicol. Appl. Pharmacol.* 187,137–146.

Margulies, M., Egholm, M., Altman, W. E., Attiya, S., Bader, J. S., Bemben, L. A., Berka, J., Braverman, M. S., Chen, Y. J., Chen, Z. T., Dewell, S. B., Du, L., Fierro, J. M., Gomes, X. V., Godwin, B. C., He, W., Helgesen, S., Ho, C. H., Irzyk, G. P., Jando, S. C., Alenquer, M. L., Jarvie, T. P., Jirage, T. P., Jirage, K. B., Kim, J. B., Knight, J. R., Lanza, J. R., Leamon, J. R., Lefkowitz, S. M., Lei, M., Li, J., Lohman, K. L., Lu, H.,

160

Makhijani, D. M., McDade, K. E., McKenna, M. P., Myers, E. W., Nickerson, E., Nobile, J. R., Plant, R., Puc, B. P., Ronan, M. T., Roth, G. T., Sarkis, G. J., Simons, G. J., Simons, J. F., Simpson, J. W., Srinivasan, M., Tartaro, K. R., Tomasz, A., Vogt, K. A., Volkmer, G. A., Wang, S. H., Wang, Y., Weiner, M. P., Yu, P., Begley, R. F. and Rothberg, J. M. (2005). Genome sequencing in microfabricated high-density picolitre reactors. *Nature* 437,376–380.

Marino, F., Winters, C. and Morgan, A. J. (1999). Heat shock protein (hsp60, hsp70, hsp90) expression in earthworms exposed to metal stressors in the field and laboratory. *Pedobiologia* 43,615–624.

Menzel, R., Bogaert, T. and Achazi, R. (2001). A systematic gene expression screen of *Caenorhabditis elegans* cytochrome P450 genes reveals CYP35 as strongly xenobiotic inducible. *Arch. Biochem. Biophys.* 395,158–168.

Menzel, R., Rodel, M., Kulas, J. and Steinberg, C. E. W. (2005). CYP35: Xenobiotically induced gene expression in the nematode *Caenorhabditis elegans. Arch. Biochem. Biophys.* 438,93–102.

Nadeau, D., Corneau, S., Plante, I., Morrow, G. and Tanguay, R. M. (2001). Evaluation for Hsp70 as a biomarker of effect of pollutants on the earthworm *Lumbricus terrestris. Cell Stress Chaperones* 6,153–163.

Natsoulis, G., El Ghaoui, L., Lanckriet, G. R. G., Tolley, A. M., Leroy, F., Dunlea, S., Eynon, B. P., Pearson, C. I., Tugendreich, S. and Jarnagin, K. (2005). Classification of a large microarray data set: Algorithm comparison and analysis of drug signatures. *Genome Res.* 15,724–736.

Newman, M. C. and Unger, M. A. (2003). *Fundamentals of Ecotoxicology*, 2nd Edition, Lewis Publishers, Boca Raton, FL.

Nicholson, J. K. and Wilson, I. D. (2003). Understanding 'global' systems biology: Metabonomics and the continuum of metabolism. *Nat. Rev. Drug Discov.* 2,668–676.

Nicholson, J. K., Connelly, J., Lindon, J. C. and Holmes, E. (2002). Metabonomics: A platform for studying drug toxicity and gene function. *Nat. Rev. Drug Discov.* 1,153–161.

Nottrot, F., Joosse, E. N. G. and Vanstraalen, N. M. (1987). Sublethal effects of iron and manganese soil pollution on *Orchesella cincta* (Collembola). *Pedobiologia* 30,45–53.

Novillo, A., Won, S. J., Li, C. and Callard, I. P. (2005). Changes in nuclear receptor and vitellogenin gene expression in response to steroids and heavy metal in *Caenorhabditis elegans. Integr. Comp. Biol.* 45,61–71.

OECD. (2004). Guideline for the testing of chemicals. Earthworm reproduction test (*Eisenia fetida/Eisenia andrei*). Rep. No. No. 222. Organisation for Economic Co-operation and Development, Geneva.

Okuno, A., Katayama, H. and Nagasawa, H. (2000). Partial characterization of vitellin and localization of vitellogenin production in the terrestrial isopod, *Armadillidium vulgare. Comp. Biochem. Physiol. B Biochem. Mol. Biol.* 126,397–407.

Oumi, T., Ukena, K., Matsushima, O., Ikeda, T., Fujita, T., Minakata, H. and Nomoto, K. (1994). Annetocin – an oxytocin-related peptide isolated from the earthworm, *Eisenia-foetida. Biochem. Biophys. Res. Commun.* 198,393–399.

Pampanin, D. M. (2004). *Integrated approach to ecosystem evaluation and control: Use of classic and new biomarkers* (Thesis). Institute of Marine Science (ISMAR), National Research Council (CNR), University of Venice. p. 330.

Parker, J. B., Leone, A. M., McMillian, M., Nie, A., Kemmerer, M., Bryant, S., Herlich, J., Yieh, L., Bittner, A., Liu, X., Wan, J. and Johnson, M. D. (2003). Toxicant class separation using gene expression profiles determined by cDNA microarray. *Toxicol. Sci.* 72,743.

Peakall, D. (1992). *Animal Biomarkers and Pollution Indicators.* Chapman and Hall, London, UK.

Peredney, C. L. and Williams, P. L. (2000). Utility of *Caenorhabditis elegans* for assessing heavy metal contamination in artificial soil. *Arch. Environ. Contam. Toxicol.* 39,113–118.

Porte, C., Janer, G., Lorusso, L. C., Ortiz-Zarragoitia, M., Cajaraville, M. P., Fossi, M. C. and Canesi, L. (2006). Endocrine disruptors in marine organisms: Approaches and perspectives. *Comp. Biochem. Physiol. C Toxicol. Pharmacol.* 143,303–315.

Posthuma, L., Hogervorst, R. F., Joosse, E. N. G. and Van Straalen, N. M. (1993). Genetic variation and covariation for characteristics associated with cadmium tolerance in natural populations of the springtail *Orchesella cincta* (L.). *Evolution* 47,619–631.

Reichert, K. and Menzel, R. (2005). Expression profiling of five different xenobiotics using a *Caenorhabditis elegans* whole genome microarray. *Chemosphere* 61,229–237.

Ricketts, H. J., Morgan, A. J., Spurgeon, D. and Kille, P. (2004). Measurement of annetocin gene expression: A new reproductive biomarker in earthworm ecotoxicology. *Ecotoxicol. Environ. Saf.* 57,4–10.

Robertson, D. G., Lindon, J. C., Nicholson, J. K. and Holmes, E. (2005). *Metabonomics in Toxicity Assessment.* CRC Press, Boca Raton, FL.

Roelofs, D., Overhein, L., de Boer, M. E., Janssens, T. K. S. and van Straalen, N. M. (2006). Additive genetic variation of transcriptional regulation: Metallothionein expression in the soil insect *Orchesella cincta. Heredity* 96,85–92.

Römbke, J. and Moser, T. (2002). Validating the enchytraeid reproduction test: Organisation and results of an international ringtest. *Chemosphere* 46,1117–1140.

Rouabahsadaoui, L. and Marcel, R. (1995). Analysis of proteinic nutrients in clitellum and cocoons albumin in *Eisenia fetida* Sav (Annelida Oligochaeta) – evidence for a vitellogenin-like glycolipoprotein. *Reprod. Nutr. Dev.* 35,491–501.

Rumpf, S., Hetzel, F. and Frampton, C. (1997). Lacewings (Neuroptera: Hemerobiidae and chrysopidae) and integrated pest management: Enzyme activity as biomarker of sublethal insecticide exposure. *J. Econ. Entomol.* 90,102–108.

SaintDenis, M., Labrot, F., Narbonne, J. F. and Ribera, D. (1998). Glutathione, glutathione-related enzymes, and catalase activities in the earthworm *Eisenia fetida andrei. Arch. Environ. Contam. Toxicol.* 35,602–614.

Sheader, D. L., Williams, T. D., Lyons, B. P. and Chipman, J. K. (2006). Oxidative stress response of European flounder (*Platichthys flesus*) to cadmium determined by a custom cDNA microarray. *Mar. Environ. Res.* 62,33–44.

Silvestre, F., Dierick, J. F., Dumont, V., Dieu, M., Raes, M. and Devos, P. (2006). Differential protein expression profiles in anterior gills of *Eriocheir sinensis* during acclimation to cadmium. *Aquat. Toxicol.* 76,46–58.

Sivasundar, A. and Hey, J. (2003). Population genetics of *Caenorhabditis elegans*: The paradox of low polymorphism in a widespread species. *Genetics* 163,147–157.

Snape, J. R., Maund, S. J., Pickford, D. B. and Hutchinson, T. H. (2004). Ecotoxicogenomics: The challenge of integrating genomics into aquatic and terrestrial ecotoxicology. *Aquat. Toxicol.* 67,143–154.

162

Soetaert, A., Moens, L. N., Van der Ven, K., Van Leemput, K., Naudts, B., Blust, R. and De Coen, W. M. (2006). Molecular impact of propiconazole on *Daphnia magna* using a reproduction-related cDNA array. *Comp. Biochem. Physiol. C Toxicol. Pharmacol.* 142,66–76.

Solassol, J., Jacot, W., Lhermitte, L., Boulle, N., Maudelonde, T. and Mange, A. (2006). Clinical proteomics and mass spectrometry profiling for cancer detection. *Expert Rev. Proteomics* 3,311–320.

Spurgeon, D. J., Ricketts, H., Svendsen, C., Morgan, A. J. and Kille, P. (2005). Hierarchical responses of soil invertebrates (earthworms) to toxic metal stress. *Environ. Sci. Technol.* 39,5327–5334.

Steenbergen, N. T. T. M., Iaccino, F., DeWinkel, M., Reijnders, L. and Peijnenburg, W. J. G. M. (2005). Development of a biotic ligand model and a regression model predicting acute copper toxicity to the earthworm *Aporrectodea caliginosa*. *Environ. Sci. Technol.* 39,5694–5702.

Steiner, G., Suter, L., Boess, F., Gasser, R., de Vera, M. C., Albertini, S. and Ruepp, S. (2004). Discriminating different classes of toxicants by transcript profiling. *Environ. Health Perspect.* 112,1236–1248.

Sterenborg, I. and Roelofs, D. (2003). Field-selected cadmium tolerance in the springtail *Orchesella cincta* is correlated with increased metallothionein mRNA expression. *Insect Biochem. Mol. Biol.* 33,741–747.

Stroomberg, G. J., Ariese, F., van Gestel, C. A. M., van Hattum, B., Velthorst, N. H. and van Straalen, N. M. (2003). Pyrene biotransformation products as biomarkers of polycyclic aromatic hydrocarbon exposure in terrestrial Isopoda: Concentration–response relationship, and field study in a contaminated forest. *Environ. Toxicol. Chem.* 22,224–231.

Sturzenbaum, S. R., Georgiev, O., Morgan, A. J. and Kille, P. (2004). Cadmium detoxification in earthworms: From genes to cells. *Environ. Sci. Technol.* 38,6283–6289.

Sturzenbaum, S. R., Parkinson, J., Blaxter, M., Morgan, A. J., Kille, P. and Georgiev, O. (2003). The earthworm Expressed Sequence Tag project. *Pedobiologia* 47,447–451.

Svendsen, C., Spurgeon, D. J., Weeks, J. M. and Hankard, P. K. (2004). A review: Lysosomal membrane stability measured by neutral red retention – is it a workable earthworm biomarker? *Ecotoxicol. Environ. Saf.* 57,20–29.

Swain, S. C., Keusekotten, K., Baumeister, R. and Sturzenbaum, S. R. (2004). *C. elegans* metallothioneins: New insights into the phenotypic effects of cadmium toxicosis. *J. Mol. Biol.* 341,951–959.

Triebskorn, R., Adam, S., Casper, H., Honnen, W., Pawert, M., Schramm, M., Schwaiger, J. and Kohler, H. R. (2002). Biomarkers as diagnostic tools for evaluating effects of unknown past water quality conditions on stream organisms. *Ecotoxicology* 11,451–465.

Van Gestel, C. A. M. and Koolhaas, J. E. (2004). Water-extractability, free ion activity, and pH explain cadmium sorption and toxicity to *Folsomia candida* (Collembola) in seven soil-pH combinations. *Environ. Toxicol. Chem.* 23,1822–1833.

Van Leeuwen, C. J. and Hermens, J. L. M. (1995). *Risk Assessment of Chemicals: An Introduction.* Kluwer, Dordrecht.

Van Straalen, N. M. and Denneman, C. A. J. (1989). Ecotoxicological evaluation of soil quality criteria. *Ecotoxicol. Environ. Saf.* 18,241–251.

Vatamaniuk, O. K., Bucher, E. A., Sundaram, M. V. and Rea, P. A. (2005). CeHMT-1, a putative phytochelatin transporter, is required for cadmium tolerance in *Caenorhabditis elegans*. *J. Biol. Chem.* 280,23684–23690.

Venter, J. C., Remington, K., Heidelberg, J. F., Halpern, A. L., Rusch, D., Eisen, J. A., Wu, D. Y., Paulsen, I., Nelson, K. E., Nelson, W., Nelson, W., Fouts, D. E., Levy, S., Knap, A. H., Lomas, M. W., Nealson, K., White, O., Peterson, J., Hoffman, J., Parsons, R., Baden-Tillson, H., Pfannkoch, C., Rogers, Y-H. and Smith, H. O. (2004). Environmental genome shotgun sequencing of the Sargasso Sea. *Science* 304,66–74.

Viant, M. R., Rosenblum, E. S. and Tjeerdema, R. S. (2003a). Identification of biomarkers for withering syndrome in red abalone using NMR-based metabonomics. *Toxicol. Sci.* 72,240.

Viant, M. R., Rosenblum, E. S. and Tjeerdema, R. S. (2003b). NMR-based metabolomics: A powerful approach for characterizing the effects of environmental stressors on organism health. *Environ. Sci. Technol.* 37,4982–4989.

Warne, M. A., Lenz, E. M., Osborn, D., Weeks, J. M. and Nicholson, J. K. (2000). An NMR-based metabonomic investigation of the toxic effects of 3-trifluoromethyl-aniline on the earthworm *Eisenia veneta*. *Biomarkers* 5,56–72.

Wasmuth, J. D. and Blaxter, M. L. (2004). Prot4EST: Translating expressed sequence tags from neglected genomes. *BMC Bioinformatics* 5,187.

Weeks, J. M. (1995). The value of biomarkers for ecological risk assessment: Academic toys or legislative tools? *Appl. Soil Ecol.* 2,215–216.

Weeks, J. M., Spurgeon, D. J., Svendsen, C., Hankard, P. K., Kammenga, J. E., Dallinger, R., Kohler, H. R., Simonsen, V. and ScottFordsmand, J. (2004). Critical analysis of soil invertebrate biomarkers: A field case study in Avonmouth, UK. *Ecotoxicology* 13,817–822.

Wharfe, J., Tinsley, D. and Crane, M. (2004). Managing complex mixtures of chemicals – a forward look from the regulators' perspective. *Ecotoxicology* 13,485–492.

Wheeler, J. R., Gimeno, S., Crane, M., Lopez-Juez, E. and Morritt, D. (2005). Vitellogenin: A review of analytical methods to detect (anti) estrogenic activity in fish. *Toxicol. Mech. Methods* 15,293–306.

Williams, T. D., Gensberg, K., Minchin, S. D. and Chipman, J. K. (2003). A DNA expression array to detect toxic stress response in European flounder (*Platichthys flesus*). *Aquat. Toxicol.* 65,141–157.

Willuhn, J., Schmittwrede, H. P., Greven, H. and Wunderlich, F. (1994). cDNA cloning of a cadmium inducible messenger RNA encoding a novel cysteine rich, nonmetallothio-nein 25 kDa protein in an Enchytraeid earthworm. *J. Biol. Chem.* 269,24688–24691.

Daphnia as an emerging model for toxicological genomics

Joseph R. Shaw[1,*], Michael E. Pfrender[2], Brian D. Eads[3], Rebecca Klaper[4], Amanda Callaghan[5], Richard M. Sibly[5], Isabelle Colson[6], Bastiaan Jansen[7], Donald Gilbert[3] and John K. Colbourne[8]

[1]*The School of Public and Environmental Affairs, Bloomington, IN 47405, USA*
[2]*Department of Biology, Utah State University, 5305 Old Main Hill Road, Logan, UT 84322, USA*
[3]*Department of Biology, Indiana University, Bloomington, IN 47405, USA*
[4]*Great Lakes WATER Institute, University of Wisconsin-Milwaukee, 600 East Greenfield Ave, Milwaukee, Wisconsin 53204, USA*
[5]*School of Biological Sciences, University of Reading, PO Box 68, Reading RG6 6AJ, UK*
[6]*Zoologisches Institut, Universität Basel, Biozentrum/Pharmazentrum, Klingelbergstrasse 50, 4056 Basel, Switzerland*
[7]*Laboratory of Aquatic Ecology, Katholieke Universiteit Leuven, Ch. De Beriostraat 32, B-3000, Leuven, Belgium*
[8]*The Center for Genomics and Bioinformatics, Indiana University, Bloomington, IN 47405, USA*

Abstract. *Daphnia* are already an established model species in toxicology. This freshwater crustacean is used commonly for environmental monitoring of pollutants around the globe and plays an important role in establishing regulatory criteria by government agencies (*e.g.*, US EPA, Environment Canada organization for Economic Cooperation and Development, Environment Agency of Japan). Consequently, daphniids represent 8% of all experimental data for aquatic animals within the toxicological databases (Denslow *et al.*, 2007). As such, their incorporation within the new field of toxicological genomics is limited only by the advancement of genomic resources. Because the development of these technologies requires the input and feedback of a large research community that extends far beyond the boundaries of any one discipline, the *Daphnia* Genomics Consortium (DGC) was formed in 2001 to: (i) provide the organizational framework to coordinate efforts at developing the *Daphnia* genomic toolbox; (ii) facilitate collaborative research and (iii) develop bioinformatics strategies for organizing the rapidly growing database. This chapter reviews the progress in establishing *Daphnia* as model species for genomic studies, with emphasis on toxicological applications. As the goals of the DGC are defined largely by extending the boundaries of current biological research in light of genomic information, this chapter first reviews *Daphnia*'s unique biological attributes that make it ideal for such an expansion of research efforts. These attributes include a long tradition of ecological, evolutionary and toxicological study, culminating in the benefits provided by emerging genomic tools.

Corresponding author: Tel.: +812-855-1392. Fax: +812-855-7802.
E-mail: joeshaw@indiana.edu (J.R. Shaw).

ADVANCES IN EXPERIMENTAL BIOLOGY
VOLUME 02   ISSN 1872-2423
DOI: 10.1016/S1872-2423(08)00005-7

166

Keywords: BioMart; BLAST; cDNA; *Ceriodaphnia dubia*; *Daphnia* Genomic Consortium; *Daphnia magna*; *Daphnia pulex*; daphniid; ecology; environment; evolution; expressed sequence tag (EST); gene; gene expression; gene-expression profiling; gene inventory; Gene Ontology (GO); Generic Model Organism Database (GMOD); genetic map; genetics; genome; genomics; microarray; model system; physical map; quantitative trait loci (QTL) analysis; toxicity test; toxicogenomics; toxicology; toxicological genomics; wFleaBase.

The role *Daphnia* plays in toxicology

Biological research using daphniids

Species of the freshwater crustacean genus *Daphnia* have been the focus of steady research by naturalists and experimental biologist for centuries (Korovchinsky, 1997). Swammerdam (1669, 1758) provided their common name, water flea, while their scientific designation, *Daphnia*, was imparted one century later by Mueller (1785). The early studies focused on functional morphology, taxonomic classification and biogeography and on their unusual mode of life. Their life cycle includes parthenogenesis, environmental sex determination and the animal's ability to produce two types of eggs (Lubbock, 1857) – one which can remain dormant for decades. During the 19th century, *Daphnia* and other cladocerans were so well characterized that Richard (1895, 1896) produced 'an historical review' with a bibliography nearing 150 titles.

Reasons for *Daphnia* receiving such early and sustained attention are due in part to their numerical abundance, their role in aquatic food webs and their geographical distribution (Edmondson, 1987). Daphniids are ecologically important (Carpenter *et al.*, 1987) as they are often the primary grazers of algae, bacteria and protozoans and the primary forage for fish (Tessier *et al.*, 2000). They inhabit a remarkable array of environments throughout the world, ranging from permanent lakes to temporary ponds, oligotrophic to eutrophic, hypersaline to freshwater and extending into the UV-rich settings of coastal dune ponds and high-alpine lakes. These radically different waters have been colonized on multiple occasions with a characteristic pattern of convergence of adaptive traits linked to specific habitats (Colbourne *et al.*, 1997). This pattern has stimulated much interest into the physiological requirements needed to persist and thrive in these environments. As a result, *Daphnia* are now recognized as a sentinel species of freshwater lakes and ponds, where their decline serves as an indicator of environmental problems (Dodson and Hanazato, 1995). Work is underway to archive the extensive literature on ecological research using *Daphnia,* which exceeds 4000 articles for the past century and the over 7,000 articles on cladocerans that have been published since 1855 (http://www.cladocera. uoguelph.ca/).

Characteristics of Daphnia *that make it useful for biological research*
Daphnia possess several characteristics that make them valuable for experimental genetic studies. These unique qualities make it possible to translate knowledge about their population structure and ecology to the study of general theories that cross biological scales and disciplines (de Bernardi and Peters, 1987). Several of these attributes revolve around their complex life cycle (Fig. 1). Most *Daphnia* species are cyclical parthenogens, therefore, capable of both clonal and sexual reproduction (Hebert, 1987). Because of clonal reproduction, their genetic background can be held constant, allowing for the maintenance of permanent intact genotypes (Hebert and Ward, 1972; Lynch and Gabriel, 1983). Then, clonal reproduction provides an effective means for comparisons of various treatments against a defined genetic background – a concept that is central to toxicological evaluations and further discussed in the

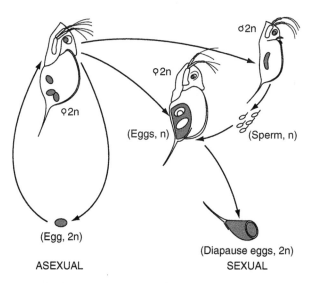

Fig. 1. Life cycle of *Daphnia. Daphnia* are cyclical parthenogens, capable of asexual and sexual reproduction. This interesting reproductive cycle provides a powerful platform for experimental genetics, which is strengthened by growing genomics resources. (From Mort (1991), used with permission from Elsevier Science Publishers LTC).

section titled *Standardization*. For *Daphnia*, sexual reproduction is environmentally induced and these cues can be transmitted in the laboratory (Olmstead and LeBlanc, 2003). During sexual reproduction, females produce sons that are genetically identical to their mothers, allowing for the development of inbred lines by selfing or genetic manipulation by out-crossing. This unusual flexibility in their breeding system makes *Daphnia* an ideal model organism for mapping and characterization of quantitative trait loci (QTL) for complex traits (discussed in section *Genetic maps and QTL analysis*). In addition, development of sexually produced diapausing eggs (*e.g.*, which are encased in ephippia) is paused by a resting stage (diapause). These embryos remain viable for decades and then provide a unique opportunity for long-term maintenance of culture stocks and for probing past populations in the field. This ability to measure evolutionary change by comparing past populations to their modern descendents is made possible by sampling buried diapausing embryos in lake or pond sediment, which also contains a chemical record of the changing environment (Hairston *et al.*, 1999; Pollard *et al.*, 2003).

Other practical attributes that make *Daphnia* a good model system for experimental investigation include the ease by which they are manipulated and maintained in the laboratory (Peters, 1987). Generation times are approximately one week in culture (20°C), which rivals that of most other model eukaryotes and makes it possible to track response throughout their ontogeny. They are maintained in relatively simple defined media (Elendt and Bias, 1990; Kilham *et al.*, 1998; US EPA, 2002) and are fed simple diets that include controlled concentrations of algae and/or bacteria.

Daphnia in molecular ecology and evolutionary biology
Many early empirical investigations that employed *Daphnia* were driven by renewed interests in Mendelian genetics, which permitted a modern understanding of evolutionary theory based on natural selection and quantitative theory of population growth (Edmondson, 1987). These early studies attempted to characterize heritable variation in such traits as male production and life history parameters, among others (Banta, 1939). At this time, there was also considerable interest in characterizing the physiochemical conditions associated with daphniid distribution/ occurrence (Edmondson, 1987). This problem was studied by chemical ecologists (and toxicologist; see section *Traditional use as toxicity test species*) that attempted to define the chemical limits tolerated by *Daphnia* and by geneticist that treated variability in chemical limits as a heritable

trait (Warren, 1900). It proved difficult, however, to identify the genetic mutations underlying these heritable traits adequately without current molecular tools (Hebert, 1987). As a result, the intersection of evolutionary theory, ecological understanding and genetic experimentation with *Daphnia* was not realized until the advent of molecular techniques in the 1970s. These techniques allowed for rapid and more precise identification of genetic variation in natural populations and resulted in a 'resurgence of *Daphnia* genetics' (De Meester, 1996; Hebert, 1974a, 1974b, 1974c; Lynch, 1983). In fact, the increase in focus attributable to the advent of molecular biology has established *Daphnia* as one of the preeminent models in ecological and evolutionary genetics (Lynch and Spitze, 1994; Mort, 1991; Schierwater *et al.*, 1994). With these techniques, the molecular phylogeny of *Daphnia* has been defined for several geographical units: North America (Colbourne and Hebert, 1996), Europe (Schwenk *et al.*, 2000), South America (Adamowicz *et al.*, 2004) and Australia (Colbourne *et al.*, 2006). This characterization of the animal's ecology, population genetics and phylogenetic history provides the necessary foundation for comparative studies on the evolutionary origin of species groups and of phenotypic diversity in *Daphnia*, including those induced by stress (*i.e.*, predation, UV, thermal, etc.). Cumulatively, this knowledge allows interpretation of molecular modifications, individual phenotypes and population-level responses in context of environmental change and represents a key advantage offered by this emerging genomic model.

Current studies across diverse disciplines
While experimental characterization of *Daphnia* has emerged from the fields of ecology, evolutionary biology and toxicology (discussed in section *Traditional use as toxicity test species*), the enhanced ability to dissect environmental influences on biological responses at multiple levels (*e.g.*, gene, cell organism and population) has proven attractive to other disciplines. Recent studies utilizing *Daphnia* have also focused on physiology (Campbell *et al.*, 2004; Glover and Wood, 2005), developmental regulation (Shiga *et al.*, 2006), innate immunity and host–parasite interactions (Ebert *et al.*, 2004; Jensen *et al.*, 2006; Little and Ebert, 2000; Little *et al.*, 2003), ageing (Dudycha, 2001, 2003; Yampolsky and Galimov, 2005) and epidemiology (Chiavelli *et al.*, 2001; Hall *et al.*, 2006). *Daphnia* are also providing the investigative platform in studies that cross disciplines and address questions fundamental to biology, such as, the causes and consequences of recombination (Nielsen, 2006; Paland and Lynch, 2006). Such expansion has occurred because the biology of

Daphnia offered unique research advantages that will only increase as genomic resources for *Daphnia* grow.

Traditional use as toxicity test species

The unique biological attributes of *Daphnia* that make it a well-suited model system for ecology and evolutionary biology also make it a useful model system for toxicology. The purpose of the following section is to highlight how *Daphnia* have been used in toxicology. In this section we do not provide detailed test methodologies/protocols, but rather illustrate the potential for expanded application of *Daphnia* as a toxicological model with the rapid growth of genomics.

Early studies
Daphnia have a long history of use in toxicological evaluations. Some of the first toxicity studies were conducted by the geneticist, Ernest Warren (1900) who defined the range of sodium chloride tolerated by *Daphnia magna* (Straus). These studies were not only the first to quantitatively establish the link between concentration, duration of exposure and organism response, which eventually was disseminated as Haber's rule (Haber, 1924; Miller et al., 2000), but they also represent one of the first intersections between toxicological principle and ecological understanding. Warren was breeding *Daphnia* to study contemporary theories of heredity, but realized that '[his] results seem to present certain features of considerable interest and of wide biological significance; they illustrate how closely the organism is knit to its external conditions of life.' This statement is one of the first to define aquatic toxicology inquiry, as most of the early studies investigated the physical and chemical conditions that governed daphniid distribution (*e.g.*, pH, Klugh and Miller, 1926; magnesium, Hutchinson, 1932).

The rapid growth in the production of synthetic compounds (*e.g.*, drugs, pesticides, munitions, etc.) during the early 1900s resulted in the rapid evolution of modern toxicology (Eaton and Klaassen, 1996). It was the early growth of the pharmaceutical industry and the advent of food and drug laws during this period that was responsible for promoting the use of *Daphnia* in bioassays (Viehoever, 1931, 1936). Toxicity tests with *Daphnia* (primarily *D. magna*) were being used to define mechanisms of action and for testing the safety and efficacy of drugs. As Arno Viehoever (1937) noted, the concept of '*D. magna* as the biological reagent – has been established in both scientific circles and in the public press (*Time*, 1937).' This assertion resulted in *Daphnia* being coined the 'diminutive drug detective' and receiving praise from Dr Mayo, Rochester, MN (Viehoever, 1937). Contaminant testing with *Daphnia* grew in a similar and parallel fashion, but without regulatory directives, was far less prevalent. It is worth noting that mercury (Breukelman, 1932) and DDT (Anderson, 1944, 1945) garnered some of the focus of these early studies. Ironically, it was environmental tragedies decades later with these very compounds that helped provide the impetus for current environmental regulations. Today *Daphnia* are one of the most used models for environmental toxicological evaluations.

Standardization
The need for defined culture conditions and standard methodologies with respect to *Daphnia* experimental biology was quickly realized, especially in light of the early emphasis on defining the limits of *Daphnia*'s external environments. Likewise, the goals of pharmaceutical and toxicological testing, which were guided by regulatory directives, favoured a reduction of test variables to increase reproducibility and facilitate data comparison (Davis, 1977; Duodoroff and Katz, 1950; Viehoever, 1937). This move towards standardization highlights a fundamental difference between *Daphnia* studies in ecology and evolutionary biology and those in toxicology. The emphasis in toxicology has traditionally focused on defining the media/chemical condition (*e.g.*, toxin, stressor, etc.) as opposed to defining the biological condition. In fact, regulatory compliance testing often employs a single clone distributed among laboratories to limit biological variability, whereas the emphasis in other fields often centres on investigating biological variability among genotypes, populations and species.

Current protocols are standardized with respect to pretest animal maintenance/care, age of test organisms, media, food, duration, ambient light and light–dark cycle, temperature and monitored endpoints. Standardized methods are also specific for species and, as previously mentioned, sometimes for clones. Without question, *D. magna* is the most common species used in toxicology followed by *Daphnia pulex* (Leydig) and *Ceriodaphnia dubia* (Richards; *e.g.*, Table 1 in Shaw *et al.*, 2006). *D. magna* is one of the largest daphniids and, as in many arenas, size has been a major factor in its widespread acceptance (Viehoever, 1937). Some studies, however, have indicated that *D. magna* is more tolerant than other species (Koivisto *et al.*, 1992; Koivisto, 1995; Shaw *et al.*, 2006), which may be problematic in regions where it is not widely distributed (*e.g.*, North America). The use of *C. dubia* is often

Table 1. The utility box for toxicological genomic investigations using *Daphnia*.

	Species	Utility	Notes	References
Arrayed and archived libraries				
Nonnormalized cDNA	*D. pulex, D. magna*	CGA, GD, GFR	Standard plus 15 experimental conditions	*Daphnia* Genomics Consortium (2007), Colbourne *et al.* (2007), Watanabe *et al.* (2005)
Normalized, full-length enriched cDNA	*D. pulex*	CGA, GD	Additional 15 experimental conditions	*Daphnia* Genomics Consortium (2007)
Genomic DNA – small inserts (avg. 8 kb)	*D. pulex*	VGS	End sequenced from 5' & 3'	*Daphnia* Genomics Consortium (2007)
Genomic DNA – large inserts (Fosmids)	*D. pulex*	VGS, VGA	End sequenced from 5' & 3'	*Daphnia* Genomics Consortium (2007)
Genomic DNA – large insert (Cosmids)	*D. pulicaria*	VGS, VGA	Gene capture by PCR screening	*Daphnia* Genomics Consortium (2007)
Genomic DNA – large insert (BACs)	*D. pulex*	VGA	Fingerprinted by RFLP	*Daphnia* Genomics Consortium (2007)
Molecular markers				
Microsatellites (SSRs)	*D. pulex* complex, *D. longispina* complex, *D. magna*	GPM	Determined from GSS of genomic libraries, ESTs and from genome sequence	Colbourne *et al.* (2004), Fox (2004), Brede *et al.* (2006), Cristescu *et al.* (2006)

Single nucleotide polymorphisms (SNPs)	*D. pulex*	GPM	From genome sequence and ESTs	*Daphnia* Genomics Consortium (2007)
Data				
Genetic map	*D. pulex*	GPM	185 SSRs in 12 linkage groups	Cristescu *et al.* (2006)
Draft genome sequence assembly	*D. pulex*	GPM, VGS, VGA	9080 scaffolds totaling 227 Mb	*Daphnia* Genomics Consortium (2007)
Draft genome sequence annotation	*D. pulex*	CGA, GD	In progress	*Daphnia* Genomics Consortium (2007)
Expressed sequence tags (ESTs)	*D. pulex, D. magna*	CGA, GD	Over 200,000 available	Colbourne *et al.* (2007), Watanabe *et al.* (2005), *Daphnia* Genomics Consortium (2007)
Infrastructure and other resources				
wFleaBase	Genus *Daphnia*		Genome database	Colbourne *et al.* (2005)
Microarrays – cDNA	*D. pulex* complex, *D. magna*	GFR	Best for comparing closely related species	Eads *et al.* (2007), Shaw *et al.* (2007), Eads *et al.* (2008), Poynton *et al.* (2007), Soetaert *et al.* (2006, 2007a, 2007b)
Microarrays – oligonucleotide	*D. pulex, D. magna*	GFR	Greater specificity and gene coverage	Watanabe *et al.* (2007), *Daphnia* Genomics Consortium (2007)

Table 1. (Continued)

	Species	Utility	Notes	References
Toxicological genomic studies	D. pulex, D. magna	GD, GFR, VGF	Exposures to cadmium, copper, fenarimol, hydrogen peroxide, ibuprofen, beta-naphthoflavone, pentachlorophenol, propiconazole, zinc	Shaw et al. (2007), Poynton et al. (2007), Soetaert et al. (2006, 2007a, 2007b), Connon et al. (2008), Heckmann et al. (2006), Soetaert et al. (2007b)
Cell lines	D. pulex	VGF	Not yet immortalized	Robinson et al. (2006)
Transformation lines	D. pulex	VGF	In progress	Robinson et al. (2006)
Daphnia stocks and mutants	Genus Daphnia	VGF	Includes mapping panels	Daphnia Genomics Consortium (2007)

Notes: CGA – candidate gene approach for finding known genes of interest; EST – expressed sequence tag; GPM – gene position mapping of predicted loci onto the genome; GD – gene discovery of novel genes; GFR – gene function and regulation predictions; GSS – genome sequence survey; Mb – mega bases; PCR – polymerase chain reaction; RFLP – restriction fragment length polymorphisms; SSR – simple sequence repeat; VGS – validation of gene structure predictions; VGA – validation of local gene arrangements; VGF – validation of gene function predictions.

preferred in lifecycle tests, because it reaches reproductive maturity (~ 3 days) about three times as quickly as *D. magna* or *D. pulex* (~ 9 days).

Regulatory tests
To date, standardized procedures have been adopted by numerous environmental protection agencies throughout the world (*i.e.*, American Public Health Association, US Environmental Protection Agency, American Society for Testing and Materials, International Standardization Organization, Environment Canada and Organization for Economic Cooperation and Development, European Commission). More information on standardized test methods is found in Cooney (1995) and Versteeg *et al.* (1997). Information from regulatory tests is used for three primary purposes: (i) criterion development (*e.g.*, establishing regulatory limits); (ii) testing chemical safety and (iii) compliance monitoring. In addition, these data are often deposited in toxicological databases, where they become primary sources for risk assessors. The inclusion of toxicogenomic endpoints has been discussed in context of all these functions (Cook *et al.*, 2007).

Advantages in using a model system in ecology and evolution for toxicology

Although the questions addressed by toxicologists, ecologists and evolutionary biologists are similar, there are differences in directives that influence the methods in which they are addressed. The advent of genomic tools, should bridge these (or some of these) differences and expand the uses of *Daphnia* in toxicology. This section discusses the advantages of using *Daphnia* to address toxicological problems, but with approaches drawn from ecology and evolutionary biology.

Cross-disciplinary nature of toxicology
The science of toxicology is inherently multidisciplinary, borrowing and improving on most all of the basic sciences to test its hypotheses. As Micheal Gallo (1996) stated, 'toxicology has drawn its strength and diversity from its proclivity to borrowing.' The need for model species that span disciplinary boundaries is only part of the case for including *Daphnia*. While *Daphnia* were incorporated in toxicological study early in its modern expansion (Warren, 1900), toxicologists traditionally have not exploited the biological attributes of *Daphnia* that other disciplines sought to expand. Rather, they have worked to reduce the contributions of biological variation and focused instead on the role of the chemical surroundings. Perhaps this difference was the result of regulatory

directives or the imprecision noted by Hebert (1987) in correlating genetic change with phenotype prior to the advent of molecular biology. Regardless of the cause for this early split in philosophy, there are substantial reasons to reconsider – at least some – toxicological problems for which *Daphnia* are used in light of modern ecological and evolutionary practices.

Guided by similar questions
Toxicology is focused on understanding response outcomes to pollutant exposure, often with the goal of defining levels that are well tolerated for the organism. Typically, this goal is accomplished by integrating responses at the level of the individual (or below, such as organ tissues or cell lines). Organisms are not passive targets of their external environment, however, and collectively their ranges of response define population-level effects. The magnitude of this impact depends on the organisms' ability to alter their tolerance limits (*i.e.*, acclimate) and over time, these limits re-structure within populations, as the genotypes that favour acclimation are selected (*i.e.*, adaptation). Understanding the dynamic responses that shape tolerance limits are often further complicated, because tolerant phenotypes can be costly to maintain resulting in ecological tradeoffs. Currently, risk assessment techniques only account for the increased fitness associated with acclimation or adaptation, without debiting their associated costs. With this situation in mind, an integration of ecology, population genetics and evolutionary biology practices would provide tools that could improve the derivation of safety limits and ultimately, risk predictions (*i.e.*, susceptibility).

The US National Institute of Environmental Health Sciences has recently suggested such an alignment of philosophies (http://www.niehs. nih.gov/external/plan2006/). This suggestion has highlighted the need for integrated research teams to investigate the complex interplay between genes and the environment in order to identify the environmental factors that influence disease risk. Two areas of research focus towards this endeavour include: (i) an 'expansion of our understanding of environmental influences on genome maintenance/stability ...' and (ii) 'concerted efforts to improve our understanding of epigenetic influences on health.' More simply stated, these objectives encompass understanding the environmental influence on the limits and underlying mechanisms of genetic adaptation and physiological acclimation. Then, a growing area of research interest in both environmental toxicology and evolutionary biology is the relationship between genome function and environment. For example, recent analyses of whole genome sequences show that there are often many more predicted genes than there are genes with known function. One possibility for this observation is that the expression, regulation and function of many genes may be highly context dependent and may only manifest in particular environments. The importance of understanding the genetic basis of interactions between genotype and environment is reflected in a renewed interest in phenotypic plasticity and its relationship to adaptive evolution (Miner *et al.*, 2005; Pigliucci, 2005; West-Eberhard, 2003). This shift in focus is also reflected by an increase in research on the relationship between environmental factors and epistatic interactions among genes that may make a substantial contribution to variation in complex traits such as disease susceptibility (Carlborg and Haley, 2004). Increasingly, the connection between genes and phenotypes is being established by combining genomic information with QTL studies. These studies focus on deciphering regulatory networks of polymorphic genes utilizing a combination of QTL analysis and microarray expression profiles (eQTLs) (Bing and Hoeschele, 2005; Carlborg *et al.*, 2005; de Koning *et al.*, 2005). The combination of these two methods, referred to as genetical genomics (de Koning *et al.*, 2005; Jansen and Nap, 2001) is a powerful approach to inferring gene transcriptional relationships (Li *et al.*, 2005) and has been utilized to demonstrate that regulation of many genes has a heritable basis (Cheung and Spielman, 2002; Hubner *et al.*, 2005; Morley *et al.*, 2004). A genetical genomic approach utilized in an organismal system in which environmental conditions can be accurately and systematically manipulated will significantly advance our understanding of the relationship between the phenotype and the underlying genotypic and environmental effects.

As previously highlighted (see section *Biological research using daphniids*), *Daphnia* possess several biological attributes to integrate disciplines and address such research needs and this situation will only be strengthened as genomic resources mature.

Change in response over time
The *Daphnia* system is poised to become a leading research model for understanding environmental influences on gene regulation and subsequent stressor induced acclimation and adaptation. One reason for this emerging utility derives from their life cycle, which during sexual reproduction produces embryos that diapause (*e.g.*, delayed development). Diapausing embryos, which encase in ephippia, are resistant to harsh environmental conditions (*e.g.*, desiccation, freezing) and represent a 'bank of genetic diversity from which an existing population can draw new genotypes (Mort, 1991).' Experimentally, sediments contain banks

of diapaused *Daphnia* that provide access to past populations, as these can be hatched from sediments several decades old (Cáceres, 1998; Hairston *et al.*, 1995; Kerfoot *et al.*, 1999). Thus, egg banks allow the past products of evolution to be resurrected and evaluated against their current descendants in a controlled setting.

The resting egg bank of *Daphnia* has been used to study the influence of natural stressors (*i.e.*, cyanobacteria, predation) on the distribution of stressor-induced phenotypes in populations and through time (Cousyn *et al.*, 2001; Hairston *et al.*, 2001). These studies have demonstrated rapid adaptive change (*i.e.*, acquisition and loss of phenotypes) in presence of stress. Current studies have extended these applications to investigate the influence of metal pollution on the structure of *Daphnia* communities (Pollard *et al.*, 2003), examining pollution and subsequent recovery. In cases in which the egg banks of interest are no longer viable, the DNA is still accessible to genetic probes for centuries (Limburg and Weider, 2002).

Benefits in applying genomic tools

Genomics add a new level of knowledge to traditional toxicology studies of model test species such as *Daphnia*. Traditional toxicity assays, although informative as to the levels of a chemical that may be toxic to a population, do not provide information on the mechanism by which a chemical has its toxic effect. In addition, many chemicals are found at sublethal levels in the environment, affecting populations by altering the general physiology, reproductive capacity or the ability of an organism to fight disease. Examining changes in gene expression provides a means to identify biochemical pathways that are altered in an organism after even a low-level chemical exposure. Genomics can provide a level of detail that is absent in general toxicity studies, indicating mode of action, differences between low- and high-dose effects, effects caused by exposures to complex mixtures, providing detailed early biomarkers or a 'canary' to indicate exposure or potential effects of an exposure (Klaper and Thomas, 2004). The potential of genomics for studies of exposure and effects in toxicology and environmental risk assessment is now recognized by the US Environmental Protection Agency (Dix *et al.*, 2006; Gallagher *et al.*, 2006). This potential is illustrated by a pilot project using *Daphnia* in a proof of concept experiment to integrate metabolomic, proteomic and genomic responses to pollutant exposures (http://www.epa.gov/heasd/edrb/comptox.htm).

Identifying mechanisms of action
One of the most promising immediate applications of genomics is in determining the mode of action of a chemical. Gene expression patterns provide clues as to the biochemical pathways that are affected by a particular toxin (Amin *et al.*, 2002). Genomic biomarkers can also distinguish effects of different chemicals. Several studies have now shown that gene expression patterns can be signatures of exposure to a particular chemical (*e.g.*, Bartosiewicz *et al.*, 2001; Merrick and Bruno, 2004). Organisms under stress may show a generalized pattern of gene expression associated with a stress response. Unique gene expression patterns, however, are also present in each of these studies. Chemicals with similar modes of action provide similar expression patterns (*e.g.*, heavy metals, Andrew *et al.*, 2003) even within a chemical class, exposures can be distinguished based upon gene expression patterns (Hamadeh *et al.*, 2002; Poynton *et al.*, 2007; Watanabe *et al.*, 2007).

Improved biomarkers
Traditional field studies for ecological risk assessments try to identify what factors are affecting a population after an insult has already occurred. Molecular biomarkers are taken from laboratory to field studies to diagnose the effects of the many different stressors an organism may be exposed to in its environment. Genomic biomarkers are likely more sensitive and more specific than other biomarkers, which could be highly valuable for field assessments. For example, gene expression patterns are already being used in field studies, to determine when various fish species have been exposed to endocrine disrupting chemicals (Larkin *et al.*, 2002) and studies demonstrate that other chemicals (such as metals) that phenotypically cause the same reproductive issues as endocrine disruptors actually have different mechanisms of action (Klaper *et al.*, 2006; Shaw *et al.*, 2007).

Cross-species interpretations
Another issue with laboratory toxicology studies is the ability to predict the response of one organism using data from another related organism. Will all *Daphnia* species, for example, have the same susceptibilities to a chemical in question? As noted by Shaw *et al.* (2006), there is often a disconnect between species studied in the laboratory and natural populations exposed in the field. Will *Daphnia* toxicity assays correctly predict what will happen to other important aquatic invertebrates? Currently, arbitrary extrapolation factors are employed to provide a conservative estimate of minimum exposure limits to protect the most

sensitive species. From the use of genomic data and biochemical pathway homology across species, direct comparisons can be made of genomic changes related to these biochemical pathways among species. If different pathways are affected when two species are exposed to the same chemical, the data could indicate that a receptor is present or triggered in one species and not in the other or the more resistant species may have additional pathways that are triggered for detoxification. Differences in levels of gene expression may indicate higher chemical sensitivity caused by physiological modification or genetic alteration.

Biosensor/predictive models
Some would argue that genomics for *Daphnia* will never be used for toxicology modelling as standard *Daphnia* toxicity tests are cheap and provide enough relevant information. We argue that because of their ecological importance and distribution, their ease of use, the resources developed through the DGC and most importantly their ability to act as sensitive sensors for other organisms, *Daphnia* present the ideal species group to use for proof-of-principle toxicogenomic modelling efforts. We already see *Daphnia* incorporated into sensor systems for freshwater for human consumption (*e.g.*, de Hoogh *et al.*, 2006), in which a change in *Daphnia* mortality levels or behaviour triggers alarms for water intake systems. In the study of de Hoogh *et al.* (2006), *Daphnia* mortality signalled that an unknown chemical had been released into the River Meuse in the Netherlands. Potential chemical suspects were identified and then traditional toxicity assays were used to determine if the suspected chemical was the cause of the *Daphnia* deaths. In this case, in particular, *Daphnia* from the monitoring device could be sampled for RNA expression and compared with a database of gene expression patterns recorded by laboratory screening of thousands of chemicals and mixtures. This direct sampling would provide information to inform what type of chemical was involved and what biochemical pathways are altered to better predict the potential impact on sensitive animal or human populations.

Daphnia *toxicogenomics*
Developing genomics resources in the model *Daphnia* is the most cost-effective and ecologically relevant investment currently proposed for models of toxicology and toxicogenomics. As discussed above, there is significant information known on their behaviour, ecology, population genetics, reproduction and physiology that can now through the efforts of the DGC be linked to genes and gene expression data. The linkage between

phenotype, specifically those related to survival and fitness and genomic biomarkers will ultimately be necessary to make genomic information relevant for toxicology and the environmental sciences. The DGC is providing a means to link genomic data to environmentally relevant phenotypic characteristics that are currently elusive for some model organisms and too expensive to explore in any context but the laboratory for many others.

Daphnia genomics initiative

Community-based approach to Daphnia *genomics*

Proliferation of the consortium approach to big science
Initiating and managing a large-scale genomics initiative and developing the genetic tools and bioinformatic infrastructure to support these efforts are challenging tasks. Perhaps the most important component is fostering a user community with the level of collaboration and shared expertise to take full advantage of developing genomic resources. One method currently employed to accomplish this objective is to build a coordinated consortium with the common goal of advancing a particular model system. The consortium approach works well by ensuring the backing of an organized group with a vested interest in utilizing and managing the vast amount of data from such a project. For instance, most of the completed and ongoing genome sequence projects have employed this approach to secure funding and to achieve success, as demonstrated by the diverse array of taxa and broad utility in terms of both basic and applied issues that have been the focus of genome projects. For large eukaryotic genomes, international cooperative efforts are a critical component of future success.

Rapidly expanding genome sequence data
An overview of current and planned genome projects recorded in NCBI's GenBank, which is by no means exhaustive, documents 338 eukaryotic and 1,471 prokaryotic genome sequence projects either completed or in progress (From NCBI Entrez Genome Project; http://www.ncbi.nlm. nih.gov/genomes/static/gpstat.html; accessed 31 October 2007). These data reveal the taxonomic focus of existing genome projects. For example, the vast majority of eukaryotic genome projects are focused on five taxonomic groups; mammalian vertebrates, insects, fungi, protists and plants (Fig. 2). Some interesting historical trends are imbedded in this array of genome

182

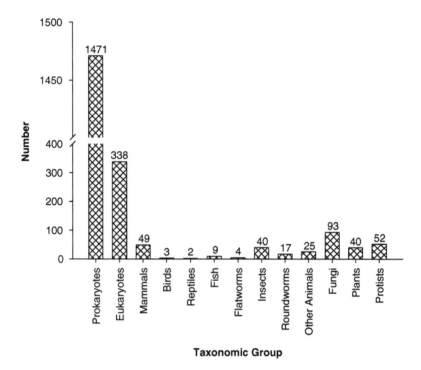

Fig. 2. Taxonomic representation of genome projects. Data compiled from NCBI EntrezGenome Project; http://www.ncbi.nlm.nih.gov/genomes/static/gpstat.html; accessed October 31, 2007.

projects. The initial projects were driven primarily by two factors: (i) genome size (smaller genomes are simpler and cheaper to sequence than larger ones) and (ii) utility as a model system for basic genetic or human health research. *Chlamydomonas, Caenorhabditis, Drosophila* and humans were among the first whole-genome sequences completed for these reasons. Subsequently the focus turned to a broader set of criteria including consideration of economic and phylogenetic utility in informing the selection of genomes. An example of the former is rice (*Oryza*) and of the latter is the tunicate (*Ciona*). Recently, the repertoire of genome projects has included organisms that are important ecological model systems such as the ever so useful plant *Arabidopsis* and models for toxicology and environmental monitoring including the zebrafish (*Danio*) and the fresh-water microcrustacean *Daphnia*. These examples highlight the changing perspectives and prioritization contributing to the current diversity of completed and ongoing genome projects and the increasing availability of genomic data for toxicological application.

Why a community consortium?
There are a number of advantages to developing a community consortium. Foremost is the increased availability of data and techniques to researchers in the science community at large. Consortia function in a number of ways to facilitate the development and utility of a model system. One primary function is to provide an open-access platform as a repository for genomic reagents and data. Experimental model systems, like *Daphnia*, are ultimately used for study in a large number of biological contexts. These diverse contexts range from developmental and evolutionary biology to toxicology and ecosystem monitoring and assessment. The result will be the generation of multiple complex data tracks. Integrating these complex biological databases requires a common web-based forum that will promote synergistic research and enhance the utility of the system as a whole. For example, the phenomenal success of *Drosophila* as a genetic model is attributed to the strength of its research community and the community's devotion at creating and maintaining shared resources, ranging from mutant stocks (http://flystocks.bio.indiana.edu/) and cell lines to vectors, clones, microarrays (http://dgrc.cgb.indiana.edu/) and a continuously improved genome sequence annotation and curated research literature (http://flybase.bio.indiana.edu/). A second function of a coordinated consortium is to provide a level of oversight for the community by openly generating quality control standards, including controlled methodologies, nomenclature, test samples and reagents such as DNA libraries and microarray probes. There are numerous examples of the recent adoption of standards for genomic data collection, such as microarray experiments (Ball and Brazma, 2006) and toxicological studies (Mattingly *et al.*, 2004). For toxicological studies in *Daphnia*, there is already a number of experimental and agency protocols that require uniformity to ensure that data collected in separate facilities and studies are comparable. Given the sensitivity of genomic assays (gene expression profiling) to test conditions and to the physiological/developmental state of isolates under investigation, these standards are required if genomic data are to be compared across experiments. A third function of a consortium and perhaps the single most important factor requiring a community of investigators is in leveraging the technical difficulties and high cost associated with creating genomics tools for a new species.

Outcomes of a consortium approach
The outcomes of a well-developed genome consortium manifest in a number of ways. A community-based approach facilitates inter-institutional and international collaboration. This approach facilitates

synergistic interdisciplinary research by promoting a common set of tools and sets a forum for the free exchange of ideas and results. Because consortia draw on the combined efforts of many individuals, they enable a scale of science that is well outside the realm of the individual investigator. Resources such as a complete genome sequence and expression arrays are costly to develop and the facilities to conduct such projects are limited. The likelihood of generating financial and technical support to conduct these large-scale projects is greatly increased by coordinated efforts and the presence of a well-organized and documented consortium. The result is a much more rapid development of the diverse set of tools required for genomic level science.

Development of the Daphnia *Genome Consortium (DGC)*
Recognizing the need for a well-developed genomic model system for ecological, evolutionary and toxicogenomic studies the *Daphnia* Genome Consortium (DGC) was initiated in the autumn of 2001 and held its first meeting the following year. This consortium aimed to develop a model system that would address these issues:

(1) A need for tractable models to study the genetic basis and evolution of organismal responses to environmental stressors, as individuals and among populations.
(2) A need for appropriate model systems to dissect the interaction between genotype and environment, *i.e.*, phenotypic plasticity.
(3) A need for model aquatic systems with a wide distribution amenable to use as biological indicators for ecosystem monitoring and risk assessment.
(4) A need for model systems to study the genetic plus environmental basis of gene regulation.

The traditional model organisms (*i.e., Escherichia, Saccharomyces, Arabidopsis, Caenorhabditis, Drosophila, Danio* and *Mus*) were selected for genomics by earlier research groups because of their utility in development, cell biology and genetics. Unfortunately, all of these systems lack significant biological context outside of the laboratory. Thus, despite the deep understanding of the molecular and developmental properties of these species, we know almost nothing about the natural environmental factors that lead to their evolution or govern their responses to environmental stressors. The inaccessibility of one or more life stages of these species in nature does not inspire confidence that this situation will change in the near future. One of the ultimate goals of biology is to understand how organisms and populations respond to and evolve in variable environments. With this goal in mind, the DGC has been developing a new model system with clear application to toxicogenomics. The microcrustacean *Daphnia*, because of the biological attributes enumerated earlier is an ideal candidate for further genomic development. The ecological diversity for this organism provides a unique opportunity to ask if independent lineages of *Daphnia* evolve to meet environmental challenges in the same way (Pfrender *et al.*, 2000).

Until recently, the main limitation of the *Daphnia* system was the lack of well-developed genetic tools, but rapid progress has been made on this front. The first international *Daphnia* Genomics Consortium (DGC) meeting was held at Indiana University in October 2002 (*Daphnia* Genomics Consortium, 2007). Consisting of diverse scientists from 17 countries, the DGC's goal is 'to develop the *Daphnia* system to the same depth of molecular, cell and developmental biological understanding as other model systems, but with the added advantage of being able to interpret observations in the context of natural ecological challenges.' *Daphnia* is now one of the best genomically characterized organisms with a deeply understood ecology. Although *D. magna* is more commonly used for toxicological research and significant tools are in place and will continue to grow for this species, *D. pulex* was first chosen for genomics because of its natural history is pertinent to a greater number of investigators. Its geographic range is vast compared with other narrowly endemic taxa in North America (Hebert and Finston, 1993, 1996, 1997). It is closely allied to a 'complex' of hybridizing species that have adapted to live in a great diversity of habitats within only the last few million years (Colbourne *et al.*, 1998). In certain lineages, the sexual phase of reproduction is altogether lost, therefore enabling comparative studies on the consequences of shuffling the genome by recombination (Paland and Lynch, 2006).

In the following sections, we outline some of the significant genomic developments in the *Daphnia* system as a direct result of adopting a consortium philosophy. The centerpiece of this effort is the generation and annotation of a draft genome sequence assembly for *D. pulex* through the combined work of the DGC and the US Department of Energy's Joint Genome Institute. In parallel, we have developed extensive cDNA libraries and sequences, microarray-based gene expression systems, a large number of polymorphic microsatellite markers to facilitate population genetic studies and genetic mapping and an open-access web portal to maintain and distribute these data (Table 1).

In the sections below, we detail our progress in each of these areas and outline progress in developing QTL mapping panels for community use. The consortium approach allowed for the coordination of multiple simultaneous efforts across varied institutions, providing extremely rapid progress in tool building and application of these emerging tools for the fields of toxicological and ecological genomics. The *Daphnia* experience is an exemplar for a coordinated community collaboration that benefits the biological scientific community at large.

The genomic toolbox for Daphnia – *linking genomic resources via the genome sequence*

The discovery and functional analysis of ecologically relevant genes is critical to the goals of toxicological genomics. At one end of a spectrum of analytical approaches, unique patterns of gene expression can be used as indicators of specific environmental stressors. At the other end, these patterns of expression implicate identifiable genes and specific genetic regulatory pathways in the response of organisms to stressors. In either case, understanding the functionality of expressed genes and their place within a network of interacting genes is highly informative. Nevertheless, gaining an understanding of gene function in a novel organism is a challenging task that requires multiple tactics and a suite of phenotypic and genomic tools (Fig. 3).

A central component of our efforts is the recent generation of the complete genome sequence of *D. pulex*. With this tool in hand, we have a complete catalogue of the coding and noncoding components of a *Daphnia* genome. To understand the functional relevance of these components, however, requires tools to link the phenotypic response of organisms with the genome, tools to examine patterns of transcriptional and translational variation and tools to systematically isolate the function of particular genes. In essence, the research community must have a set of methodologies that implicate particular genes in the genetic basis of organismal response and then another set to verify the functional role of these genes.

There is a number of approaches to associate gene function with the catalogue represented by the genome sequence. Establishing a direct connection between phenotypic variation and physical locations with in the genome that influence this variation can be accomplished using a QTL approach. This methodology utilizes a large set of recombinant individuals and a genetic map based on recombination frequencies to make the link. There is also a need to establish the transcriptional and

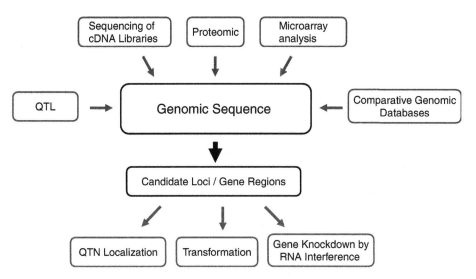

Fig. 3. Strategies for applying the *Daphnia* genomic tool box for functional analysis of toxicologically relevant genes and gene networks. A series of tools (blue) are available to probe the genome (black), via candidate gene, global expression profiling or genetic mapping approaches. These are validated using standard molecular biological approaches (red). (Modified from material presented by Michael Lynch to the Joint Genome Institute on behalf of the *Daphnia* Genomics Consortium, April 28, 2003; (See color figure 5.3 in color plate section).

translational responses of organism to environmental stressors. A first step towards a complete understanding of translation response is to develop a rich collection of cDNA libraries representing as large a fraction of the coding portion of the genome as possible. *Daphnia* is an ideal organism for this task as a common genetic background can be exposed to any number of environmental perturbations and the resulting transcriptional response captured for analysis. Once a library of expressed genes is developed the use of microarray chips to assay gene expression patterns and 2D gels or mass spectrophotometry to assay protein production provides tremendous insight. Finally, a comparative approach can take advantage of the growing understanding of gene function in other phylogenetically related species. This latter approach will be among the most informative in the short run.

Currently the leading model systems for the study of gene function and genome structure in *Daphnia* are a growing number of insect systems. The number of complete genome sequences available for species in this group is rapidly expanding, with over 40 projects completed or in progress (From NCBI Entrez Genome Project; http://www.ncbi.nlm. nih.gov/genomes/static/gpstat.html; accessed 31 October 2007). Functional information derived from these organisms combined with a bioinformatic analysis of the *Daphnia* genome will be a valuable starting point to develop a functional annotation of *Daphnia*. The transfer of information will also flow in the opposite direction. At present there is a conspicuous lack of a relevant outgroup for these numerous insect systems with comparable genomic level infrastructure in the form of a complete genome sequence and available tools for gene expression and QTL mapping studies. Crustaceans, in particular *Daphnia*, are a logical candidate as an outgroup for comparative genomic studies. The close relationship of the crustacea and insects is clearly supported by both molecular and morphological studies (Averof and Akam, 1995a, 1995b; Boore *et al.*, 1998; Friedrich and Tautz, 1995; Nardi *et al.*, 2003; Regier and Shultz, 1997; Regier *et al.*, 2005).

The approaches listed above all serve to increase our understanding of how *Daphnia* respond to environmental stressors and in essence allows for the development of candidate loci and gene regions linked to an organism's response. To increase this understanding to another level, a clear demonstration of the functional relationships among these genes and gene regions, will require the development of techniques to systematically assay genetic variants in natural populations (QTNs) and to knock out genes or transform them into novel genetic backgrounds. In the following sections, we outline the current development of these tools in the *Daphnia* system and suggest priorities for future expansion.

The utility box for toxicological genomic investigations using *Daphnia* is summarized in Table 1.

cDNA sequencing projects

The high-throughput sequencing and analysis of transcribed genes, archived in cDNA libraries, is a powerful method of discovering genes for toxicological studies using *Daphnia*. The straightforward approach of creating cDNA libraries from extracted mRNA within selected tissues, then picking a large number of clones at random for sequencing, was originally proposed to characterize new genes and to facilitate the identification of coding regions in genomic sequences (Adams *et al.*, 1991). cDNA sequencing continues to serve as a necessary component of whole-genome sequencing projects, by improving the annotation of the *D. pulex* genome sequence, for instance. The growing number of *Daphnia* cDNA libraries – which are created under a variety of environmental stressors and life-stages – are also providing insights into the physiological, developmental and cellular responses of animals to toxicants in two ways. First, by clustering the sequences to represent unique transcripts among libraries, classes of genes with shared putative functions are identified. Gene transcripts with similar functions are either enriched or missing from certain libraries compared with others, then producing clues about the general mode of action and biological effects of specific toxicants. Second, both the cDNA clones and their sequences are reagents for the fabrication of microarrays. This tool for monitoring the simultaneous expression patterns of thousands of genes under controlled experimental conditions is valuable, only if the DNA from genes of interest is present on the array to observe the experimental results from cDNA hybridization. Genes of greatest interests to researchers are those whose expression is specific to chemical challenges. These genes, therefore, are most likely to be absent from an array created from a limited diversity of cDNA libraries. This section describes the *Daphnia* cDNA library production and EST sequencing projects that are currently underway to support genomic-level investigations and we summarize some preliminary findings.

Daphnia pulex cDNA libraries and EST sequencing
The *D. pulex* cDNA libraries are key genomic tools for the overall DGC efforts at mounting the freshwater crustacean *Daphnia* as a model system for ecological and toxicological genomics. At present, 37 libraries are constructed and partially sequenced from three different isolates of *D. pulex*, *araneta* strain (Table 2). The diversity of conditions and developmental stages represented among these libraries stems from the input of a research community involved in a variety of research programs. The list includes conditions to discover stress response genes, of both natural (*e.g.*, UV, starvation, hypoxia, predation, salt, infection) and anthropogenic causes (metals, acidification, nanoparticles, depleted calcium). Other libraries contrast genes expressed in males compared with females, juveniles compared with adults and at high compared with low doses of environmental toxicants. In many cases, the libraries reveal genes associated with phenotypic plastic responses to changes in the environment, such as the parthenogenetic production of males and haemoglobin synthesis under the control of a juvenoid hormone and hypoxia (Rider *et al.*, 2005) or the modulation of the carapace into defence structures against predators (Tollrian, 1993, 1995). The libraries may also bring to light shared symptoms or physiological outcomes from diverse environmental hardships. In total, over 135,000 cDNA clones are

190

Table 2. The *Daphnia pulex* cDNA libraries supporting genomic research activities.

Daphnia pulex library ID	Date created	Condition, developmental stage	No. Arrayed clones	No. Sequenced clones	No. Nuclear ESTs	No. Clusters[a]	No. Clones[a]	% Clone diversity	% Organelle ESTs
Nonnormalized libraries									
Log52	26-Jun-03	Unchallenged, mixed	10,000	1,648	1,414	804	1,435	56	9.0
Log50-1	18-Mar-04	Hypoxia, adult	2,304	2,304	3,355	1,039	1,823	57	10.6
Log50-2	18-Mar-04	Hypoxia, juvenile	3,840	3,840	5,567	1,524	3,033	50	12.7
Log50-3	1-Apr-04	Low dose UV exposure, mixed	2,304	2,304	2,620	1,013	1,433	71	23.6
Log50-4[b]	1-Apr-04	High-dose UV exposure, mixed	2,304	384	450	188	243	77	22.9
Log50-5	5-Mar-04	Unchallenged, juvenile	1,536	1,152	1,580	553	827	67	16.7
Log50-6	19-Jul-04	Low-dose cadmium, mixed	2,688	2,688	4,048	1,209	2,170	56	12.8
Log50-7	19-Jul-04	Low-dose arsenic, mixed	4,224	4,224	6,370	1,867	3,399	55	14.6
Log50-8	19-Jul-04	Low-dose zinc, mixed	4,224	4,224	6,817	1,535	3,709	41	6.5
Log50-9	19-Jul-04	High-dose mixed metals, mixed	4,608	4,608	7,185	1,863	3,770	49	14.4
Log50-10[b]	22-Jun-04	Unchallenged, mixed	3,840	384	390	159	232	69	14.6
Log50-11[b]	18-Dec-03	Unchallenged, mixed	4,992	384	405	167	238	70	30.0
Log50-12	8-Sep-04	Invertebrate (*Chaoborous*) predation, adult	4,608	4,608	6,542	2,034	3,511	58	17.7
Log50-13	8-Sep-04	Food starvation, juvenile	2,304	2,304	2,826	924	1,378	67	18.3
Log50-14	8-Sep-04	Food starvation, adult	2,304	2,304	2,684	860	1,291	67	23.1
Log50-15[b]	8-Sep-04	Microcystis fed, juvenile	2,304	384	307	150	175	86	44.8

Log50-16[b]	8-Sep-04	Microcystis fed, adult	2,304	384	368	164	208	79	39.2
Log50-17	8-Sep-04	Fish predation, juvenile	3,840	3,840	4,750	1,548	2,638	59	22.5
Log50-18[b]	8-Sep-04	Fish predation, adult	2,300	384	425	177	249	71	20.3
Log50-19[b]	8-Sep-04	Methyl farnesoate hormone, juvenile	2,300	384	413	170	227	75	31.7
Log50-20	8-Sep-04	Methyl farnesoate hormone, adult	3,840	3,840	4,833	1,323	2,604	51	11.8
Total			76,808	51,718	72,179				
Normalized libraries									
Log50-21	22-Jun-04	Unchallenged, mixed	3,840	5,376	8,962	3,413	4,762	72	0.7
Chosen One-1[b]	7-Feb-06	Females, juvenile	384	384	211	98	121	96	8.0
Chosen One-2	7-Feb-06	Females, adult	3,456	3,456	5,313	2,252	2,821	84	1.5
Chosen One-3	7-Feb-06	Males, adult	4,224	4,224	5,425	2,168	2,883	83	21.7
Chosen One-4	10-Apr-06	Low-dose nickel, mixed	4,224	4,224	6,484	2,865	3,599	86	0.5
Chosen One-5	10-Apr-06	Low-dose copper, mixed	4,224	4,224	6,852	2,963	3,685	85	1.3
Chosen One-6	10-Apr-06	Acid stress pH 6.0, mixed	3,840	3,840	6,626	2,870	3,514	86	0.5
Chosen One-7	10-Apr-06	High salinity, mixed	3,840	3,840	6,121	2,645	3,275	85	2.6
Chosen One-8	4-Apr-06	Fullerene nanoparticle, mixed	4,224	4,224	5,643	2,428	3,044	84	14.1
Chosen One-9	11-May-06	Bacterial infection, mixed	3,456	3,456	5,639	2,553	2,935	90	1.1
Chosen One-10	2-May-06	High-dose mixed metals, mixed	3,840	3,840	4,398	2,030	2,452	91	3.5

Table 2. (Continued)

Daphnia pulex library ID	Date created	Condition, developmental stage	No. Arrayed clones	No. Sequenced clones	No. Nuclear ESTs	No. Clusters[a]	No. Clones[a]	% Clone diversity	% Organelle ESTs
Chosen One-11	2-May-06	Low-dose mixed metals, mixed	3,456	3,456	5,407	2,447	2,967	89	0.8
Chosen One-12	11-May-06	Low-dose monomethylarsinic III, mixed	4,224	4,224	6,274	2,768	3,387	87	4.5
Chosen One-13	10-May-06	Titanium dioxide nanoparticle, mixed	4,224	4,224	5,742	2,490	3,037	86	3.4
Chosen One-14	10-May-06	Microcystis fed, mixed	3,072	3,072	4,734	2,052	2,522	86	8.1
Chosen One-15	10-May-06	Calcium starvation, mixed	3,840	3,840	5,309	2,278	2,887	84	11.1
Total			58,368	59,904	89,140				

Notes: cDNA clones were sequenced from both ends, except for library Log52 that was sequenced only from the 5′ end. Clone diversity is calculated by dividing the # of clusters (including clusters of 1 EST) by the # of clones. This estimate is inflated, especially for nonnormalized libraries, by ignoring clones containing organelle transcripts (6–45% of ESTs are mitochondrial, depending on library). By contrast, the normalized libraries typically contain between <1% and 10% organelle ESTs.

[a]These numbers are of clusters and clones of nuclear genes only.

[b]Libraries failing stringent quality control checks and were, therefore, excluded from high-throughput EST sequencing.

archived within 384 well plates, providing over 161,000 ESTs of nuclear genes. Mining of these sequence data, with reference to the biological conditions of the animals when the gene transcripts were sampled, uncovers regulatory genetic pathways specific to how *Daphnia* cope with environmental challenges (study in progress).

Isolate Log52 is the source of the first cDNA library, created to produce experimental *D. pulex* microarrays. This work is also a pilot study that aimed to improve protocols for fabricating subsequent libraries, which are enriched for full-length cDNA and optimized for gene discovery. A detailed characterization of this initial library assured high-quality cDNA resources for *Daphnia* (Colbourne *et al.*, 2007). Of 1,648 sequenced clones, only 9% contained mitochondrial genes. The average molecular weight of cDNA inserts within the large size fraction was 847 bp, while 64–68% of the cDNAs were full-length or close to full-length. With few exceptions, this level of quality was met and often exceeded in the 20 nonnormalized Log50 libraries – whose average insert sizes range between 575 and 819 bp – and in the 15 normalized 'Chosen One' libraries, with average insert sizes between 819 and 1504 bp (unpublished data).

Alternative splice variants of abundantly transcribed genes are more likely to be detected in standard libraries created without normalization. Moreover, data on the relative number of specific transcripts sampled from libraries that represent an array of experimental conditions may be indicative of differentially expressed loci, which possibly deserve further study (Audic and Claverie, 1997). The benefits in sampling from nonnormalized cDNA libraries, however, are offset by the cost of sequencing redundant clones. Normalization procedures reduce the number of redundant copies of gene transcripts, therefore increasing the gene discovery rate during high-throughput EST sequencing. Sequencing from normalized *D. pulex* libraries resulted in a 15% decrease in the average number of EST contaminants from mitochondrial genes and gained an average of 20% genes discovered (Table 2).

The gene inventory of Daphnia *compared with model insects*
An important use for the large cDNA sequence data is in identifying *Daphnia* genes whose functions may be inferred from their sequence similarity to well-studied loci in genetic model species. This method is the candidate gene approach to uncovering genes of interest for toxicology. Although crustaceans and insects have divergent evolutionary histories for some 600 million years, both *Daphnia* and the model insect *Drosophila* (fruitfly) are members of a monophyletic group called

Pancrustacea (Boore *et al.*, 1998). Consequently, these two species, plus related taxa, are expected to share ancestral genes that are central to their biology and development. A fraction of the gene inventory of *Daphnia* is also expected to be uniquely crustacean or specific to daphniids, given the numerous lineage-specific adaptations associated with their distinct ecologies and life histories.

A recent analysis of the Log52 cDNA sequences provides some insight about the level of sequence conservation between *Daphnia* and its fellow arthropods. Of 787 assembled gene sequences, $\sim 68\%$ matched to a similar sequence in at least one insect proteome (Colbourne *et al.*, 2007). By also comparing to the nematode genome, 21% of the genes are either derived within *Pancrustacea* or lost within nematodes. These results suggest that the elaborate functional genetic database for the fruitfly (Drysdale *et al.*, 2005) may be a valuable resource for making predictions about the biological function for a majority of *Daphnia* genes. Understandably, the link between gene sequence similarity and function can be tenuous, partly because of lineage-specific expansions and extinctions of ancestral loci or the invention of new genes. For example, within the Log52 cDNA sequence dataset, 13 sequences encode genes with putative homologues in insects that specifically bind charged atoms like metals. These include a lineage-specific expansion of the *D. pulex* ferritin genes. In contrast to flies that have three ferritins, *Daphnia* has six or more loci that probably code protein subunits. Expression data from microarray experiments suggest that at least three gene duplicates have modified functions, based on their different transcriptional responses to metals and their sex-biased expression (Colbourne *et al.*, 2007). Further experiments are required to determine which locus, if any, has retained the ancestral gene function.

Community resource

The *D. pulex* and *D. magna* EST sequences are mapped to specific clones within the archived cDNA libraries (Table 1), which can, therefore, be retrieved by researchers for their experiments. As the pace of gene discovery quickens with high-throughput and computational methods, so will the demand for reagents to validate predicted functions, through detailed gene-by-gene investigations. At present, however, the cDNA sequences are playing a vital role in the annotation of the newly assembled *D. pulex* genome sequence. They facilitate the delineations of intron/exon boundaries, mark the positions of transcribed and untranslated regions and help to identify regulatory regions of the genome. Their predicted gene translations enable more accurate similarity searched against protein databases and are more useful for pattern matching and comparison of *Daphnia* functional proteomics data. Moreover, the full set of assembled cDNA sequences were used to design 10,000 oligonucleotides that are unique to single loci in the genome, for printing microarrays to detect the transcriptional signatures of exposures to toxicants.

Gene expression profiling

The use of microarray gene expression profiling for toxicology research has been a key development linking molecular approaches with traditional toxicology studies. Large-scale sequencing efforts have resulted in the creation of several different microarray platforms for toxicologenomics research in *Daphnia* and studies using these arrays are beginning to be published (Connon *et al.*, 2008; Poynton *et al.*, 2007; Shaw *et al.*, 2007; Soetaert *et al.*, 2006, 2007a, 2007b; Watanabe *et al.*, 2007). Data from these experiments are being used for a variety of purposes, with special interest in the environmental toxicology community for biomonitoring, which has been called 'canary on a chip' (Klaper and Thomas, 2004). Other uses for mRNA abundance data include gene function discovery (Hughes *et al.*, 2000; Shaw *et al.*, 2007), genetic regulatory network analysis (Tavazoie *et al.*, 1999), defining mechanisms of toxicity and clearance (Waring *et al.*, 2001a, 2001b) and evaluating the effects of genetic and environmental variation on transcript levels (discussed in section *Toxicogenomics database – a role for wFleaBase*). Harnessing the power of transcriptional profiling will require solutions to a number of vexing problems, including general challenges for gene expression analysis, as well as, those unique to the ecological aspects involved in this work. A brief description of options for expression profiling in *Daphnia* will be followed by a summary of their current uses and a discussion of the advantages and challenges of conducting microarray studies with daphniids.

Microarray platforms for Daphnia
The number of options available for comparative expression profiling on a genome-wide scale continues to increase, from cDNA amplicon arrays or custom oligonucleotide arrays to high-throughput pyrosequencing technologies (*e.g.*, Illumina®, Solexa®; see Cook *et al.*, 2007 for review). Newer pyro-technologies have a significant advantage in that transcripts are sequenced directly, so there is no ascertainment bias caused by absence of (potentially unknown) sequences from an array.

While there are ongoing efforts to apply pyrosequencing technologies to *Daphnia*, these approaches are still too costly and in the research and development stage (Darren Bauer, personal communication). By comparison, the ease and comparatively minimal costs of microarrays makes it highly probably that this technology will be a workhorse for many years to come. With this in mind, several groups have developed microarrays for probing *Daphnia*'s expressed genome (*i.e., D. magna*, Connon *et al.*, 2008; Poynton *et al.*, 2007; Soetaert *et al.*, 2006, 2007a, 2007b; Watanabe *et al.*, 2007; *D. pulex*, Shaw *et al.*, 2007). All of these arrays relied on cDNA libraries for source material, but differed with respect to design.

One microarray format that has been utilized for *Daphnia* research followed the approach described by Gracey *et al.* (2001), which involved arraying unknown PCR-amplified cDNA clones randomly picked from a collection of high-quality cDNA libraries. These blind arrays were used to identify differentially regulated targets that were then sequenced for characterization/annotation. One of the major challenges associated with this approach lies on the analysis end, as there is an unknown and uneven amount of replication on the array. The observation of repetitive annotations, however, suggests that they are not the product of chance events, which can be addressed statistically using permutation tests to estimate the likelihood that random processes would place highly represented annotations on a list of significant annotations. This microarray platform has been used for *D. magna* and *D. pulex* and proved successful as a gene discovery tool differentiating male- and female-specific responses (Eads *et al.*, 2007, 2008) and following exposure to pollutants (*i.e.*, metals, ordinance related compounds; Poynton *et al.*, 2007; Shaw *et al.*, 2007).

The other array format that has been utilized with *Daphnia* involved spotting known cDNA amplicons. These were derived from: (i) a large EST project that sequenced through a cDNA library from unexposed mixed aged organisms (Watanabe *et al.*, 2005); (ii) suppressive subtractive hybridization (SSH) between adults and juveniles (Soetaert *et al.*, 2006) and (iii) SSH on populations exposed to selected stressors (*i.e.*, cadmium, lufenuron, pH, hardness, kerosene and ibuprofen; Connon *et al.*, 2008). SSH, which identifies genes expressed in one population, but not in the other, was used as a means of establishing condition-specific targets (Diachenko *et al.*, 1996; Lisitsyn *et al.*, 1993). Theoretically, using stressor-specific SSH to create microarrays max-imises the specificity of the response and reduces the need to have all genes represented. Retrospect analysis suggested, however, that the SSH derived spots actually underperformed random nonspecific cDNA in identifying genes regulated by the conditions from which the SSH probes were generated. This does bring into question the utility of using SSH and, if it is used, suggests that a series of SSH treatments should be used rather than just one. This microarray platform has been used for *D. magna* to identify differentially regulated genes following exposure to fenarimol, propiconazole, ibuprofen and cadmium (Connon *et al.*, 2008; Heckmann *et al.*, 2006; Soetaert *et al.*, 2006; Soetaert *et al.*, 2007a, 2007b).

The declining cost of oligonucleotide synthesis has increased their attractiveness as a microarray platform, largely because of the time and effort required to generate cDNA for spotting. Such an approach has been applied by Watanabe *et al.* (2007) for *D. magna*. This approach, however, requires careful attention be paid to probe design in order to minimize cross-hybridization of related sequences, so splice variants and closely related multigene families can be easily distinguished. Such intricate design is now possible for *D. pulex* because of the complete genome sequence and the diverse EST project (Table 1). A robust set of oligonucleotide (70mers) probes is now available for this species. One challenge by using oligonucleotide arrays is their lower signal in proportion to their length. Such arrays are thus susceptible to signal loss caused by sequence mismatches especially in distantly related study populations. It remains to be examined how broad the oligonucleotide set that is available for *D. pulex* can be applied within the *D. pulex* complex of species.

Benefits of using Daphnia *in microarray studies*
Transcriptional profiling has already begun to provide important insights into the biology of *Daphnia*. First, because the animals can be bred clonally, separation of genetic and random environmental components (*e.g.*, subtle changes in food concentration from beaker to beaker) is possible. In addition, transgenerational effects can be tested or removed because of the quick generation times and large numbers of progeny produced in culture. Investigators can, therefore, examine how the norm of reaction (*i.e.*, the range of phenotypes a given genotype is able to produce, given variable environmental conditions) of gene expression changes from one clonal genotype to another. This approach will have unique power in *Daphnia* for the dissection of the genetic basis of gene expression, sometimes called genetical genomics (Jansen and Nap, 2001). A current problem in the field is how to account for the effect of common descent, sometimes called 'phylogenetic inertia' (Blomberg and Garland, 2002), on gene expression. Recent work in this area using a traditional

phylogenetic comparative method (Whitehead and Crawford, 2006) demonstrates an approach to control for the effects of phylogenetic distance (assuming neutral drift to be the dominant component of population-level differences) and shows some patterns of gene expression to be under natural selection. More work in this area is clearly warranted, because studies at the population-level are an integral part of an ecological approach to toxicology. Another important attribute of *Daphnia* is the ability to bring field samples directly into the lab and rear them in a common garden environment. In some cases, extinct populations can even be 'resurrected' by hatching dormant ephippia from sediment core samples, providing an unparalleled opportunity to examine microevolutionary patterns. Although longitudinal studies (*e.g.*, over time) of natural populations have not yet been reported, with the proliferation of the genomics utility box they will probably be an important component of toxicology research in *Daphnia*. Together, these features provide virtually limitless opportunities to study the interplay of genetics and environmental conditions in toxicological assays.

Many of the challenges facing transcriptional profiling in environmental toxicology are not unique to the field, including a need for appropriate quality control and analytical rigour. There are, however, some issues arising from the use of natural populations and ecological stressors that demand extra attention. Foremost among these are the avoidance or characterization that confounds biological variation, such as cryptic infection by parasites (see Ford and Fernandes, 2005) or regional variation in water chemistry affecting baseline culture conditions. Thorough genetic characterization of study populations is also critical, as is the attention to age, developmental stage or reproductive status of the animals that are being assayed. Finally, integration of expression data with toxicological or other types of data will be an important determinant of the utility of microarray data. In this area, the DGC has made considerable progress in creating databases and internet tools for community use. Developments in this area are highlighted later in this chapter (see section *Toxicogenomics database – a role for wFleaBase*).

Genetic maps and QTL analysis

The availability of genomic information greatly facilitates the mapping and characterization of genes responsible for complex phenotypes. Typically, complex traits are under the control of multiple interacting loci, the segregation thereof leading to a continuous distribution of phenotypes. Such traits are also termed quantitative traits. The genomic location of quantitative trait loci or QTLs, can be estimated using various methodologies, all of which are based on the cosegregation of genetic markers with known location on a genetic map and the QTLs in question. One of the most powerful methods is line-cross mapping: by crossing two lines with different phenotypes of interest and divergent genotypes at markers loci, one can determine what part of the genome cosegregate with the QTLs (Lynch and Walsh, 1998). This methodology has been previously used in toxicity studies in rodents (Mcclearn *et al.*, 1993) and plants (Dong *et al.*, 2006). With high-resolution mapping, this method can identify putative candidate genes, which can then be functionally characterized (Mackay, 2001).

QTL panels in Daphnia
With their long history as model organisms in ecotoxicological studies and the availability of increasing amount of genomic information, *Daphnia* provide an ideal system to genetically characterize QTLs for resistance and responses to environmental toxicity. We now have at our disposal the genomic sequence of *D. pulex*, as well as an EST (expressed sequence tags) database of *D. pulex and D. magna*. Moreover, an extensive number of polymorphic markers are available for *Daphnia* species from the direct cloning and development of simple sequence repeat motifs (SSRs) (Table 1). For example, microsatellite markers have been developed for *D. pulicaria*, many of which cross-amplify in other *Daphnia* species (Colbourne *et al.*, 2004). In addition, bioinformatic analyses of the available genome and EST sequences from *D. pulex* will yield many additional SSR loci and single nucleotide polymorphisms (SNPs) useful for fine resolution mapping. The tools are then available to relatively quickly develop genetic markers in many *Daphnia* species. An important advantage of *Daphnia* as a model system for QTL analysis is the ability to clonally propagate individual *Daphnia* genotypes by parthenogenesis. Not only does this property of *Daphnia* increase the power of scoring population-level phenotypes such as toxicity tolerance by allowing replication of genotypes across environments, but also recombinant lines resulting from crosses can be kept in the laboratory for extended periods and repeatedly used for mapping of a wide variety of traits. Cyclical parthenogenesis, however, also has a drawback, as this phenomenon makes line crossing more difficult. Environmental trigger-ing of sexuality varies greatly among lines and the production of sexual eggs is linked to diapause, which is often difficult to break in the laboratory. Performing sexual crosses in *Daphnia* most often leads to substantial losses in numbers. As a result, segregation ratios may be

subject to a strong bias. Careful choice of parental clones will help to alleviate this potential source of bias.

A genetic linkage map has now been produced for *D. pulex* based on 129 recombinant F_2 lines (Cristescu *et al.*, 2006). The map comprises 185 microsatellite markers distributed over 12 linkage groups, with an average inter-marker distance of 7 cM. Notably, a substantial number of the markers (more than 20%) show evidence of a homozygote deficiency in the F_2 generation, probably because these markers are linked to recessive deleterious alleles. Indeed, genetic load seems to be high in *Daphnia* populations (De Meester and De Jager, 1993), with levels of inbreeding depression being among the highest ever reported (Haag *et al.*, 2002). This general pattern of high inbreeding depression indicates that many strains of *Daphnia* may harbour a large load of deleterious recessive alleles, which can lead to patterns of segregation distortion in line-cross mapping panels. Line-cross designs that minimize the impact of deleterious recessive alleles should be utilized in *Daphnia*. There are two possible strategies to reduce the impact of deleterious recessive alleles on subsequent QTL analyses. One method is to construct highly inbred lines to purge the genetic load and select the highest fitness inbreds for use as the parental generation. These lines are then crossed to produce an F_1 generation and then selfed to produce recombinant F_2s (Fig. 4A). This approach requires several rounds of sexual reproduction to inbreed the parental lines and adds a considerable amount of time and effort to the initial phase of mapping panel construction. An alternative strategy to avoid the consequences of fixing deleterious alleles is to conduct two initial crosses using four different outbred parental clones (four-grandparent design, Bradshaw *et al.*, 1998). The resulting F_1s are then crossed reciprocally to produce F_2 generation individuals (Fig. 4B). This design has the advantage of avoiding fixing deleterious recessive alleles in the recombinant F_2 lines. In addition, greater phenotypic diversity can be incorporated in the F_2 mapping panel. The major limitation to the four-grandparent design is the larger number of markers required to distinguish chromosomal segments from each parental line. Given the possibility of long-term maintenance of recombinant F_2's through asexual reproduction, the additional effort required to create highly inbred parental lines may be well warranted and below we describe our efforts to construct QTL panels using the four grand parent design.

We are at present developing resources for QTL analysis in *D. magna* and *D. pulex*. In order to minimize segregation of deleterious recessive alleles in the F_2 generation, inbred clones will be used as parentals in the crosses, then hopefully purging the experimental system of deleterious

A)

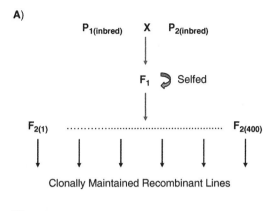

B)

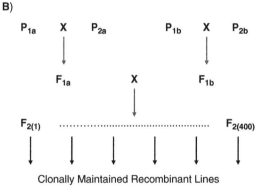

Fig. 4. Alternative crossing designs for QTL panel construction. (A) Inbred parental lines and selfed F_1. (B) Four-grandparent design with outcrossed F_1s.

alleles. In addition, to avoid biased segregation of markers linked to genes involved in diapause, it is necessary to use hatching conditions that will not favour one of the parental clones. These precautions should allow us to produce a high-quality recombinant mapping populations that will be available to the whole community of *Daphnia* researchers. For *D. magna*, we are developing markers based on simple sequence repeat motifs from two sources: (i) genomic DNA libraries enriched for repetitive sequences yielded more than 60 polymorphic microsatellite markers and (ii) to supplement these markers, we identified repetitive sequences in a *D. magna* EST database (Watanabe *et al.*, 2005). Utilizing an EST database allows us to develop markers closely linked to coding regions, increasing the probability to detect QTLs (Coulibaly *et al.*, 2005; Vasemagi *et al.*, 2005). A total of 330 EST sequences containing tandemly repeated DNA have been identified and are currently being tested for polymorphism in *D. magna*. In *D. pulex*, there are in excess of 500 polymorphic SSR loci available (Colbourne *et al.*, 2004) and markers linked to virtually every open reading frame can be developed from the complete genome sequence. Availability of these QTL resources in the future will make it possible to efficiently map QTLs involved in a variety of toxicological and ecologically relevant traits, including toxicological responses. There is a strong interest in evolutionary toxicological studies on *Daphnia* examining patterns of genetic adaptation to local ambient pollution levels (Lopes *et al.*, 2004). Both the identification of QTLs and the analysis of gene expression profiles using cDNA microarrays (Soetaert *et al.*, 2006) are promising avenues for linking genetic adaptation to specific genes. QTL panels from *D. magna* and *D. pulex* will provide a comparative basis for examining the generality of mapping results and will help bridge the existing gap between a model for toxicology (*D. magna*) and a resource-rich model for ecological and evolutionary genomic studies (*D. pulex*). The shared use of common mapping populations for the analysis of different traits will also allow the investigation of correlations among complex phenotypes.

Currently, the genomic tools available for *D. pulex* are more advanced than for *D. magna*. Given the importance of *D. magna* as a prime model organism in toxicological studies and the investment of many research groups in developing genomic resources for this species, there would be enormous benefit for the community to have the genomic tools (including a complete genome sequence) for this organism advanced as quickly as possible. In fact, the *D. magna* genome project has been launched by the DGC (http://dgc.cgb.indiana.edu/display/magna/Home).

Toxicogenomics database – a role for wFleaBase

Toxicological genomics research has a new resource in the *Daphnia* genome and the new genome database wFleaBase (http://wFleaBase.org/; Colbourne *et al.*, 2005) provides useful access to it. New genome sequencing projects and communities are facing large informatics tasks for incorporating, curating and annotating and disseminating sequence and annotation data. Biologists should now expect rapid access to new genomes, including basic annotations from well-studied model organisms and predictions to locate potential new genes, to make sense of them. Expertise from existing genome projects can be leveraged into building such tools. The Generic Model Organism Database (GMOD; Stein *et al.*, 2002) project has this goal, to fully develop and extend a genome database tool set to the level of quality needed to create and maintain new genome databases. The wFleaBase database, which is constructed on the GMOD platform, provides scientists with rapid access to this *Daphnia* genome, facilitating new discoveries and understanding for sciences such as toxicological genomics.

Genome database components
wFleaBase is built with common GMOD database components and open source software shared with other genome databases. Use of common components facilitates rapid construction and interoperability. The GMOD ARGOS replicable genome database template (www.gmod.org/argos/) provides a tested set of integrated components. The genome access tools of GMOD – GBrowse (Stein *et al.*, 2002), BioMart (Durinck *et al.*, 2005) and BLAST (Altschul *et al.*, 1997) – are available for searching the *D. pulex* genome. The GMOD Chado relational database schema (www.gmod.org/chado/) is used for managing an extensible range of genome information. Middleware in Perl and Java were added to bring together BLAST, BioMart, sequence reports, searches and other bioinformatics programs for public access. Another aid to integrating and mining these data is GMOD Lucegene (www.gmod.org/lucegene/), that forms a core component for rapid data retrieval by attributes, GBrowse data retrieval and databank partitioning for Grid analyses. wFleaBase operates on several Unix computers including Apple Macintosh OSX and Intel Linux and is portable enough to run on laptop computers for field studies. Genome maps include homologies to nine eukaryote proteomes, marker genes, microsatellite and EST locations and gene predictions. The assemblies and predicted genes can be searched by BLAST and linked to genome maps. BioMart provides searches of the full genome annotation sets, allowing selections of genome regions with and without specific features.

Genome annotations at wFleaBase produced by several groups are provided for map viewing and data mining, including contributions listed in Table 1. Gene predictions with SNAP (Korf, 2004) have been generated to locate new as well as known genes. SNAP guided by protein homology evidence is one of the better *ab initio* predictors when: (i) new genes are sought and (ii) there are no close relatives with an experimentally verified genome annotation. SNAP works well on the range of eukaryote genomes (plant to animal, small to big) with minimal homology data. A drawback is that SNAP overpredicts genes, but such aggressiveness can prove useful in identifying gene-like features.

The TeraGrid project (www.teragrid.org) is part of a shared cyber infrastructure for sciences, funded primarily by NSF. TeraGrid provides

204

collaborative, cost-effective scientific computing infrastructure much in the same way the GMOD initiative is building common tools for genome databases. The TeraGrid system is particularly suitable for genome assembly, annotation, gene finding and phylogenetic analyses. TeraGrid computers have been employed to annotate and validate the assembly of *D. pulex*. Results include homologies to nine eukaryote proteomes, gene predictions, marker genes, microsatellite and EST locations. Proteome comparisons included 217,000 proteins drawn from source genome databases, Ensembl and NCBI, for human, mouse, zebrafish, fruitfly, mosquito, bee, worm, mustard weed and yeast.

Database uses
wFleaBase provides a resource to biologists interested in comparing *Daphnia* to known genomes, finding novel and known genes, genome structure and evolution and gene function associations. These known genes provide useful access to this new genome for many researchers interested in locating a particular gene or gene family. The known gene matches also offer searches and cataloguing gene contents by known functions. Figure 5 summarizes known model organism genes found in *Daphnia* and two insects.

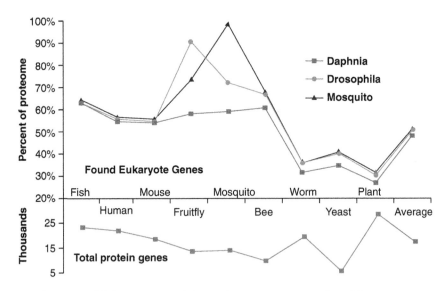

Fig. 5. Percent of full gene sets of nine eukaryote genomes found in new genomes *Daphnia pulex, Drosophila virilis,* with out-group *Anopheles gambia.* Lower line shows count of protein gene sets. *D. virilis* has 90% similarity to model fruitfly *D. melanogaster and A. gambiae* has 100% similarity to itself. (See color figure 5.5 in color plate section).

BLAST searches in wFleaBase
Searching for known gene sequences or gene families with BLAST remains one of the best ways to probe genomes. For example, one might be interested in cytochrome P450 genes and want to search for their presence in the *Daphnia* genome. Using the mouse P450 gene, MGI:88607 or GenBank:NP_067257, which is involved in haeme and iron ion binding, monooxygenase and oxidoreductase activity, a BLAST search against predicted proteins with BlastP returned several high-scoring matches. The best was to 'scaffold_3-snapho.108.' The BLAST report contained usual statistics and alignment values and also provided visualization by linking to a genome map view of the match. As a check of this predilation, a search of the chromosome 'scaffold' DNA with tBlastN returned several matches and the best was also at scaffold_3. The genome map view of this BLAST search linked to the same predicted gene, 'scaffold_3-snapho.108,' which is shown in Fig. 6. Two *Daphnia*

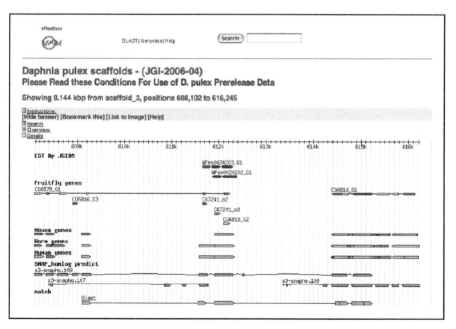

Fig. 6. Cytochrome P450 gene located on *Daphnia pulex* genome. This GBrowse map view at wFleaBase is returned from a BLAST search for mouse gene MGI:88607 (GenBank:NP_067257) and matches well the predicted *Daphnia* gene s3-snapho.108. Two *Daphnia* EST matches part of this gene and homologous genes from fruitfly, mouse, worm and human match. (See color figure 5.6 in color plate section).

ESTs are also found at this location and the source material can be readily accessed for follow-up studies of this gene.

Genome maps
Maps of the genome form the core, with BLAST searches, of discovery tools for bioscientists. Maps including available annotations from several groups are provided using GBrowse. The BLAST result reports include hyperlinks from each alignment match to the respective genome map, as well as to sequence and GFF annotation results. As seen in Fig. 6, this tool allows one to view evidence of common gene predictions and features in homologous regions.

Gene functions and biological processes
To provide an assessment of possible gene gain and loss among *Daphnia*, gene matches to Gene Ontology (GO) categories by species were tabulated and provided at section genome-summaries/gene-GO-function-association, which are discussed in functional detail in Colbourne *et al.* (2007). These may indicate species differences in functional categories when statistically significant deviations are indicated. While low counts, suggestive of missing genes, may be caused by divergence of genes, extra gene matches more strongly suggest categories in which species differ. Among the interesting effects, *Daphnia* may have higher gene counts than insects or *Caenorhabditis elegans* for catalytic activity (GO:0003824), hydrolase activity (GO:0016787), peptidase activity (GO:0008233) and transferase activity (GO:0016740). There is also a suggestion of lower gene numbers for receptor activity, protein binding and enzyme regulator activity.
    The gene matches were high-scoring segment pair groupings and include various events: gene duplications, alternate splice exons within genes, new genes that appear composed of exons from other genes, as well as computational artifacts. Detailed evidence pages provide links to GBrowse genome map views showing all secondary high-scoring segment pair groupings. Proteome sources in this analysis were those organism with extensive GO annotations: Dmel fruitfly, mouse, *C. elegans* worm and yeast. GO-Slim groupings are used for Biological Process, Molecular Function, Cell Location (125 categories).

Genome data mining
An emerging trend among bioscientists and bioinformaticians is to use data mining of large subsets of genome data, often focused on summary information for a range of common attributes. These data are used in spreadsheets and simple databases or analyses. The Ensembl project has

produced BioMart (Durinck *et al.*, 2005), used at wFleaBase for searches of the full *Daphnia* genome by various attributes of homologous genes, such as known toxicogenomic genes. One can retrieve tables of matching gene locations, Ids and annotations or retrieve the genome sequences for these in bulk Fasta format, which is suitable for BLAST and sequence tools. With BioMart, one can select or exclude genome regions with the available annotations and download tables or sequences of the selection set. For instance, one could select the regions of the *Daphnia* genome in which there are gene homologs similar to fish but fruitfly homologs are missing or regions with gene predictions in which there are no known homologs.

Future additions

Additions planned for this *Daphnia* genome database include experimental data from toxicology studies, such as gene expression in environmentally stressed populations. Metabolic pathways and cell cycle views of identified genes will be added to this database. These will provide an aid to quickly classify genes that belong to common and toxicologically important biochemical processes. Future growth of this database as a toxicogenomic resource will benefit much from use and contributions of data from the scientific community.

Concluding remarks

Daphnia are without question an established model species for toxicological studies. This sentinel species of inland aquatic habitats is among the most well-studied animals for toxicological testing and for ecological research. Their position in the emerging field of toxicological genomics will progress concomitantly with the development of genomic resources. In general, these resources are expected to provide reliable diagnostic tools that determine whether a defined ecosystem (or individual) is exposed to toxic agents (Andrew *et al.*, 2003), whether substances are harmful (Hayes and Bradfield, 2005) and whether a population is susceptible to certain chemicals (Pedra *et al.*, 2004). Genomic tools offer the promise of providing data that are more sensitive, more quickly obtained, less expensive and better able to detect all bioreactive compounds, including parent compounds and their metabolites, than traditional monitoring tools. When linked to well-defined biological endpoints – including cellular, physiological, life-history and population-level responses – they also promise to reveal insights into the modes of toxicity that are shared among varying classes of compounds and across species. For *Daphnia*, genomic technologies benefit from anchoring to abundant existing knowledge about the response of species to toxicological and ecological challenges. These challenges can at times have important and unexpected manifestations by their combined actions. For example, research suggests that certain toxicants negatively affect the ability of populations to respond appropriately to predation threat (Hunter and Pyle, 2004) and to the normal cycles of population growth and reproduction (Oda *et al.*, 2005). *Daphnia* also provide the experimental platform to probe past populations and integrate genomic response through evolutionary time scales.

Genomic resources are rapidly becoming available for *Daphnia* research by the coordinated efforts of the DGC. Over the past five years, the genomic utility box (Table 1) for *Daphnia* has experienced rapid growth. The box started with a set of molecular markers for genetic mapping and identifying genes of interest by QTL analyses; a pilot microarray platform to identify synchronized gene responses to experimental conditions; and cDNA libraries and sequences used to uncover the existence of conserved genes of interest that have already been characterized in other model systems. The utility box for *Daphnia* now includes a fully sequenced genome and draft annotation. The genome sequence provides a resource that will greatly facilitate research to identify major genes and their interactions (Fig. 3) that can account for the successes or failures of individuals at coping with chemicals and of populations at adapting to these challenges. Given the complexities of genome-wide studies, however and the scope of toxicological research, which often necessitates validation of results by detailed functional studies in the lab and field, integration would benefit from an approach similar to studies in development, cell and molecular biology. As a starting point, this involves building a research community around a common set of model species. Research using *Daphnia* is at present focused on two species. Expanding this research to include comparative studies among many species relies on creating the maximum set of resources for both *D. magna* and *D. pulex*. From this foundation, studies using other daphniids or even more distantly related species are made easier by comparing results to a reference model system for toxicological genomics. This strategy should lead to a marked improvement in our ability to extrapolate toxicological responses and biological outcomes across animal populations, through ecological food webs and from animal models to humans – a goal that is made possible by the general evolutionary conservation of genes and their interactions within biochemical pathways.

Acknowledgements

We thank Luc De Meester and Dieter Ebert for helpful comments on the manuscript. cDNA libraries were created by Darren Bauer and Zack Smith in the laboratories of Kelley Thomas and John Colbourne. Sequencing was produced by Erika Lindquist and Jeff Boore at the Joint Genome Institute. Kelley Thomas and Michael Lynch are key contributors of the *Daphnia* Genomics Consortium. Support for the *Daphnia* genomic utility box was provided in part by the U.S. National Science Foundation and by the DOE Joint Genome Institute under the auspices of the U.S. Department of Energy's Office of Science, Biological and Environmental Research Program and by the University of California, Lawrence Livermore National Laboratory under Contract No. W-7405-Eng-48, Lawrence Berkeley National Laboratory under Contract No. DE-AC02-05CH11231, Los Alamos National Laboratory under Contract No. W-7405-ENG-36, and in collaboration with the *Daphnia* Genomics Consortium, http://daphnia.cgb.indiana.edu. wFleaBase was developed at the Genome Informatics Lab of Indiana University with support to Don Gilbert from the National Science Foundation and the National Institute of Health. Other sources of support include NSF DEB-022183 (JKC, JRS); NIEHS-SBRP ES ES07373-07 (JRS); NSF DEB-021212487 (MEP); a USU Center for Integrated BioSystems research grant (MEP); NERC NER/D/S/2002/00413 (RMS, AC); The School of Public and Environmental Affairs, Indiana University, and The Center for Genomics and Bioinformatics at Indiana University, which is supported in part by the METACyt Initiative of Indiana University, funded in part through a major grant from the Lilly Endowment, Inc. Our work benefits from and contributes to the *Daphnia* Genomics Consortium.

References

Adamowicz, S. J., Hebert, P. D. N. and Marinone, M. C. (2004). Species diversity and endemism in the *Daphnia* of Argentina: a genetic investigation. *Zool. J. Linn. Soc. Lond.* 140,171–205.

Adams, M. D., Kelley, J. M., Gocayne, J. D., Dubnick, M., Polymeropoulos, M. H., Xiao, H., Merril, C. R., Wu, A., Olde, B., Moreno, R. F. *et al.* (1991). Complementary-DNA sequencing – expressed sequence tags and Human Genome Project. *Science* 252,1651–1656.

Altschul, S. F., Madden, T. L., Schaffer, A. A., Zhang, J. H., Zhang, Z., Miller, W. and Lipman, D. J. (1997). Gapped BLAST and PSI-BLAST: A new generation of protein database search programs. *Nucleic Acids Res.* 25,3389–3402.

Amin, R., Hamadeh, H. K., Bushel, P. R., Bennett, L., Afshari, C. A. and Paules, R. S. (2002). Genomic interrogation of mechanism(s) underlying cellular responses to toxicants. *Toxicology* 27,555–563.

Anderson, B. (1944). The toxicity thresholds of various substances found in industrial wastes as determined by the use of *Daphnia magna*. *Sewage Work J.* 16,1156–1165.

Anderson, B. G. (1945). The toxicity of DDT to *Daphnia*. *Science* 102,539.

Andrew, A., Warren, A. J., Barchowsky, A., Temple, K. A., Klei, L., Soucy, N. V., O'Hara, K. A. and Hamilton, J. W. (2003). Genomic and proteomic profiling of responses to toxic metals in human lung cells. *Environ. Health Perspect.* 111, 825–838.

Audic, S. and Claverie, J. M. (1997). The significance of digital gene expression profiles. *Genome Res.* 7,986–995.

Averof, M. and Akam, M. (1995a). Hox genes and the diversification of insect and crustacean body plans. *Nature* 376,420–423.

Averof, M. and Akam, M. (1995b). Insect-crustacean relationships: Insights from comparative developmental and molecular studies. *Philos. Trans. R. Soc.* B347, 293–303.

Ball, C. A. and Brazma, A. (2006). MGED standards: Work in progress. *Omics* 10, 138–144.

Banta, A. (1939). *Studies on the physiology, genetics and evolution of some cladocera*. Carnegie Institution, Washington, DC.

Bartosiewicz, M., Jenkins, D., Penn, S., Emery, J. and Buckpitt, A. (2001). Unique gene expression patterns in liver and kidney associated with exposure to chemical toxicants. *J. Pharmacol. Exp. Ther.* 297,895–905.

Bing, N. and Hoeschele, I. (2005). Genetical genomics analysis of a yeast segregant population for transcription network inference. *Genetics* 170,533–542.

Blomberg, S. P. and Garland, T. (2002). Tempo and mode in evolution: Phylogenetic inertia, adaptation and comparative methods. *J. Evolution. Biol.* 15,899–910.

Boore, J. L., Lavrov, D. V. and Brown, W. M. (1998). Gene translocation links insects and crustaceans. *Nature* 392,667–668.

Bradshaw, H. D., Otto, K. G., Frewen, B. E., McKay, J. K. and Schemske, D. W. (1998). Quantitative trait loci affecting differences in floral morphology between two species of monkeyflower (*Mimulus*). *Genetics* 149,367–382.

Brede, N., Thielsch, A., Sandrock, C., Spaak, P., Keller, B., Streit, B. and Schwenk, K. (2006). Microsatellite markers for European *Daphnia*. *Mol. Ecol. Notes* 6,536–539.

Breukelman, J. (1932). Effect of age and sex on resistance of daphnids to mercuric chloride. *Science* 76,302.

Cáceres, C. E. (1998). Interspecific variation in the abundance, production and emergence of *Daphnia* diapausing eggs. *Ecology* 79,1699–1710.

Campbell, A. K., Wann, K. T. and Matthews, S. B. (2004). Lactose causes heart arrhythmia in the water flea *Daphnia pulex*. *Comp. Biochem. Physiol. B* 139,225–234.

Carlborg, O. and Haley, C. S. (2004). Epistasis: Too often neglected in complex trait studies? *Nat. Rev. Genet.* 5,618–625.

Carlborg, O., De Koning, D. J., Manly, K. F., Chesler, E., Williams, R. W. and Haley, C. S. (2005). Methodological aspects of the genetic dissection of gene expression. *Bioinformatics* 21,2383–2393.

211

Carpenter, S. R., Kitchell, J. F., Hodgson, J. R., Cochran, P. A., Elser, J. J., Elser, M. M., Lodge, D. M., Kretchmer, D., He, X. and Vonende, C. N. (1987). Regulation of lake primary productivity by food web structure. *Ecology* 68,1863–1876.

Cheung, V. and Spielman, R. S. (2002). The genetics of variation in gene expression. *Nat. Genet.* 32,522–555.

Chiavelli, D. A., Marsh, J. W. and Taylor, R. K. (2001). The mannose-sensitive hemagglutinin of *Vibrio cholerae* promotes adherence to zooplankton. *Appl. Environ. Microbiol.* 67,3220–3225.

Colbourne, J. K. and Hebert, P. D. N. (1996). The systematics of North American *Daphnia* (*Crustacea: Anomopoda*): A molecular phylogenetic approach. *Philos. Trans. R. Soc. Lond. B* 351,349–360.

Colbourne, J. K., Crease, T. J., Weider, L. J., Hebert, P. D. N., Dufresne, F. and Hobaek, A. (1998). Phylogenetics and evolution of a circumarctic species complex (*Cladocera: Daphnia pulex*). *Biol. J. Linn. Soc. Lond.* 65,347–365.

Colbourne, J. K., Eads, B. D., Shaw, J. R., Bohuski, E., Bauer, D. and Andrews, J. (2007). Sampling *Daphnia*'s expressed genes: Preservation, expansion and invention of crustacean genes with reference to insect genomes. *BMC Genomics* 8,217.

Colbourne, J. K., Hebert, P. D. N. and Taylor, D. J. (1997). Evolutionary origins of phenotypic diversity in *Daphnia*. In *Molecular Evolution and Adaptive Radiation* (eds T. J. Givnish and K. J. Sytsma), pp. 163–188, Cambridge University Press, London.

Colbourne, J. K., Robison, B., Bogart, K. and Lynch, M. (2004). Five hundred and twenty-eight microsatellite markers for ecological genomic investigations using *Daphnia*. *Mol. Ecol. Notes* 4,485–490.

Colbourne, J. K., Singan, V. R. and Gilbert, D. G. (2005). wFleaBase: The *Daphnia* genome database. *BMC Bioinformatics* 6,45.

Colbourne, J. K., Wilson, C. C. and Hebert, P. D. N. (2006). The systematics of Australian Daphnia and Daphniopsis (Crustacea: Cladocera): A shared phylogenetic history transformed by habitat-specific rates of evolution. *Biol. J. Linn. Soc.* 89,469–488.

Connon, R., Hooper, H. L., Lim, F. L., Moore, D. J., Watanabe, H., Soetart, A., Cook, K., Sibly, R. M., Orphanides, G., Maund, S. J., Hutchinson, T. H., Moggs, J., De Coen, W., Iguchi, T. and Callaghan, A. (2008). Linking molecular and population stress responses in *Daphnia magna* exposed to cadmium. *Environ. Sci. Technol.* 42,2181–2188.

Cook, J., Denslow, N. D., Iguchi, T., Linney, E. A., Miracle, A., Shaw, J. R., Viant, M. R. and Zacharewski, T. R. (2007). 'Omics' approaches in the context of environmental toxicology. In *Genomic Approaches for Cross-Species Extrapolation in Toxicology* (eds R. DiGiulio and W. H. Benson), Taylor and Francis, Washington, DC.

Cooney, J. (1995). Effects-toxicity testing. In *Fundamentals of Aquatic Toxicology* (ed. G. Rand), 2nd Edition, pp. 71–102, Taylor and Francis, Washington, DC.

Coulibaly, I., Gharbi, K., Danzmann, R. G., Yao, J. and Rexroad, C. E. (2005). Characterization and comparison of microsatellites derived from repeat-enriched libraries and expressed sequence tags. *Anim. Genet.* 36,309–315.

Cousyn, C., De Meester, L., Colbourne, J. K., Brendonck, L., Verschuren, D. and Volckaert, F. (2001). Rapid, local adaptation of zooplankton behavior to changes in predation pressure in the absence of neutral genetic changes. *Proc. Natl. Acad. Sci. USA* 98,6256–6260.

Cristescu, M. E., Colbourne, J. K., Radivojac, J. and Lynch, M. (2006). A microsatellite-based genetic linkage map of the water flea, *Daphnia pulex*: On the prospect of crustacean genomics. *Genomics* 88,415–430.

Davis, J. (1977). Standardization and protocols of bioassays – their role and significance for monitoring, research and regulatory usage. In *Proceedings of the 3rd Aquatic Toxicity Workshop*, (eds E. P. W. R. Parker, P. G. Wells and G. F. Westlake), Environmental Protection Service, Technical Report No. EPS-5-AR-77-1, Halifax, NS.

Daphnia Genomics Consortium (2007). http://daphnia.cgb.indiana.edu (Accessed on October 26).

de Bernardi, R. and Peters, R. H. (1987). Why *Daphnia*. In '*Daphnia*' *Memorie dell'Istituto Italiano di Idrobiologia Dr Marco De Marchi* (eds R. H. Peters and R. de Bernardi), Vol. 45, Consiglio Nazionale Delle Ricerche Instituto Italiano Di Idrobiologia, Verbania Pallanza.

De Hoogh, C., Wgenvoort, A. J., Jonker, F., Van Leerdam, J. A. and Hogenboom, A. C. (2006). HPLC-DAD and Q-TOF MS techniques identify cause of *Daphnia* biomonitor alarms in the river Meuse. *Environ. Sci. Technol.* 40,2678–2685.

de Koning, D. J., Carlborg, O. and Haley, C. S. (2005). The genetic dissection of immune response using gene-expression studies and genome mapping. *Vet. Immunol. Immunopathol.* 105,343–352.

De Meester, L. (1996). Local genetic differentiation and adaptation in freshwater zooplankton populations: Patterns and processes. *Ecoscience* 3,385–399.

De Meester, L. and De Jager, H. (1993). Hatching of *Daphnia* sexual eggs. I. Intraspecific differences in the hatching responses of *D. magna* eggs. *Freshwater Biol.* 30,219–226.

Denslow, N., Colbourne, J. K., Dix, D., Freedman, J. H., Helbing, C. C., Kennedy, S. and Williams, P. L. (2007). Selection of surrogate animal species for comparative toxicogenomics. In *Genomic Approaches for Cross-Species Extrapolation in Toxicology* (eds R. DiGiulio and W. H. Benson), Taylor and Francis, Washington, DC.

Diachenko, L. B., Ledesma, J., Chenchik, A. A. and Siebert, P. D. (1996). Combining the technique of RNA fingerprinting and differential display to obtain differentially expressed mRNA. *Biochem. Biophys. Res. Commun.* 219,824–828.

Dix, D., Gallagher, K., Benson, W. H., Groskinsky, B. L., McClintock, J. T., Dearfield, K. L. and Farland, W. H. (2006). A framework for the use of genomics data at the EPA. *Nat. Biotechnol.* 24,1108–1111.

Dodson, S. I. and Hanazato, T. (1995). Commentary on effects of anthropogenic and natural organic-chemicals on development, swimming behavior and reproduction of *Daphnia*, a key member of aquatic ecosystems. *Environ. Health Perspect.* 103,7–11.

Dong, Y. J., Ogawa, T., Lin, D. Z., Koh, H. J., Kamiunten, H., Matsuo, M. and Cheng, S. H. (2006). Molecular mapping of quantitative trait loci for zinc toxicity tolerance in rice seedling (*Oryza sativa* L.). *Field Crops Res.* 95,420–425.

Drysdale, R. A., Crosby, M. A. and the Fly Consortium. (2005). FlyBase: Genes and gene models. *Nucleic Acids Res.* 33,D390–D395.

Duodoroff, P. and Katz, M. (1950). Critical review of literature on the toxicity of industrial wastes and their components to fish. I. Alkalies, acids and inorganic gases. *Sewage Ind. Wastes* 22,1432–1458.

Dudycha, J. L. (2001). The senescence of *Daphnia* from risky and well tolerated habitats. *Ecol. Lett.* 4,102–105.

Dudycha, J. L. (2003). A multienvironment comparison of senescence between sister species of *Daphnia*. *Oecologia* 135,555–563.

Durinck, S., Moreau, Y., Kasprzyk, A., Davis, S., De Moor, B., Brazma, A. and Huber, W. (2005). BioMart and Bioconductor: A powerful link between biological databases and microarray data analysis. *Bioinformatics* 21,3439–3440.

Eads, B., Colbourne, J. K., Bohuski, E. and Andrews, J. (2007). Profiling sex-biased gene expression during parthenogenetic reproduction in *Daphnia pulex*. *BMC Genomics* 8,464.

Eads, B. D., Andrews, J. and Colbourne, J. K. (2008). Ecological genomics in *Daphnia*: Stress responses and environmental sex determination. *Heredity* 100,184–190.

Eaton, D. and Klaassen, C. D. (1996). Principles of toxicology. In *Casarett and Doull's Toxicology: The Basic Science of Poisons* (ed. C. Klaassen), 5th Edition, McGraw-Hill, New York.

Ebert, D., Carius, H. J., Little, T. and Decaestecker, E. (2004). The evolution of virulence when parasites cause host castration and gigantism. *Am. Nat.* 164,S19–S32.

Edmondson, W. (1987). *Daphnia* in experimental ecology: Notes on historical perspectives. In '*Daphnia*' *Memorie dell'Istituto Italiano di Idrobiologia Dr Marco De Marchi* (eds R. H. Peters and R. de Bernardi), Vol. 45, Consiglio Nazionale Delle Ricerche Instituto Italiano Di Idrobiologia, Verbania Pallanza.

Elendt, B. and Bias, W. R. (1990). Trace nutrient deficiency in *Daphnia magna* cultured in standard medium for toxicity testing: Effects of the optimization of culture conditions on life history parameters of *Daphnia magna*. *Water Res.* 24,1157–1167.

Ford, A. T. and Fernandes, T. F. (2005). Better the devil you know? A precautionary approach to using amphipods and daphnids in endocrine disruptor studies. *Environ. Toxicol. Chem.* 24,1019–1021.

Fox, J. A. (2004). New microsatellite primers for *Daphnia galeata mendotae*. *Mol. Ecol. Notes* 4,544–546.

Friedrich, M. and Tautz, D. (1995). Ribosomal DNA phylogeny of the extant arthropod classes and the evolution of myriapods. *Nature* 376,165–167.

Gallagher, K., Benson, W. H., Brody, M., Fairbrother, A., Hasan, J., Klaper, R., Lattier, D., Lundquist, S., McCarroll, N., Miller, G., Preston, J., Sayre, P., Smith, B., Street, A., Troast, R., Vu, B., Reiter, L., Farland, W. and Dearfield, K. (2006). Genomics: Applications, challenges and opportunities for the U.S. Environmental Protection Agency. *Hum. Ecol. Risk Assess.* 12,572–590.

Gallo, M. (1996). History and scope of toxicology. In *Casarett and Doull's Toxicology: The Basic Science of Poisons* (ed. C. Klaassen), 5th Edition, McGraw-Hill, New York.

Glover, C. N. and Wood, C. M. (2005). Physiological characterisation of a pH- and calcium-dependent sodium uptake mechanism in the freshwater crustacean, *Daphnia magna*. *J. Exp. Biol.* 208,951–959.

Gracey, A. Y., Troll, J. V. and Somero, G. N. (2001). Hypoxia-induced gene expression profiling in the euryoxic fish *Gillichthys mirabilis*. *Proc. Natl. Acad. Sci. USA* 98, 1993–1998.

Haag, C. R., Hottinger, J. W., Riek, M. and Ebert, D. (2002). Strong inbreeding depression in a *Daphnia* metapopulation. *Evolution* 56,518–526.

Haber, F. (1924). Zur geschichte des gaskrieges (On the history of gas warfare). In *Funf Vortrage aus den Jahren 1920–1923 (Five Lectures from the Years 1920–1923)*, pp. 76–92, Springer, Berlin.

214

Hairston, N., Van Brunt, R. A., Kearns, C. M. and Engstrom, D. R. (1995). Age and survivorship of diapausing eggs in a sediment egg bank. *Ecology* 76,1706–1711.

Hairston, N. G., Holtmeier, C. L., Lampert, W., Weider, L. J., Post, D. M., Fischer, J. M., Cáceres, C. E., Fox, J. A. and Gaedke, U. (2001). Natural selection for grazer resistance to toxic cyanobacteria: Evolution of phenotypic plasticity? *Evolution* 55,2203–2214.

Hairston, N. G., Lampert, W., Cáceres, C. E., Holtmeier, C. L., Weider, L. J., Gaedke, U., Fischer, J. M., Fox, J. A. and Post, D. M. (1999). Lake ecosystems – rapid evolution revealed by dormant eggs. *Nature* 401,446.

Hall, S. R., Tessier, A. J., Duffy, M. A., Huebner, M. and Cáceres, C. E. (2006). Warmer does not have to mean sicker: Temperature and predators can jointly drive timing of epidemics. *Ecology* 87,1684–1695.

Hamadeh, H., Bushel, P. R., Jayadey, S., Martin, K., DiSorbo, O., Sieber, S., Bennett, L., Tennant, R., Stoll, T., Barrett, J. C., Blanchard, K., Paules, R. S. and Afshari, C. A. (2002). Gene expression analysis reveals chemical-specific profiles. *Toxicol. Sci.* 67,232–240.

Hayes, K. and Bradfield, R. (2005). Advances in toxicogenomics. *Chem. Res. Toxicol.* 18,403–414.

Hebert, P. (1987). Genetics of *Daphnia*. In '*Daphnia*' *Memorie dell'Istituto Italiano di Idrobiologia Dr Marco De Marchi* (eds R. H. Peters and R. de Bernardi), Vol. 45, Consiglio Nazionale Delle Ricerche Instituto Italiono Di Idrobiologia, Verbania Pallanza.

Hebert, P. and Ward, R. D. (1972). Inheritance during parthenogenesis in *Daphnia magna*. *Genetics* 71,639–642.

Hebert, P. D. N. (1974a). Ecological differences between genotypes in natural populations of *Daphnia magna*. *Heredity* 33,327–337.

Hebert, P. D. N. (1974b). Enzyme variability in natural populations of *Daphnia magna* II. Genotypic frequencies in permanent populations. *Genetics* 77,323–334.

Hebert, P. D. N. (1974c). Enzyme variability in natural populations of *Daphnia magna* III. Genotypic frequencies in intermittent populations. *Genetics* 77,335–341.

Hebert, P. D. N. and Finston, T. L. (1993). A taxonomic reevaluation of North American *Daphnia* (*Crustacea: Cladocera*). I. The *Daphnia similis* complex. *Can. J. Zool.* 71,908–925.

Hebert, P. D. N. and Finston, T. L. (1996). A taxonomic reevaluation of North American *Daphnia* (*Crustacea: Cladocera*). II: New species in the *Daphnia pulex* group from the south-central United States and Mexico. *Can. J. Zool.* 74,632–653.

Hebert, P. D. N. and Finston, T. L. (1997). A taxonomic reevaluation of North American *Daphnia* (*Crustacea: Cladocera*): III. The *D. catawba* complex. *Can. J. Zool.* 75,1254–1261.

Heckmann, L. H., Connon, R., Hutchinson, T. H., Maund, S. J., Sibly, R. M. and Callaghan, A. (2006). Expression of target and reference genes in *Daphnia magna* exposed to ibuprofen. *BMC Genomics* 7,175.

Hubner, N., Wallace, C. A., Zimdahl, H., Petretto, W., Schultz, H., Maciver, F., Muller, M., Hummel, O., Monti, J., Zidek, V., Musiolova, A., Kren, V., Causton, H., Game, M., Born, G., Schmidt, S., Muller, A., Cook, S. A., Kurtz, T. W., Wittaker, J., Pravenec, M. and Aitman, T. J. (2005). Integrating transcriptional profiling and linkage analysis for identification of genes underlying disease. *Nat. Genet.* 37,243–253.

Hughes, T. R., Marton, M. J., Jones, A. R., Roberts, C. J., Stoughton, R., Armour, C. D., Bennett, H. A., Coffey, E., Dai, H. Y., He, Y. D. D. *et al.* (2000). Functional discovery via a compendium of expression profiles. *Cell* 102,109–126.

Hunter, K. and Pyle, G. (2004). Morphological responses of *Daphnia pulex* to *Chaoborus americanus* kairomone in the presence and absence of metals. *Environ. Toxicol. Chem.* 23,1311–1316.

Hutchinson, G. (1932). Experimental studies in ecology. I. The magnesium tolerance of *Daphniidae* and its ecological significance. *Int. Rev. Ges. Hydrobiol. Hydrogr.* 28,90–108.

Jansen, R. C. and Nap, J. P. (2001). Genetical genomics: The added value from segregation. *Trends Genet.* 17,388–391.

Jensen, K. H., Little, T., Skorping, A. and Ebert, D. (2006). Empirical support for optimal virulence in a castrating parasite. *PLoS Biol.* 4,1265–1269.

Kerfoot, W. C., Robbins, J. A. and Weider, L. J. (1999). A new approach to historical reconstruction: Combining descriptive and experimental paleolimnology. *Limnol. Oceanogr.* 44,1232–1247.

Kilham, S. S., Kreeger, D. A., Lynn, S. G., Goulden, C. E. and Herrera, L. (1998). COMBO: A defined freshwater culture medium for algae and zooplankton. *Hydrobiologia* 377,147–159.

Klaper, R. and Thomas, M. A. (2004). At the crossroads of genomics and ecology: The promise of a canary on a chip. *BioScience* 54,403–412.

Klaper, R., Rees, C., Carvan, M., Weber, D., Drevnick, P. and Sandheinrich, M. (2006). Gene expression links to endocrine function and reproduction decline after mercury exposure in fathead minnows. *Environ. Health Perspect.* 114,1337–1343.

Klugh, A. and Miller, H. C. (1926). The hydrogen ion concentration range of *Daphnia magna*. *Trans. R. Soc. Can.* 20,225–227.

Koivisto, S. (1995). Is *Daphnia magna* an ecologically representative zooplankton species in toxicity tests. *Environ. Pollut.* 90,263–267.

Koivisto, S., Ketola, M. and Walls, M. (1992). Comparison of 5 cladoceran species in short-term and long-term copper exposure. *Hydrobiologia* 248,125–136.

Korf, I. (2004). Gene finding in novel genomes. *BMC Bioinformatics* 5,59.

Korovchinsky, N. M. (1997). On the history of studies on cladoceran taxonomy and morphology, with emphasis on early work and causes of insufficient knowledge of the diversity of the group. *Hydrobiologia* 360,1–11.

Larkin, P., Folmar, L. C., Hemmer, M. J., Poston, A. J., Lee, H. S. and Denslow, N. D. (2002). Array technology as a tool to monitor exposure of fish to xenoestrogens. *Mar. Environ. Res.* 54,395–399.

Li, H. Q., Lu, L., Manly, K. F., Chesler, E. J., Bao, L., Wang, J. T., Zhou, M., Williams, R. W. and Cui, Y. (2005). Inferring gene transcriptional modulatory relations: A genetical genomics approach. *Hum. Mol. Genet.* 14,1119–1125.

Limburg, P. A. and Weider, L. J. (2002). 'Ancient' DNA in the resting egg bank of a microcrustacean can serve as a palaeolimnological database. *Proc. R. Soc. Lond. B Biol. Sci.* 269,281–287.

Lisitsyn, N., Lisitsyn, N. and Wigler, M. (1993). Cloning the differences between two complex genomes. *Science* 259,946–951.

Little, T. J. and Ebert, D. (2000). The cause of parasitic infection in natural populations of *Daphnia* (*Crustacea: Cladocera*): The role of host genetics. *Proc. R. Soc. Lond. B Biol. Sci.* 267,2037–2042.

Little, T. J., O'Connor, B., Colegrave, N., Watt, K. and Read, A. F. (2003). Maternal transfer of strain-specific immunity in an invertebrate. *Curr. Biol.* 13,489–492.

Lopes, I., Baird, D. J. and Ribeiro, R. (2004). Genetic determination of tolerance to lethal and sublethal copper concentrations in field populations of *Daphnia longispina*. *Arch. Environ. Contam. Toxicol.* 46,43–51.

Lubbock, J. (1857). An account of the two methods of reproduction in *Daphnia* and of the structure of the ephippium. *Philos. Trans. R. Soc. Lond.* 8,352–354.

Lynch, M. (1983). Ecological genetics of *Daphnia pulex*. *Evolution* 37,358–374.

Lynch, M. and Gabriel, W. (1983). Phenotypic evolution and parthenogenesis. *Am. Nat.* 122,745–764.

Lynch, M. and Spitze, K. (1994). Evolutionary genetics of *Daphnia*. In *Ecological Genetics* (ed. L. A. Real), pp. 109–128, Princeton University Press, Princeton, NJ.

Lynch, M. and Walsh, B. (1998). *Genetics and Analysis of Quantitative Traits*. Sinauer Associates, Inc., Sunderland.

Mackay, T. F. C. (2001). Quantitative trait loci in *Drosophila*. *Nat. Rev. Genet.* 2,11–20.

Mattingly, C. J., Colby, G. T., Rosenstein, M. C., Forrest, J. N. and Boyer, J. L. (2004). Promoting comparative molecular studies in environmental health research: An overview of the comparative toxicogenomics database (CTD). *Pharmacogenomics J.* 4,5–8.

McClearn, G. E., Jones, B., Blizard, D. A. and Plomin, R. (1993). The utilization of quantitative trait loci in toxicogenetics. *J. Exp. Anim. Sci.* 35,251–258.

Merrick, B. and Bruno, M. E. (2004). Genomic and proteomic profiling for biomarkers and signature profiles of toxicity. *Curr. Opin. Mol. Ther.* 6,600–607.

Miller, F., Schlosser, P. M. and Janszen, D. B. (2000). Haber's rule: A special case in a family of curves relating concentration and duration of exposure to a fixed level of response for a given endpoint. *Toxicology* 149,21–34.

Miner, B. G., Sultan, S. E., Morgan, S. G., Padilla, D. K. and Relyea, R. A. (2005). Ecological consequences of phenotypic plasticity. *Trends Ecol. Evol.* 20,685–692.

Morley, M., Molony, C. M., Weber, T. M., Delvin, J. L., Ewens, K. G., Spielman, R. S. and Cheung, V. G. (2004). Genetic analysis of genome-wide expression variation in human gene expression. *Nature* 430,743–747.

Mort, M. A. (1991). Bridging the gap between ecology and genetics: The case of freshwater zooplankton. *Trends Ecol. Evol.* 6,41–45.

Mueller, O. (1785). Entomostraca seu insecta testacea, quae in aquis Daniae et Norvegiae reperit, descriptsit et inconibus illustravit. Lipsiae et Harniae.

Nardi, F., Spinsanti, G., Boore, J. L., Carapelli, A., Dallai, R. and Frati, F. (2003). Hexapod origins: Monophyletic or paraphyletic? *Science* 299,1887–1889.

Nielsen, R. (2006). Evolution – Why sex? *Science* 311,960–961.

Oda, S., Tatarazako, N., Watanabe, H., Morita, M. and Iguchi, T. (2005). Production of male neonates in *Daphnia magna* (*Cladocera, Crustacea*) exposed to juvenile hormones and their analogs. *Chemosphere* 61,1168–1174.

Olmstead, A. W. and LeBlanc, G. A. (2003). Insecticidal juvenile hormone analogs stimulate the production of male offspring in the crustacean *Daphnia magna*. *Environ. Health Perspect.* 111,919–924.

Paland, S. and Lynch, M. (2006). Transitions to asexuality result in excess amino acid substitutions. *Science* 311,990–992.

Pedra, J., McIntyre, L., Scharf, M. and Pittendrigh, B. (2004). Genome-wide transcription profile of field- and laboratory-selected dichlorodiphenyltrichloroethane (DDT)-resistant *Drosophila*. *Proc. Natl. Acad. Sci. USA* 101,7034–7039.

Peters, R. H. (1987). *Daphnia* culture. In '*Daphnia*' *Memorie dell'Istituto Italioano di Idrobiologia Dr Marco De Marchi* (eds R. H. Peters and R. de Bernardi), Vol. 45, Consiglio Nazionale Delle Ricerche Instituto Italiono Di Idrobiologia, Verbania Pallanza.

Pfrender, M. E., Spitze, K. and Lehman, N. (2000). Multilocus genetic evidence for rapid ecologically based speciation in *Daphnia*. *Mol. Ecol.* 9,1717–1735.

Pigliucci, M. (2005). Evolution of phenotypic plasticity: Where are we going now? *Trends Ecol. Evol.* 20,481–486.

Pollard, H. G., Colbourne, J. K. and Keller, W. (2003). Reconstruction of centuries-old *Daphnia* communities in a lake recovering from acidification and metal contamination. *Ambio* 32,214–218.

Poynton, H. C., Varshavsky, J. R., Chang, B., Cavigiolio, G., Chan, S., Holman, P. S., Loguinov, A. V., Bauer, D. J., Komachi, K., Theil, E. C., Perkins, E. J., O Hughes, O. and Vulpe, C. D. (2007). *Daphnia magna* exotoxicogenomics provides mechanistic insights into metal toxicity. *Environ. Sci. Technol.* 41,1044–1050.

Regier, J. C. and Shultz, J. W. (1997). Molecular phylogeny of the major arthropod groups indicates polyphyly of crustaceans and a new hypothesis for the origin of hexapods. *Mol. Biol. Evol.* 14,902–913.

Regier, J. C., Shultz, J. W. and Kambic, R. E. (2005). Pancrustacean phylogeny: Hexapods are terrestrial crustaceans and maxillopods are not monophyletic. *Proc. R. Soc. Lond. B Biol. Sci.* 272,395–401.

Richard, J. (1895, 1896). Revision des Cladoceres. *Ann. Sci. Nat. Zool. Paleon.* 18,279–389.

Rider, C. V., Gorr, T. A., Olmstead, A. W., Wasilak, B. A. and Leblanc, G. A. (2005). Stress signaling: Coregulation of hemoglobin and male sex determination through a terpenoid signaling pathway in a crustacean. *J. Exp. Biol.* 208,15–23.

Robinson, C. D., Lourido, S., Whelan, S. P., Dudycha, J. L., Lynch, M. and Isern, S. (2006). Viral transgenesis of embryonic cell cultures from the freshwater micro-crustacean *Daphnia*. *J. Exp. Zool.* 305A,62–67.

Schierwater, B., Ender, A., Schwenk, K., Spaak, P. and Streit, B. (1994). The evolutionary ecology of *Daphnia*. In *Molecular Ecology and Evolution: Approaches and Applications* (eds B. S. B. Schierwater, G. P. Wagner and R. DeSalle), pp. 495–508, Birkhauser Verlag, Basel, Switzerland.

Schwenk, K., Posada, D. and Hebert, P. D. N. (2000). Molecular systematics of European *Hyalodaphnia*: The role of contemporary hybridization in ancient species. *Proc. R. Soc. Lond. B Biol. Sci* 267,1833–1842.

Shaw, J. R., Colbourne, J. K., Davey, J. C., Glaholt, S. P., Hampton, T. H., Chen, C. Y., Folt, C. L. and Hamilton, J. W. (2007). Gene response profiles for *Daphnia pulex* exposed to cadmium reveal a novel crustacean metallothionein. *BMC Genomics* 8,477.

Shaw, J. R., Dempsey, T. D., Chen, C. Y., Hamilton, J. W. and Folt, C. L. (2006). Comparative toxicity of cadmium, zinc and mixtures of cadmium and zinc to daphnids. *Environ. Toxicol. Chem.* 25,182–189.

Shiga, Y., Sagawa, K., Takai, R., Sakaguchi, H., Yamagata, H. and Hayashi, S. (2006). Transcriptional readthrough of Hox genes Ubx and Antp and their divergent after transcriptional control during crustacean evolution. *Evol. Dev.* 8,407–414.

Soetaert, A., Moens, L. N., Van der Ven, K., van Leemput, K., Naudts, B., Blust, R. and De Coen, W. M. (2006). Molecular impact of propiconazole on *Daphnia magna* using a reproduction-related cDNA array. *Comp. Biochem. Physiol. C* 142,66–76.

218

Soetaert, A., Van der Ven, K., Moens, L. N., Vandenbrouck, T., van Remortel, P. and De Coen, W. M. (2007a). *Daphnia magna* and ecotoxicogenomics: Gene expression profiles of the antiecdysteroidal fungicide fenarimol using energy-, molting- and life stage-related cDNA libraries. *Chemosphere* 67,60–71.

Soetaert, A., Vandenbrouck, T., van der Ven, K., Maras, M., van Remortel, P., Blust, R. and De Coen, W. J. (2007b). Molecular responses during cadmium-induced stress in *Daphnia magna*: Integration of differential gene expression with higher-level effects. *Aquat. Toxicol.* 83,212–222.

Stein, L. D., Mungall, C., Shu, S. Q., Caudy, M., Mangone, M., Day, A., Nickerson, E., Stajich, J. E., Harris, T. W., Arva, A. *et al.* (2002). The generic genome browser: A building block for a model organism system database. *Genome Res.* 12,1599–1610.

Swammerdam, J. (1669). Historia insectorum generalis, of te algemeene verhandeling van de bloedeloose dierkens. t'Utrecht.

Swammerdam, J. (1758). *The Book of Nature; or The History of Insects:* Reduced to Distinct Classes, Confirmed by Particular Instances, Displayed in the Anatomical Analysis of Many Species and Illustrated with Copperplates. John HIll, London.

Tavazoie, S., Hughes, J. D., Campbell, M. J., Cho, R. J. and Church, G. M. (1999). Systematic determination of genetic network architecture. *Nat. Genet.* 22,281–285.

Tessier, A. J., Leibold, M. A. and Tsao, J. (2000). A fundamental trade-off in resource exploitation by *Daphnia* and consequences to plankton communities. *Ecology* 81, 826–841.

Tollrian, R. (1993). Neckteeth formation in *Daphnia pulex* as an example of continuous phenotypic plasticity – morphological effects of *Chaoborus* kairomone concentration and their quantification. *J. Plankton Res.* 15,1309–1318.

Tollrian, R. (1995). *Chaoborus* crystallinus predation on *Daphnia pulex*: Can induced morphological changes balance effects of body size on vulnerability? *Oecologia* 101,151–155.

U.S.E.P.A. (2002). *Methods for measuring the acute toxicity of effluents and receiving waters to freshwater and marine organisms*, 5th Edition, US EPA, Office of Water, Washington, DC. 275pp.

Vasemagi, A., Nilsson, J. and Primmer, C. R. (2005). Expressed sequence tag-linked microsatellites as a source of gene-associated polymorphisms for detecting signatures of divergent selection in Atlantic salmon (*Salmo salar* L.). *Mol. Biol. Evol.* 22, 1067–1076.

Versteeg, D. J., Stalmans, M., Dyer, S. D. and Janssen, C. (1997). *Ceriodaphnia* and *Daphnia*: A comparison of their sensitivity to xenobiotics and utility as a test species. *Chemosphere* 34,869–892.

Viehoever, A. (1931). Transparent life. *Am. J. Pharm.* 103,252–278.

Viehoever, A. (1936). *Daphnia* – the biological reagent. *J. Am. Pharm. Assoc.* 25,112–117.

Viehoever, A. (1937). The development of *Daphnia magna* for the evaluation of active substances. *Am. J. Pharm.* 109,360–366.

Waring, J. F., Ciurlionis, R., Jolly, R. A., Heindel, M. and Ulrich, R. G. (2001a). Microarray analysis of hepatotoxins in vitro reveals a correlation between gene expression profiles and mechanisms of toxicity. *Toxicol. Lett.* 120,359–368.

Waring, J. F., Jolly, R. A., Ciurlionis, R., Lum, P. Y., Praestgaard, J. T., Morfitt, D. C., Buratto, B., Roberts, C., Schadt, E. and Ulrich, R. G. (2001b). Clustering of hepatotoxins based on mechanism of toxicity using gene expression profiles. *Toxicol. Appl. Pharm.* 175,28–42.

Warren, E. (1900). On the reaction of *Daphnia magna* to certain changes in its environment. *Q. J. Microsc. Sci.* 43,199–224.

Watanabe, H., Takahashi, E., Nakamura, Y., Oda, S., Tatarazako, N. and Iguchi, T. (2007). Development of a *Daphnia magna* DNA microarray for evaluating the toxicity of environmental chemicals. *Environ. Toxicol. Chem.* 26,669–676.

Watanabe, H., Tatarazako, N., Oda, S., Nishide, H., Uchiyama, I., Morita, M. and Iguchi, T. (2005). Analysis of expressed sequence tags of the water flea *Daphnia magna. Genome* 48,606–609.

West-Eberhard, M. (2003). *Developmental Plasticity and Evolution.* Oxford University Press, Oxford.

Whitehead, A. and Crawford, D. L. (2006). Neutral and adaptive variation in gene expression. *Proc. Natl. Acad. Sci. USA* 103,5425–5430.

Yampolsky, L. Y. and Galimov, Y. R. (2005). Evolutionary genetics of aging in *Daphnia. Zh. Obshch. Biol.* 66,416–424.

Whole genome microarray analysis of the expression profile of *Escherichia coli* in response to exposure to *para*-nitrophenol

Angela Brown[1],*, Jason R. Snape[2], Colin R. Harwood[3] and Ian M. Head[1]

[1]*School of Civil Engineering and Geosciences, Newcastle University, Devonshire Terrace, Newcastle upon Tyne, NE1 7RU, UK*
[2]*AstraZeneca, Brixham Environmental Laboratory, Freshwater Quarry Brixham, Devon, TQ5 8BA, UK*
[3]*Institute for Cell and Molecular Biosciences, Newcastle University, Framlington Place, Newcastle upon Tyne, NE2 4HH, UK*

Abstract. In ecotoxicology, standard biological assays are used to determine the effects of chemicals on microorganisms. One assay, the microbial multiplication inhibition test, measures the degree of growth inhibition of a population of microorganisms when exposed to a chemical. Whilst this test indicates crude inhibitory effects of chemicals on cells, it offers no insight as to why the cells are inhibited. Genomic array technology was used to investigate the effects of a nitroaromatic compound, *para*-nitrophenol (PNP), on *Escherichia coli* K12-MG1655 cells. Global changes in gene expression showed exposure to PNP caused *E. coli* cells to prematurely enter stationary phase, as shown by downregulation of genes involved in protein synthesis (*rpl*, *rps*, *rpm*). Genes of the *emrRAB* operon, which confers resistance to compounds that uncouple oxidative phosphorylation, were upregulated in cells in response to PNP exposure. PNP also induced the *marRAB* operon and *dps* gene, which bestow resistance to oxidative stress. A compound structurally similar to PNP, dinitrophenol (DNP) a protonophore that uncouples oxidative phosphorylation, has previously been shown to induce the *marRAB* operon. Like DNP, we suggest that PNP uncouples oxidative phosphorylation within *E. coli* cells. The upregulated *marRAB* and *emrRAB* genes also confer antibiotic resistance and efflux mechanisms, respectively, within *E. coli*. A downregulation of genes encoding porins, for the transport of solutes, in the outer membrane of cells (*ompA*, *ompC*, *ompF* and *ompT*), indicated that PNP also affected cell membrane constituents. In addition, *rpoE*, which encodes a sigma factor involved in cell envelope stress response, was upregulated in cells following PNP exposure. Genes that conferred resistance to low pH (*hdeA*, *hdeB*) were upregulated in cells that were exposed to PNP. Furthermore, the acidic nature of the PNP medium may have activated a pH-inducible gene, *inaa*, which (as with *marRAB* operon) bestows antibiotic stress resistance in *E. coli*.

Keywords: ecotoxicology; *para*-nitrophenol; genome array; gene; expression; stress response; antibiotic resistance; growth inhibition; *Escherichia coli*; stationary phase; downregulation; upregulation; oxidative phosphorylation; toxicant; operon; oxidative stress; protonophore; uncoupler; efflux; translation; protein synthesis; pH adaptation;

Corresponding author: Tel.: +44(0)191 2464885. Fax: +44(0)191 2464961.
E-mail: angela.brown1@ncl.ac.uk (A. Brown).

ADVANCES IN EXPERIMENTAL BIOLOGY
VOLUME 02   ISSN 1872-2423
DOI: 10.1016/S1872-2423(08)00006-9

porins; cell membrane; cell envelope; nitroaromatic; genomics; pollutant; lethal concentration; effective concentration.

Introduction

Para-nitrophenol (PNP), a nitroaromatic compound, is used in the manufacturing of analgesics (acetaminophen), pesticides, dyes, explosives and in the processing of leather (Lei *et al.*, 2003). Release of PNP into the aquatic and terrestrial environments occurs through industrial effluents or in the soil as a hydrolytic product of some organophosphorus pesticides (parathion and methyl-parathion) (Shimazu *et al.*, 2001). Although these by-products are considerably less toxic than the pesticides themselves, their fate in the environment is of particular concern. Currently, organophosphates are one of the most widely used classes of pesticide in industrialised countries. For example, over 40 million kilograms of organophosphate pesticides are applied annually in the United States (Shimazu *et al.*, 2001). PNP was found in approximately 8% of the sites on the National Priorities List identified by the U.S. Environmental Protection Agency in 2001 (Lei *et al.*, 2003) and, because of the toxic effect PNP has on biological systems, it has been classified as a priority pollutant (Shimazu *et al.*, 2001). PNP is carcinogenic, mutagenic, cytotoxic and embryotoxic to mammals (Lei *et al.*, 2003). The compound is highly water soluble, therefore runoff from the soil into surrounding waters (groundwaters, streams, lakes and rivers) may exert dangerous impact upon aquatic environments. PNP is used as a reference or test compound in biodegradability and ecotoxicology tests. PNP is readily biodegraded by bacterial cells at concentrations less than $50 \, mg \, ml^{-1}$, but is toxic/inhibitory to cells at higher concentrations (Thouand *et al.*, 1995).

In order to ensure specific chemicals such as PNP or industrial effluents do not adversely affect environmental compartments upon their release, a number of microbiology based ecotoxicology assays exist. These assays determine the effects of a chemical on either a mixed preparation of microorganisms, such as activated sludge, or a pure microbiological culture such as *Escherichia coli*, *Pseudomonas putida* (British standard BS EN ISO 10712, 1995) or *Photobacterium phosphoreum*. The degree of chemical toxicity to bacteria is normally estimated by measuring the effects on either viability or growth. LC_{50} or EC_{50} measurements are often calculated, which determine the 'lethal concentration' or 'effective concentration' of a chemical at which 50% of the microbial population are killed or growth inhibited, respectively, in a given time. Generally, the lower the LC_{50}/EC_{50} value, the more toxic is the chemical, whereas if the LC_{50}/EC_{50} value is large, the toxicity of the chemical is regarded as low. Whilst this type of assay typically measures holistic parameters such as inhibition of population growth, it offers no insight into why the bacteria are inhibited.

To obtain a greater understanding of the biological effects of chemically induced toxicity, the effects of toxicants on gene expression in an organism can be determined using genome array technology. With the elucidation of gene sequences, it is now possible to produce genome arrays that represent all of the genes in any given organism. This situation allows the simultaneous analysis of expression patterns for thousands of genes. The theory behind this approach is that exposure to toxicants will alter gene expression levels transiently within the cells, as the cells respond to the direct and indirect influences of the toxicant on cell structure and function. Messenger RNA abundance will increase from toxicant-induced genes and decrease from toxicant-repressed genes. These changes are then quantified using gene array technology. The resultant data are mined in an attempt to identify any specific pathways or unique stressor-induced signatures activated by the toxicant. Furthermore, because gene expression responds to low levels of toxicant exposure, genome array analysis may provide toxicity information that would otherwise be apparent only with chronic exposure.

Gene expression profiling has previously been utilised to assess the response of toxicants that affect human cells (Kramer et al., 2004) and this technology was termed 'toxicogenomics' (Iannaccone, 2001; Rockett and Dix, 2000). As the name suggests, this technology encompasses genomics together with mammalian toxicology. Consequently, Snape et al. (2004) proposed the term 'ecotoxicogenomics' to describe the application of genomic technologies (transcriptomics, proteomics and metabolomics) in ecotoxicology and environmental studies. The authors go on to define ecotoxicogenomics as 'the study of gene and protein expression in nontarget organisms that are important in responses to environmental toxicant exposures'.

The first E. coli genome was sequenced in 1997, the non-pathogenic E. coli K12-MG1655 (Blattner et al., 1997). The completion of the E. coli K12-MG1655 genome project permitted the development of new tools for genome-wide analysis. Consequently, global gene expression profiling in E. coli, using array-based technology, has been used to examine gene expression in rich and minimal medium (Tao et al., 1999), alkali stress (Bordi et al., 2003), osmotic stress (Weber and Jung, 2002), chemical stress (Allen *et al.*, 2006; Kershaw *et al.*, 2005; Phadtare *et al.*, 2002) and environmental stress (Rozen *et al.*, 2002). Transcriptional response to a range of chemical stressors in different species of bacteria have also been studied, for example, *Pseudomonas aeruginosa* (Chang *et al.*, 2005a, 2005b; Salunkhe *et al.*, 2005), *Staphylococcus aureus* (Chang *et al.*, 2006), *Corynebacterium glutamicum* (Jakob *et al.*, 2007), *Enterococcus faecalis* (Aakra *et al.*, 2005) and *Bacillus subtilis* (Nguyen *et al.*, 2007). The ability of bacterial cells to rapidly adapt to their changing environment is crucial to their survival and growth. The general stress response (a known phenomena, Hengge-Aronis, 1999) is induced in response to a number of different stress conditions, for example, starvation, changes in pH, osmolarity and temperature. This response is usually accompanied by reduced growth rate or entry of cells into stationary phase. Physiological changes to the cell include multiple stress resistance, accumulation of storage compounds, changes in cell envelope composition and altered morphology. Bacterial gene expression studies have also demonstrated that, in addition to previously recognised stress responses, previously unknown stress-related gene expression patterns have been discovered, thus providing a more comprehensive picture of the stress response (Zheng *et al.*, 2001). These studies also show the enormous potential of global gene expression profiling to identify rapidly multiple genes that are involved in the response to environmental stimuli and to elucidate the biological effects of toxic chemicals at the molecular level.

A single-species/single-toxicant (*E. coli* exposed to PNP) approach was used to identify specific toxicant-induced gene expression profiles amongst complex gene expression data sets (a comparison of 36 gene arrays, each containing 4,290 genes was conducted). PNP was chosen as a test chemical because of its status as a priority pollutant, toxicity to mammals and the fact that it is used as a reference chemical in ecotoxicological biodegradability and inhibition tests. Furthermore, the mode of action of PNP is unclear and global gene expression analysis may provide some insights into the manner in which PNP exerts its toxic effects on *E. coli* cells. Whilst other investigators have chosen to work with a compound with known mode of action (Phadtare *et al.*, 2002; Shaw *et al.*, 2003), this study examined the potential for gene expression profiling to deliver this information, with little or no knowledge of the mode of action of the toxicant. Allen *et al.* (2006) determined the transcriptional response of *E. coli* to a biocide (polyhexamethylene biguanide) whose mode of action was thought to be known and, interestingly, the array-based experiments provided a new dimension to the mechanism of action based on the way the biocide interacted with nucleic acids.

The hypothesis of the study was to use gene array technology to investigate the mode of action of PNP, a reference compound in ecotoxicology, on *E. coli* cells in order to study the effect of this toxicant on bacterial gene expression.

Methods

Expression profiling experiments conformed to MIAME (Minimum Information About a Microarray Experiment) guidelines (Brazma, 2001; Brazma *et al.*, 2001).

Determination of EC50 for PNP in E. coli

In order to determine concentrations of PNP that showed different degrees of growth inhibition for use in gene expression profiling experiments, a standard microbial growth inhibition test was performed (adapted from British Standard BS EN ISO 10712, 1995). The test, with PNP as test compound and *E. coli* K12-MG1655 as test species, determined an EC_{50} (effective concentration at which 50% of the population were inhibited by PNP) of $56\,mg\,l^{-1}$. Additionally, viable cell counts and measurement of the dimensions of cells exposed to a range of PNP concentrations were performed, in order to determine any effects of PNP that were not detected by gross changes in growth. These data were used to inform appropriate concentrations of PNP and sampling times for subsequent gene expression analyses (data not shown).

Growth conditions

Triplicate cultures of *E. coli* K12-MG1655 cells were grown in minimal medium (50 ml, Prytz *et al.*, 2003) batch cultures in 250 ml Erlenmeyer flasks at 27°C, with shaking (150 rpm, Gallenkamp Orbital Incubator, Sanyo-Gallenkamp Plc, Loughborough, UK). Bacterial growth was monitored hourly by measuring attenuance at 600 nm on an Ultrospec III spectrophotometer (Pharmacia/LKB Biotech Ltd., Cambridge, UK). At mid-logarithmic growth phase ($D_{600} = 0.3$), PNP was added to triplicate cultures at concentrations of 0 (control), 10, 32 and $56\,mg\,l^{-1}$ and hourly monitoring of D_{600} resumed. At 1, 3 and 4 h after the addition of PNP, cells (10 ml) were harvested from each flask.

Cell lysis

Cells (10 ml) were pipetted directly into 5 ml of ice-cold killing buffer (20 mM Tris/HCl, pH 7.5; 5 mMMgCl$_2$; 20 mMNaN$_3$) and centrifuged (Beckman Avanti™ J-25, Beckman Coulter Inc, C.A., USA) (8,000 rpm, 3 min, 4°C). Cell pellets were resuspended in supernatant and pipetted directly into a Teflon vessel with disruption ball (Sartorius Ltd, Surrey, UK) containing liquid nitrogen. Cells were physically lysed by bead beating (Mikro-Dismembrator S, B. Braun Biotech International, Sartorius Ltd, Surrey, UK) (2,600 rpm, 2 min). Cell powders were resuspended in 2–3 ml prewarmed (60°C) lysis solution, (4 M guanidine thiocyanate; 0.025 M sodium acetate, pH 5.2; 0.5% *N*-laurylsarcosinate; 1 M β-mercaptoethanol) and placed on ice until complete lysis occurred ($\sim$ 1 min).

Nucleic acid extraction

Two rounds of acid phenol extraction (aqua-phenol/chloroform/isoamyl alcohol (25:24:1)) were performed on cell lysates, followed by further extraction with chloroform/isoamyl alcohol (96:4, Sigma-Aldrich Company Ltd, Dorset, UK). RNA was precipitated with 3 M sodium acetate (pH 5.2) and isopropanol at –70°C for 1 h. RNA was resuspended in diethylpyrocarbonate (DEPC)-treated water (50 μl) on ice. On-column DNase I digestion was performed using the RNeasy® RNase-free DNase kit (Qiagen® Ltd, West Sussex, UK), according to manufacturer's instruction. RNA was stored at –70°C prior to labelling.

cDNA synthesis of probes

RNA (4 μg) and cDNA labelling primers (Sigma-Genosys Inc, Cambridgeshire, UK) were mixed. Primers were annealed (90°C for 2 min, 80°C for 5 min, 70°C for 5 min, 60°C for 5 min, 50°C for 5 min and 42°C for 20 min) (Omn-E, Hybaid Ltd, Middlesex, UK). Following primer annealing, 5 × first strand buffer (Invitrogen Ltd, Paisley, UK), 0.1 mM DTT (Invitrogen), sterile dH2O, each of dCTP, dGTP and dTTP (final concentration 0.33 mM each) (Roche Diagnostics GmbH, Mannheim, Germany), [α-^{33}P]-dATP (74–111 GBq m mol^{-1}) (Amersham Plc, Buckinghamshire, UK) and Superscript II reverse transcriptase (400 U) (Invitrogen Ltd, Paisley, UK) were added to a final volume of 30 μl. Reverse transcription was performed at 42°C for 2.5 h. Unincorporated nucleotides were removed using Micro Bio-Spin

Chromatography Columns (Bio-Rad Laboratories Ltd, Hertfordshire, UK), according to manufacturer's instructions. Any remaining RNA within the cDNA was hydrolysed with 0.5 M EDTA and 3 M NaOH at 65°C for 30 min, then at room temperature for 15 min. Alkali was neutralised with 1 M Tris–HCl, pH 8 and 2 N HCl. cDNA was precipitated with 3 M sodium acetate and ethanol (96–100%, v/v) at − 80°C for 1 h, washed in 70% (v/v) ethanol and resuspended in DEPC-treated water (100 μl) at 65°C for 30 min. Approximate percentage incorporation of radionucleotides into cDNA was determined with a Geiger counter (Mini 5.10 series EP15, Mini instruments Ltd, Essex, UK), using the formula:

$$\frac{\text{incorporated nucleotides}}{(\text{incorporated nucleotides} + \text{unincorporated nucleotides})} \times 100$$

Hybridisation

DNA macroarrays (Panorama *E. coli* gene arrays) were produced by Sigma-GenoSys Biotechnologies, Inc. Each DNA array consisted of a 12- by 24-cm positively charged nylon membrane on which 10 ng each of all 4,290 PCR-amplified ORF-specific DNA fragments were robotically printed in duplicate ($n = 4,608$, including genes ($n = 4,290$), genomic DNA ($n = 24$) and blanks ($n = 294$) spotted onto the array in duplicate). In this study, eight DNA macroarrays were used for the hybridisation of labelled cDNA samples (AH018A, AH018B, AL009A, AL009B, AW004A, AW004B, AO033A and AO033B). Thirty-six hybridisations were performed with labelled cDNA generated from 36 individual RNA samples. RNA samples were extracted from 36 individual cultures that were exposed to three concentrations of PNP (10, 32 and 56 mg l^{-1}, plus an untreated control (0 mg l^{-1})) and harvested 1, 3 and 4 h after PNP addition (12 samples). Three biological repeats of the whole experiment were performed (36 samples).

Arrays were rinsed in 2 × SSPE (1 × SSPE is 0.18 M NaCl, 10 mM NaH$_2$PO$_4$ and 1 mM EDTA, pH 7.7) at room temperature for 10 min and prehybridised in hybridisation solution (5 × Denhardts solution, 5 × SSC, 0.5% SDS and 100 μg/ml denatured salmon sperm DNA) at 65°C for 2 h. Subsequently, radiolabelled cDNA samples (100 μl each) were added to 5 ml of hybridization solution and hybridized to arrays for 20 h at 65°C. Arrays were washed (2 × SSC, 0.1% SDS) for 5 min at room temperature, then for 20 min at 65°C, with a final wash step (0.1% SSC, 0.1% SDS) for 60 min at 65°C. Arrays were wrapped in plastic foodwrap and exposed to an imaging screen (35 × 43 cm, Fuji-BAS MS-type, Raytek Scientific Ltd, Sheffield, UK) for 48 h.

Data analysis

The exposed imaging screen was scanned (resolution 50 μm) on a STORM 840 PhosphorImager (Molecular Dynamics, Buckinghamshire, UK). Image files (.GEL) were used in ArrayVision™ (GE Healthcare Life Sciences, Buckinghamshire, UK) to quantify signal intensities from each spot ($n = 4{,}608$) on each array ($n = 36$). Background was calculated from the perimeter of the array and subtracted from raw signal intensity values. Each ORF-specific spot was present in duplicate on an array and intensities were averaged for analysis. A gene expression visualisation and statistical analysis package, GeneSpring® (Silicon Genetics, Redwood City, CA, USA), was used to compare changes in global gene expression profiles.

Array stripping

Arrays were stripped for reuse by washing twice at 100°C with stripping solution (5 mM sodium phosphate (pH 7.5), 0.1% SDS). Stripped arrays were exposed for 24 h to check for residual label.

Data reproducibility

Reproducibility is a vital factor in the interpretation of array data (Beissbarth *et al.*, 2000; Hess *et al.*, 2001) and a comprehensive study of reproducibility was conducted as part of this project. As a quality control measure, signal intensity values from duplicate spots on the same array were compared. The analysis compared the $\log_{10}$ signal intensity of one spot to its duplicate spot on the same array and included all genes, blank spots and genomic DNA ($n = 4{,}608$), the analysis was performed on all 36 arrays. Spearman's rho correlations were calculated in SPSS (Version 11.0) to assess the relationship between duplicate spots. High levels of reproducibility were observed between duplicate spots for all 36-gene arrays (mean: 0.970 ± 0.028 (range 0.877–0.996), significant at the 0.01 level).

Reproducibility was also assessed at a biological replicate level (array experiments were performed in triplicate at each PNP concentration and time point). Pairwise comparisons of the $\log_{10}$ signal intensities of all spots ($n = 4{,}290$, genes only) on an array were conducted for all possible pairs of data within a triplicate set. Spearman's rho correlations were calculated to indicate relationships between $\log_{10}$ signal intensities of the three biological replicate data sets. The levels of reproducibility observed between triplicate array data sets were (mean: 0.846 ± 0.17 (range 0.374–0.967), significant at the 0.01 level). From this, two array hybridizations were identified as having produced poor quality data and these data sets were omitted from further analysis, thus improving levels of reproducibility between replicate data sets (mean: 0.903 ± 0.045 (range 0.802–0.967), significant at the 0.01 level).

Data normalisation

All data were normalized to ensure signal intensities from different macroarray batches were comparable to one another. In GeneSpring®, a standard (default) one-colour experiment normalisation method was applied in which intensity values below a cutoff value of 0.0001 were set to 0.0001. This normalization was followed by 'Per Chip' normalization in which each intensity value was divided by the 50.0th percentile (median) of all measurements in that sample. This normalization controls chip-wide variations in intensity. Finally, 'Per Gene' normalization was executed, in which each gene was divided by the median of its measurements across all samples. Following normalization, signal intensities across triplicate experiments were averaged, resulting in an individual set of signal intensity values for each of the four PNP concentrations at each of the three time points. $\log_{10}$ ratios were calculated between conditions and indicated whether gene expression was higher/lower under one condition or the other or remained unchanged.

Differential expression

A multistep statistical approach was used to identify genes that were significantly differentially expressed in at least one condition in the total data set. The statistical approach was designed to eliminate false positives from consideration (Arfin *et al.*, 2000). In GeneSpring®, statistically significant genes had to pass two filters, firstly, $\log_{10}$ ratios had to be ≥ 2 standard deviations from the mean of the $\log_{10}$ ratios in one out of 12 conditions (untreated cells and cells exposed to $10\,\mathrm{mg\,l^{-1}}$ PNP, $32\,\mathrm{mg\,l^{-1}}$ PNP and $56\,\mathrm{mg\,l^{-1}}$ PNP at 1, 3 and 4 h). This analysis revealed 577 differentially expressed genes. The second filter evaluated all possible pairwise comparisons of the $\log_{10}$ ratios using a Student's t-test, which determined the probability that the expression ratios were significant. From the literature, it was decided that only those genes with $\log_{10}$ ratio t-test p-values of ≤ 0.0002 (99.9998% significance) (Chang et al., 2002) in at least one out of the 12 conditions would be considered further. Therefore, the 577 genes that passed the first filter were further filtered on confidence; in addition the Benjamini and Hochberg multiple testing correction (Benjamini and Hochberg, 1995) was applied, which corrected for false positive results. The second filter revealed 423 genes that were classified as significantly differentially expressed. The 423 genes were subsequently grouped using K-means clustering (five groups, Gene-Spring® software default).

A less stringent procedure to identify differentially expressed genes was also adopted in this study. A fold change approach (at least ± 2-fold) compared the expression values of genes from PNP-treated cells with untreated control cells. Normalized signal intensities of genes from untreated cells (0 mg l^{-1} PNP) were compared with those of PNP-treated cells (10, 32 and 56 mg l^{-1} PNP, respectively). Genes were considered differentially expressed when $\log_{10}$ ratios were at least ± 2-fold that of the value in untreated cultures.

Functional groups

Genes considered significantly differentially expressed were further classified into functional groups based on the classifications of Blattner et al. (1997) and Riley (1998). Stress-response genes were identified in Rocha et al. (2002) and Keseler et al. (2005) (http://ecocyc.org/). Differentially expressed genes were compared with known stress-response genes to provide an unbiased indication of the probable significance of relationships when attempting to find functional classifications among genes identified as differentially expressed. Only gene membership to the sets was considered.

Results and discussion

Viable cell counts

Viable cell counts were determined in cultures exposed to a range of PNP concentrations (Untreated, 10, 32 and 56 mg l^{-1} PNP), in order to determine any effects of PNP that were not detected by gross changes in growth characteristics. No significant differences were noted between the mean number of viable cells across the concentration series at 1 h or 3 h after the addition of PNP. At 4 h, untreated cells showed a significantly larger number of viable cells compared with cells exposed to $32\,mg\,l^{-1}$ and $56\,mg\,l^{-1}$ PNP. As expected no significant difference in viable cell numbers was shown in cells exposed to $10\,mg\,l^{-1}$ PNP compared with untreated cells (data not shown).

Cell morphology

Cell morphology was also investigated in cultures exposed to a range of PNP concentrations (untreated, 10, 32 and $56\,mg\,l^{-1}$ PNP). At 4 h following mid-log phase, untreated control cells were entering stationary phase and the morphological changes followed what are normal changes within the bacterial growth cycle. Untreated control cells became shorter and narrower at 4 h, consistent with cellular adaptation to stationary phase (data not shown). It is well established that *E. coli* cells when entering stationary phase undergo a number of morphological and physiological changes, one of which is that the cells become smaller and rounder (Farewell *et al.*, 1998; Nyström, 1995). At 3 h after the addition of PNP, PNP-treated cells displayed morphological changes compared with untreated cells; the cells were shown to be shorter and narrower than untreated cells (data not shown). This finding suggests that cells exposed to PNP were entering stationary phase prematurely, as they showed similar morphological characteristics to the untreated cells entering stationary phase at 4 h. Cells exposed to stress, often undergo morphological changes to adapt to their changing environment (Hengge-Aronis, 1999) and this change was indicated in PNP-treated cells.

Part A: Differential expression
When applying stringent statistical analyses to gene expression profiling, there is a danger that important biologically significant genes will be missed. Therefore, it is not unusual to apply a range of methods to analyse gene expression data (for example, Chang *et al.*, 2002). Two approaches to gene expression data analysis were employed during this study, a stringent statistical approach and a less stringent and at least ± 2-fold change in log_{10} ratio values approach (Part B). The stringent statistical approach yielded 423 genes that were significantly differentially expressed in at least one out of the 12 conditions.

Cluster analysis of differentially expressed genes

The 423 genes identified as showing differential expression using a stringent statistical approach were grouped using K-means cluster analysis (five clusters) to further filter genes that exhibited similar levels of expression (Fig. 1). Cluster 1 contained 21 genes with expression values in the range ~ 1 to <0.1, cluster 2 with 175 genes with expression values in the range 1 to ~ 80. Cluster 3 contained 161 genes and showed expression values in the range ~ 3 to ~ 100. Cluster 4 contained 53 genes, the majority of which showed downregulation in response to increasing concentrations of PNP, with expression values ranging from ~ 10 to ~ 0.3. Finally cluster 5 contained 13 genes and showed upregulation in response to increasing concentrations of PNP, the range of expression values was ~ 0.8 to 10.

Comparison of gene clusters with stress-response subsets and functional classifications

Subsequently, the genes within each of the five clusters were statistically compared with stress-response genes and functional classifications, in order to identify any stress responses or genes grouped according to physiological role within the cells that were effected by PNP. The probability was calculated based on the probability of the clusters of genes and a functional group/stress-response subset of genes sharing at least as many common elements as they did by chance. Interestingly, the genes in cluster 4 ($n = 53$), which showed downregulation of expression in response to increasing concentrations of PNP, showed a statistical similarity with genes involved in translation ($p = 8.08E\text{-}16$), suggesting translation genes were downregulated in response to increasing concentrations of PNP. The genes of cluster 2 ($n = 175$) showed significant relationships with those genes involved with phage, plasmids and transposons ($p = 0.001$) within the cell and also genes involved in translation ($p = 1.16E\text{-}07$). No statistically similar gene classifications were observed with the genes belonging to clusters 1, 3 and 5, indicating that differentially expressed genes from those clusters are spread over a range of different functional groups, rather than many of the genes in each cluster belonging to the same functional classification.

Effect of PNP on genes involved in translation

The genes in cluster 4 ($n = 53$) that showed downregulation in expression in response to increasing concentrations of PNP and that showed a

233

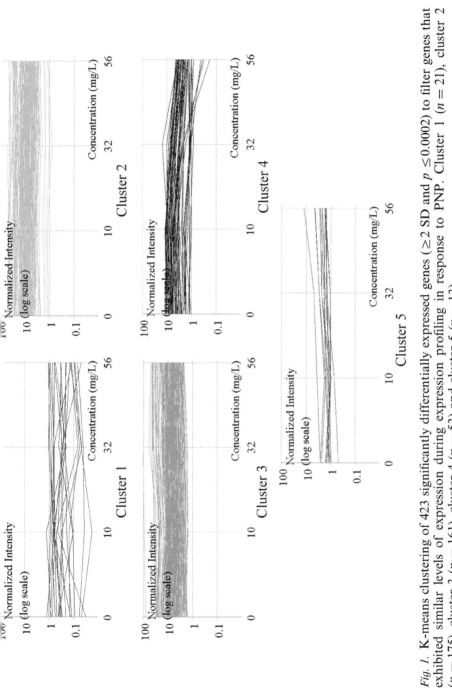

Fig. 1. K-means clustering of 423 significantly differentially expressed genes (≥2 SD and *p* ≤0.0002) to filter genes that exhibited similar levels of expression during expression profiling in response to PNP. Cluster 1 (*n* = 21), cluster 2 (*n* = 175), cluster 3 (*n* = 161), cluster 4 (*n* = 53) and cluster 5 (*n* = 13).

statistical similarity with genes involved in translation were further investigated. Twenty genes involved in translation were identified, of these genes, nine encoded 50S ribosomal L proteins (*rpl* genes) of which there are 24 associated genes within *E. coli* cells. Six genes encoded 30S ribosomal subunit proteins (*rps* genes) of which there are 21 genes and two genes encoded 50S ribosomal subunit proteins (*rpm* genes) of which there are 10 associated genes within *E. coli* cells. The three remaining genes (*infA*, *prfB* and *slyD*) encode translation and modification proteins.

The nine genes encoding 50S ribosomal L proteins (*rpl* genes) (Fig. 2A) showed similar levels or slight upregulation in expression levels in cells exposed to 10 mg l^{-1} PNP compared with untreated cells. Cells exposed to 32 mg l^{-1} PNP showed an upregulation of four genes (*rplN*, *rplC*, *rplF* and *rplS*) and a downregulation of five genes (*rplY*, *rplD*, *rplB*, *rplJ* and *rplX*) involved in translation compared with untreated control cells. Finally, cells exposed to 56 mg l^{-1} PNP showed a downregulation of all nine *rpl* genes when compared with the expression level in untreated cells.

The six genes encoding for 30S ribosomal subunit proteins (*rps* genes) showed a similar trend to those of the *rpl* genes (Fig. 2B). In cells exposed to 10 mg l^{-1} PNP similar levels or upregulation of expression levels were shown in comparison to levels within untreated cells. Cells exposed to 32 mg l^{-1} PNP exhibited upregulation of two genes (*rpsB and rpsD*) and a downregulation of four genes (*rpsN*, *rpsJ*, *rpsB and rpsT*) involved in translation, compared with untreated control cells. Finally, cells exposed to 56 mg l^{-1} PNP downregulated all six *rps* genes compared with untreated cells. The remaining differentially expressed translation associated genes (*rpmBG*, *infA*, *slyD* and *prfB*) showed a similar trend (Fig. 2C).

Similarly, Brocklehurst and Morby (2000) showed a pattern of diminished gene expression in genes encoding proteins involved in translation in a study to assess the effects of metal-ion tolerance on *E. coli* gene expression. An increasing range of metal-ion concentrations of zinc sulphate, cadmium sulphate, cobalt sulphate and nickel sulphate were shown to increase the tolerance of *E. coli* to the metal ions. However, all of the metal-ion tolerant strains showed diminished gene expression in translation genes. Brocklehurst and Morby (2000) suggest the gradual increase in the number of downregulated translation genes in response to increasing concentrations of toxicant may be a specific stress response (to PNP in this study) within the cells or it may merely be a function of lower growth rate in the presence of higher concentrations of a toxicant.

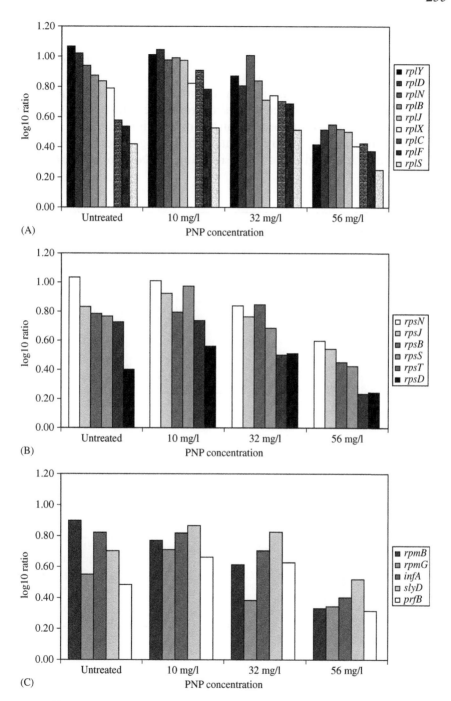

Fig. 2. Effect of PNP concentration on expression levels of genes involved in translation; (A) *rpl* genes; (B) *rps* genes; and (C) *rpm* genes plus *infA*, *slyD*, *prfB*.

236

Effect of PNP on stress-response genes

Genes associated with the general stress response and pH adaptation (*dps, hdeA, hdeB*) were activated in response to PNP (Fig. 3A). *dps* encodes for Dps, which is a highly abundant protein in stationary phase *E. coli* and is required for the starvation response. The protein has been shown to be involved in protection from multiple stresses, including oxidative stress caused by treatment with hydrogen peroxide. Dps has been shown to bind tightly to DNA forming a DNA-protein crystal that

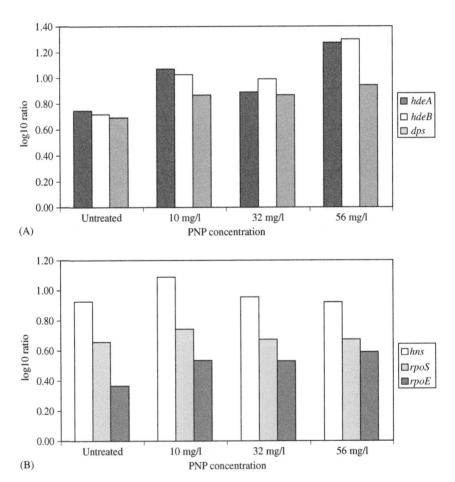

Fig. 3. Upregulation of genes involved in (A) pH adaptation and (B) the general and cell envelope stress responses in *E. coli* cells exposed to PNP.

protects the DNA from damage (Martinez and Kolter, 1997). It is, therefore, not unexpected that *dps* is upregulated in cells exposed to PNP or that expression of the gene increases with increasing concentration of PNP (Fig. 3A), either as a result of PNP as a stressor causing damage to the cells or as a response to the growth inhibition caused by PNP within the cells.

The *hdeA* gene encodes for HdeA, a periplasmic protein that plays a role in resistance to low pH. HdeA dimers have been shown to dissociate at low pH and the monomer is involved in protecting proteins from acid-induced denaturation (Gajiwala and Burley, 2000). HdeB (encoded by *hdeB*) is a protein related to an acid resistance protein of *Shigella flexneri* and it is predicted to have structural similarity to HdeA (Gajiwala and Burley, 2000), indicating that it too may play a role in adaptation of the cells to low pH. During a standard microbial growth inhibition test, the pH of PNP measured 4.31 in the test flasks at time 0. It is, therefore, perhaps not surprising that pH adaptation genes (*hdeA* and *hdeB*) were upregulated in cells exposed to PNP compared with untreated control cells (Fig. 3A).

A slightly elevated level of gene expression was shown in the gene encoding the global regulator involved in the general stress response, sigma^S, in PNP-treated compared with untreated cells (*rpoS*, Fig. 3B). The product of the *rpoS* gene was originally seen as the key regulator for stationary phase cells (Nyström, 1995), but it is now seen as a central component of the general stress response of *E. coli* (Hengge-Aronis, 1999). It was not unexpected, therefore, that expression of this gene was elevated in response to PNP treatment. Interestingly, another sigma factor gene was also expressed at elevated levels in PNP-treated cells compared with untreated cells (although not significantly). The gene *rpoE* encodes for SigmaE (σ^E), which is involved in the cell envelope stress response (Raivio and Silhavy, 1999). The level of this gene was elevated in all three concentrations of PNP compared with those of untreated control cells (Fig. 3B). The σ^E pathway has been shown to respond to high temperatures and ethanol and it has been suggested that σ^E may be induced by overexpression or misfoldings of outer membrane proteins (OMPs) (Raivio and Silhavy, 1999) (Fig. 4).

OmpA encodes for the OMP, OmpA. This protein is believed to be a nonspecific diffusion channel, allowing various small solutes to cross the outer membrane (Sugawara and Nikaido, 1992). It is also believed to be involved in shape stabilisation within *E. coli* cells. It is 325 amino acids long and is one of the most abundant proteins in the outer membrane of *E. coli* (Karp *et al.*, 2002: EcoCyc). OmpC (encoded by *ompC*) and

238

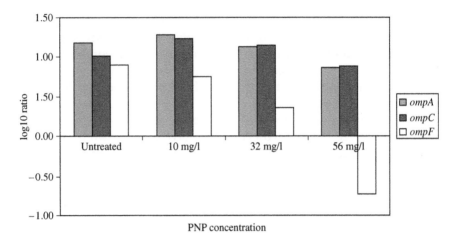

Fig. 4. Downregulation of genes encoding outer membrane proteins in response to cells exposed to PNP.

OmpF (encoded by *ompF*) are outer membrane porins involved in the transport of ions and solutes. OmpC forms a pore allowing ions and other hydrophilic solutes to cross the outer membrane of the cells, whereas OmpF allows the passage of solutes such as sugars, ions and amino acids (Nikaido and Varra, 1985). A transcriptional regulatory protein, OmpR, is a member of the two-component regulatory system, EnvZ/OmpR. It has been suggested that OmpR regulates transcription of the membrane porin genes, *ompC* and *ompF* (Karp *et al.*, 2002: EcoCyc). Gene expression levels of *ompA* and *ompC* appeared similar in untreated cells and cells exposed to $10\,\text{mg}\,\text{l}^{-1}$ PNP (Fig. 4), which possessed similar growth characteristics. In cells exposed to $56\,\text{mg}\,\text{l}^{-1}$ PNP gene expression levels of *ompA* and *ompC* were slightly reduced compared with those in untreated control cells (Fig. 4). A marked reduction in the gene expression level of the *ompF* gene was shown in cells exposed to increasing concentrations of PNP compared with those in untreated control cells (Fig. 4). This finding suggests that higher concentrations of PNP may be affecting cell membrane constituents or affecting the permeability of the cell, which elicits a downregulation effect on certain *omp* genes.

Part B: Growth characteristics and differential expression
The less stringent and at least ±2-fold approach yielded 307 (untreated versus $10\,\text{mg}\,\text{l}^{-1}$ PNP), 568 (untreated versus $32\,\text{mg}\,\text{l}^{-1}$ PNP) and 585

(untreated versus $56 \, \text{mg} \, l^{-1}$ PNP) differentially expressed genes, respectively, after 1 h exposure. These gene expression differences were reflected in the bacterial growth characteristics, whereby *E. coli* cells exposed to $10 \, \text{mg} \, l^{-1}$ PNP for 1 h showed similar growth characteristics to untreated cells. Cells exposed to $32 \, \text{mg} \, l^{-1}$ and $56 \, \text{mg} \, l^{-1}$ PNP for 1 h showed increasing levels of inhibition compared with untreated cells (data not shown).

Functional classification of differentially expressed genes

Following the identification of genes that were differentially expressed by at least ± 2-fold in cells exposed to PNP compared with untreated cells, the genes were divided into functional groups in order to make biological inferences about the data and to gain an understanding of the processes within the cell that may be affected by PNP exposure (Fig. 5). The largest number of genes that were differentially expressed in PNP-treated cells belonged to the hypothetical/unknown functional group and the number

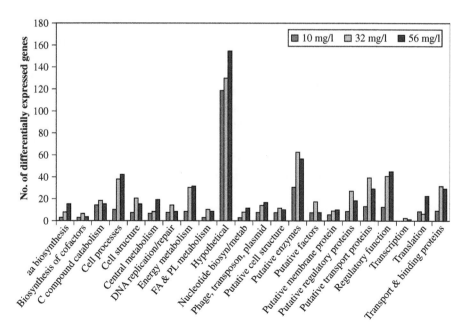

Fig. 5. Genes (differentially expressed by at least ± 2-fold in cells exposed to PNP for 1 h compared with untreated cells) classified by functional group.

of genes increased with increasing concentration of PNP ($10\,\mathrm{mg\,l^{-1}}$, 120 genes; $32\,\mathrm{mg\,l^{-1}}$, 130 genes and $56\,\mathrm{mg\,l^{-1}}$, 150 genes). This finding suggested that many adaptive changes that might occur as a result of PNP exposure are as yet uncharacterised. This situation can make the interpretation of whole genome expression data difficult, although it may provide a route towards identifying the role of unknown genes in genomes. Generally, within eight of the functional groups larger numbers of differentially expressed genes were shown in response to 32 and $56\,\mathrm{mg\,l^{-1}}$ PNP, compared with $10\,\mathrm{mg\,l^{-1}}$ PNP. The eight groups were cell processes, cell structure, energy metabolism, putative enzymes, putative regulatory proteins, putative transport proteins, genes with a regulatory function and genes involved in translation.

Comparison of differentially expressed genes (at least ±2-fold) with stress-response subsets and functional classifications

Genes showing at least ±2-fold differential expression were also statistically compared with stress-response genes and genes divided into functional classification, in order to identify any stress responses or functional groupings of genes within the cells that were affected by PNP exposure. Differentially expressed genes from cells exposed to $32\,\mathrm{mg\,l^{-1}}$ PNP and $56\,\mathrm{mg\,l^{-1}}$ for 1 h showed a statistical similarity to genes involved in cell processes ($p = 0.03$ and $p = 0.003$, respectively). Differentially expressed genes from cells exposed to $56\,\mathrm{mg\,l^{-1}}$ PNP for 1 h also showed similarity to genes involved in pH adaptation ($p = 0.01$). Differentially expressed genes from cells exposed to $10\,\mathrm{mg\,l^{-1}}$ PNP showed no similarities to any functional group classification or stress-response genes.

Effect of PNP exposure (1 h) on drug sensitivity/resistance genes

Many genes belonging to the cell processes functional group were shown to be differentially expressed (at least ±2-fold) following 1 h exposure to all three concentrations of PNP (Fig. 5). Individual genes and operons within this group, therefore, which the array data indicated were upregulated in response to PNP, were investigated further.

emrRAB operon

Genes of the *emrRAB* operon were upregulated in response to PNP treatment compared with untreated cells (Fig. 6A). The *emrRAB* operon is involved in multidrug resistance in *E. coli and* consists of *emrR*, which encodes the transcriptional regulator of the operon (EmrR), the *emrA* gene which encodes for a membrane fusion protein (EmrA) and *emrB*, which encodes a multidrug efflux protein (EmrB) (Karp *et al.*, 2002: EcoCyc). There is evidence to suggest that the gene products of the operon may function to assemble a multidrug resistance pump that provides an efflux pathway across both the inner and outer membranes of the cells (Lomovskaya and Lewis, 1992, Lomovskaya *et al.*, 1995). The *emrRAB* pump was discovered during an investigation to identify the genes involved in *E. coli* adaptation to protonophores, which uncouple oxidative phosphorylation. Expression of both *emrA* and *emrB* were both necessary to confer resistance to the uncouplers (carbonyl cyanide m-chlorophenylhydrazine, CCCP and tetrachlorosalicylanilide, TCS) and to two antibiotics, nalidixic acid and thiolactomycin (Lomovskaya and Lewis, 1992). It is possible, therefore, that cells exposed to PNP for 1 h are acting to reduce the effects of PNP as indicated by an upregulation of *emrRAB* genes in this study (Fig. 6A). Another study investigated the regulation of *emrR*, the first gene of the operon and discovered that the same compounds (CCCP, TCS, certain antibiotics) induced expression of the *emr* genes (Lomovskaya *et al.*, 1995). The data from the current study suggest that PNP may also induce the operon, as evidenced by the upregulation of *emrRAB* genes in PNP-treated cells compared with untreated cells (Fig. 6A). Interestingly, dinitrophenol (DNP) is a protonophore and uncouples oxidative phosphorylation by transporting protons across the membrane (Pinchot, 1967). This finding may suggest that PNP also acts as an uncoupler of oxidative phosphorylation within *E. coli* cells as it is structurally similar to DNP and has a similar effect on *emr* gene expression.

marRAB operon and associated genes

The EmrR protein is homologous to a regulatory protein, MarR, which is a repressor of the *mar* (*m*ultiple *a*ntibiotic *r*esistance) locus that confers multiple-antibiotic resistance in *E. coli* cells (Lomovskaya *et al.*, 1995). MarR is encoded by the *marR* gene, which was upregulated in cells exposed to PNP for 1 h compared with untreated cells (Fig. 6B). The *marRAB* operon also encodes MarA, a transcriptional regulator and MarB, a multiple antibiotic resistance protein, both of which were also upregulated in response to PNP (Fig. 6B). The activation of the *marRAB* operon results in enhanced resistance to both oxidative stress (Pomposiello *et al.*, 2001) and multiple antibiotics (Chollet *et al.*, 2002)

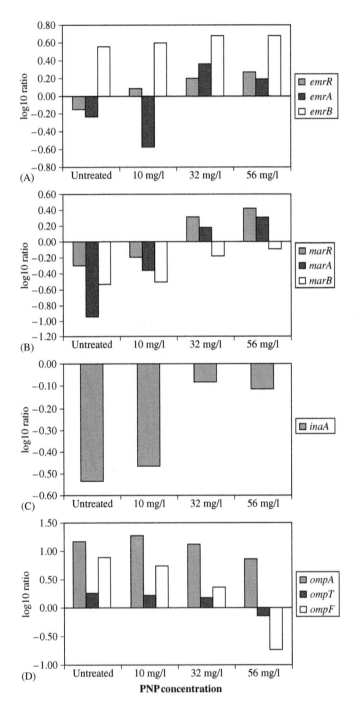

Fig. 6. Upregulation of gene expression in genes involved in drug resistance and sensitivity; (A) *emrRAB*, (B) *marRAB* and (C) *inaa* in PNP-treated cells compared with untreated cells and (D) downregulation of gene expression in genes encoding outer membrane proteins (*ompA*, *ompF*, *ompT*). Note difference in scale.

within the cell and it is, therefore, perhaps not surprising that the genes of this operon were upregulated in cells in response to PNP. The *marRAB* operon has been shown to respond to a variety of compounds, including antibiotics (tetracycline, chloramphenicol), oxidative stress agents (salicylate) and organic solvents. Interestingly, 2,4-dinitrophenol (a structurally similar compound to PNP) that uncouples oxidative phosphorylation, has been shown to induce expression of the *marRAB* operon (Alekshun and Levy, 1997). This observation is further evidence to support the suggestion that PNP may cause uncoupling of oxidative phosphorylation within *E. coli* cells.

Macroarray analysis has previously shown the differential expression of over 60 genes in response to the constitutive expression of MarA (the product of *marA*) in *E. coli* (Barbosa and Levy, 2000). A gene known to respond to MarA, *inaa*, was also upregulated in response to growth in all three concentrations of PNP (Fig. 6C). The *inaa* gene encodes a pH-inducible protein of unknown function that is involved in the multiple-antibiotic-resistance stress response.

Porins in the outer membrane of *E. coli* restrict the passage of large molecules and small hydrophobic substances, including antibiotics. These porins are encoded by the *omp* genes. The product of *mar* locus is also known to repress the synthesis of OmpF (Lomovskaya and Lewis, 1992) consistent with the observation that the gene encoding OmpF (*ompF*) was downregulated in cells exposed to PNP treatment (Fig. 6D). This finding may possibly be accounted for by the increase in expression of the *marA* gene in PNP-treated cells (Fig. 6B).

It has been suggested that the physiological function of the *marRAB* operon is to simultaneously induce a decrease in antibiotic uptake by altering the porin content of the outer membrane and to increase antibiotic release from the cell by activating efflux mechanisms (Alekshun and Levy, 1997; Chollet *et al.*, 2002). The data from this study indicate that PNP induces a multiple-antibiotic-resistance stress response in *E. coli* cells, as evidenced by a downregulation of genes encoding porins in the outer membrane of cells (*ompA, ompC, ompF* and *ompT*) (Figs. 5 and 7D) and an upregulation of genes involved in conferring antibiotic resistance (*marRAB*, Fig. 6B) and efflux mechanisms (*emrRAB*, Fig. 6A). A similar response was previously demonstrated during a study that investigated whether growth in salicylate induced antibiotic resistance by activating the *mar* operon in *E. coli*. This study indicated that growth on salicylate resulted in increased antibiotic resistance and also decreased amounts of OmpF (Cohen *et al.*, 1993).

244

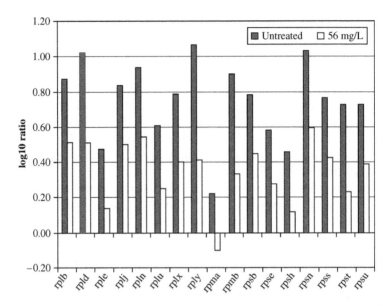

Fig. 7. Downregulation of ribosomal subunit protein (translation) genes in cells in response to 56 mg l⁻¹ PNP.

Effect of PNP exposure (1 h) on translation genes

One manifestation of growth inhibition is an effect on translational genes (Phadtare *et al.*, 2002). In cells exposed to 56 mg l⁻¹ PNP, 17 translational genes showed reduced expression levels compared with those of untreated cells (Fig. 7). Of these genes, eight encoded for 50S ribosomal L proteins (*rpl* genes) of which there are 24 associated genes within *E. coli*. Seven genes encoded for 30S ribosomal subunit proteins (*rps* genes) of which there are 21 genes and two genes encoded for 50S ribosomal subunit proteins (*rpm* genes) of which there are 10 associated genes within *E. coli*. A downregulation of translation genes and genes encoding outer membrane proteins (*omp* genes) in *E. coli* in response to PNP treatment was also apparent following the stringent statistical analysis (Figs. 2 and 4, respectively) indicating that these genes are both statistically and biologically important.

When using both a statistical and a two-fold approach to determine differential expression, all of the PNP-treated cultures showed a similar pattern of diminished gene expression, with particular bias towards genes whose products were involved in translation (Figs. 2 and 7, respectively). This bias may be caused by PNP causing the cells to prematurely enter stationary or survival phase. Viable cell counts showed significantly larger numbers of cells in untreated cultures compared with those exposed to $32\,mg\,l^{-1}$ and $56\,mg\,l^{-1}$ PNP for 4 h, confirming this. Additionally, morphological changes were apparent in cells exposed to PNP for 3 h. PNP-treated cells were shown to be shorter and narrower than untreated cells, a physiological change associated with cells entering stationary phase (Farewell *et al.*, 1998; Nyström, 1995).

Acknowledgements

GeneSpring™ licenses were kindly provided by the Natural Environmental Research Council (NERC) Environmental Bioinformatics Centre (NEBC).

References

Aakra, A., Vebø, H., Snipen, L., Hirt, H., Aastveit, A., Kapur, V., Dunny, G., Murray, B. E. and Nes, I. F. (2005). Transcriptional response of *Enterococcus faecalis* V583 to erythromycin. *Antimicrob. Agents Chemother.* 49,2246–2259. Erratum in: *Antimicrob. Agents Chemother.* 49,3989.

Alekshun, M. N. and Levy, S. B. (1997). Regulation of chromosomally mediated multiple antibiotic resistance: The *mar* regulon. *Antimicrob. Agents Chemother.* 41,2067–2075.

Allen, M. J., White, G. F. and Morby, A. P. (2006). The response of *Escherichia coli* to exposure to the biocide polyhexamethylene biguanide. *Microbiology* 152,989–1000.

Arfin, S. M., Long, A. D., Ito, E., Richle, M. M., Paegle, E. S. and Hatfield, G. W. (2000). Global gene expression profiling in *Escherichia coli* K12: The effects of integration host factor. *J. Biol. Chem.* 275,29672–29684.

Barbosa, T. M. and Levy, S. B. (2000). Differential expression of over 60 chromosomal genes in *Escherichia coli* by constitutive expression of MarA. *J. Bacteriol.* 182, 3467–3474.

Beissbarth, T., Fellenberg, K., Brors, B., Arribas-Prat, R., Boer, J. M., Hauser, N. C., Scheideler, M., Hoheisel, J. D., Schütz, G., Poustka, A. and Vingron, M. (2000). Processing and quality control of DNA array hybridization data. *Bioinformatics* 16,1014–1022.

Benjamini, Y. and Hochberg, Y. (1995). Controlling the False Discovery Rate: A practical and powerful approach to multiple testing. *J. R. Stat. Soc.* 57,289–300.

Blattner, F. R., Plunkett III, G., Bloch, C. A., Perna, N. T., Burland, V., Riley, M., Collado-Vides, J., Glasner, J. D., Rode, C. K., Mayhew, G. F., Gregor, J., Davis, N. W., Kirkpatrick, H. A., Goeden, M. A., Rose, D. J., Mau, B. and Shao, Y. (1997). The complete genome sequence of *Escherichia coli* K-12. *Science* 277,1453–1462.

Bordi, C., Théraulaz, L., Méjean, V. and Jourlin-Castelli, C. (2003). Anticipating an alkaline stress through the Tor phosphorelay system in *Escherichia coli*. *Mol. Microbiol.* 48,211–223.

246

Brazma, A. (2001). On the importance of standardisation in life sciences (Editorial). *Bioinformatics* 17,113–114.

Brazma, A., Hingamp, P., Quackenbush, J., Sherlock, G., Spellman, P., Stoeckert, C., Aach, J., Ansorge, W., Ball, C. A., Causton, H. C., Gaasterland, T., Glenisson, P., Holstege, F. C. P., Kim, I. F., Markowitz, V., Matese, J. C., Parkinson, H., Robinson, A., Sarkans, U., Schulze-Kremer, S., Stewart, J., Taylor, R., Vilo, J. and Vingron, M. (2001). Minimum Information About a Microarray Experiment (MIAME) – toward standards for microarray data. *Nat. Genet.* 29,365–371.

British Standard BS EN ISO 10712. (1995). Water quality – *Pseudomonas putida* growth inhibition test (*Pseudomonas* cell multiplication inhibition test). BSI British Standards, London, UK.

Brocklehurst, K. R. and Morby, A. P. (2000). Metal-ion tolerance in *Escherichia coli*: Analysis of transcriptional profiles by gene-array technology. *Microbiology* 146, 2277–2282.

Chang, D-E., Smalley, D. J. and Conway, T. (2002). Gene expression profiling of *Escherichia coli* growth transitions: An expanded stringent response model. *Mol. Microbiol.* 45,289–306.

Chang, W., Small, D. A., Toghrol, F. and Bentley, W. E. (2005a). Microarray analysis of *Pseudomonas aeruginosa* reveals induction of pyocin genes in response to hydrogen peroxide. *BMC Genom.* 6,115.

Chang, W., Small, D. A., Toghrol, F. and Bentley, W. E. (2005b). Microarray analysis of toxicogenomic effects of peracetic acid on *Pseudomonas aeruginosa*. *Environ. Sci. Technol.* 39,5893–5899.

Chang, W., Toghrol, F. and Bentley, W. E. (2006). Toxicogenomic response of *Staphylococcus aureus* to peracetic acid. *Environ. Sci. Technol.* 40,5124–5131.

Chollet, R., Bollet, C., Chevalier, J., Malléa, M., Pagès, J-M. and Davin-Regli, A. (2002). *mar* operon involved in multidrug resistance of *Enterobacter aerogenes*. *Antimicrob. Agents Chemother.* 46,1093–1097.

Cohen, S. P., Levy, S. B., Foulds, J. and Rosner, J. L. (1993). Salicylate induction of antibiotic resistance in *Escherichia coli*: Activation of the mar operon and a mar-independent pathway. *J. Bacteriol.* 175,7856–7862.

Farewell, A., Kvint, K. and Nyström, T. (1998). UspB, a, new, σ^S-regulated gene in *Escherichia coli* which is required for stationary-phase resistance to ethanol. *J. Bacteriol.* 180,6140–6147.

Gajiwala, K. S. and Burley, S. K. (2000). HDEA, a periplasmic protein that supports acid resistance in pathogenic enteric bacteria. *J. Mol. Biol.* 295,605–612.

Hengge-Aronis, R. (1999). Interplay of global regulators and cell physiology in the general stress response of *Escherichia coli*. *Curr. Opin. Microbiol.* 2,148–152.

Hess, K. R., Zhang, W., Baggerly, K. A., Stivers, D. N. and Coombes, K. R. (2001). Microarrays: Handling the deluge of data and extracting reliable information. *Trends Biotechnol.* 19,463–468.

Iannaccone, P. M. (2001). Toxicogenomics: The call of the wild chip. *Environ. Health Perspect.* 109,A8–A11.

Jakob, K., Satorhelyi, P., Lange, C., Wendisch, V. F., Silakowski, B., Scherer, S. and Neuhaus, K. (2007). Gene expression analysis of *Corynebacterium glutamicum* subjected to long-term lactic acid adaptation. *J. Bacteriol.* 189,5582–5590.

Karp, P. D., Riley, M., Saier, M., Paulsen, I. T., Collado-Vides, J., Paley, S. M., Pellegrini-Toole, A., Bonavides, C. and Gama-Castro, S. (2002). The EcoCyc database. *Nucleic Acids Res.* 30,56–58.

Keseler, I. M., Collado-Vides, J., Gama-Castro, S., Ingraham, J., Paley, S., Paulsen, I. T., Peralta-Gil, M. and Karp, P. D. (2005). EcoCyc: A comprehensive database resource for *Escherichia coli*. *Nucleic Acids Res.* 33,D334–D337.

Kershaw, C. J., Brown, N. L., Constantinidou, C., Patel, M. D. and Hobman, J. L. (2005). The expression profile of *Escherichia coli* K-12 in response to minimal, optimal and excess copper concentrations. *Microbiology* 151,1187–1198.

Kramer, J. A., Pettit, S. D., Amin, R. P., Bertram, T. A., Car, B., Cunningham, M., Curtiss, S. A., Davis, J. W., Kind, C., Lawton, M., Naciff, J. M., Oreffo, V., Roman, R. J., Sistare, F. D., Stevens, J., Thompson, K., Vickers, A. E., Wild, S. and Afshari, C. A. (2004). Overview of the application of transcription profiling using selected nephrotoxicants for toxicology assessment. *Environ. Health Perspect.* 112,460–464.

Lei, Y., Mulchandani, P., Chen, W., Wang, J. and Mulchandani, A. (2003). A microbial biosensor for *p*-nitrophenol using *Arthrobacter* sp. *Electroanalysis* 15,1160–1164.

Lomovskaya, O. and Lewis, K. (1992). *emr*, an *Escherichia coli* locus for multidrug resistance. *Proc. Natl. Acad. Sci. USA* 89,8938–8942.

Lomovskaya, O., Lewis, K. and Matin, A. (1995). EmrR is a negative regulator of the *Escherichia coli* multidrug resistance pump EmrAB. *J. Bacteriol.* 177,2328–2334.

Martinez, A. and Kolter, R. (1997). Protection of DNA during oxidative stress by the nonspecific DNA-binding protein Dps. *J. Bacteriol.* 179,5188–5194.

Nikaido, H. and Varra, M. (1985). Molecular basis of bacterial outer membrane permeability. *Microbiol. Rev.* 49,1–32.

Nguyen, V. D., Wolf, C., Mäder, U., Lalk, M., Langer, P., Lindequist, U., Hecker, M. and Antelmann, H. (2007). Transcriptome and proteome analyses in response to 2-methylhydroquinone and 6-brom-2-vinyl-chroman-4-on reveal different degradation systems involved in the catabolism of aromatic compounds in *Bacillus subtilis*. *Proteomics* 7,1391–1408.

Nyström, T. (1995). The trials and tribulations of growth arrest. *Trends Microbiol.* 3,131–136.

Phadtare, S., Kato, I. and Inouye, M. (2002). DNA microarray analysis of the expression profile of *Escherichia coli* in response to treatment with 4,5-dihydroxy-2-cyclopenten-1-one. *J. Bacteriol.* 184,6725–6729.

Pinchot, G. B. (1967). The mechanism of uncoupling of oxidative phosphorylation by 2,4-dinitrophenol. *J. Biol. Chem.* 242,4577–4583.

Pomposiello, P. J., Bennik, M. H. J. and Demple, B. (2001). Genome-wide transcriptional profiling of the *Escherichia coli* responses to superoxide stress and sodium salicylate. *J. Bacteriol.* 183,3890–3902.

Prytz, I., Sandén, A. M., Nyström, T., Farewell, A., Wahlström, A., Förberg, C., Pragai, Z., Barer, M., Harwood, C. and Larsson, G. (2003). Fed-batch production of recombinant β-galactosidase using the universal stress promoters *uspA* and *uspB* in high cell density cultivations. *Biotechnol. Bioeng.* 83,595–603.

Raivio, T. L. and Silhavy, T. J. (1999). The σ^E and Cpx regulatory pathways: Overlapping but distinct envelope stress responses. *Curr. Opin. Microbiol.* 2,159–165.

248

Riley, M. (1998). Genes and proteins of *Escherichia coli* K-12 (GenProtEC). *Nucleic Acids Res.* 26,54.

Rocha, E. P. C., Matic, I. and Taddei, F. (2002). Over-representation of repeats in stress response genes: A strategy to increase versatility under stressful conditions? *Nucleic Acids Res.* 30(9),1886–1894.

Rockett, J. C. and Dix, D. J. (2000). DNA arrays: Technology, options and toxicological applications. *Xenobiotica* 30,155–177.

Rozen, Y., LaRossa, R. A., Templeton, L. J., Smulski, D. R. and Belkin, S. (2002). Gene expression analysis of the response by *Escherichia coli* to seawater. *Antonie Leeuwenhoek* 81,15–25.

Salunkhe, P., Töpfer, T., Buer, J. and Tümmler, B. (2005). Genome-wide transcriptional profiling of the steady-state response of *Pseudomonas aeruginosa* to hydrogen peroxide. *J. Bacteriol.* 187,2565–2572.

Shaw, K. J., Miller, N., Liu, X., Lerner, D., Wan, J., Bittner, A. and Morrow, B. J. (2003). Comparison of the changes in global gene expression in *Escherichia coli* induced by four bactericidal agents. *J. Mol. Microbiol. Biotechnol.* 5,105–122.

Shimazu, M., Mulchandani, A. and Che, W. (2001). Simultaneous degradation of organophosphorus pesticides and *p*-nitrophenol by a genetically engineered *Moraxella* sp. with surface-expressed organophosphorus hydrolase. *Biotechnol. Bioeng.* 76, 318–324.

Snape, J. R., Maund, S. J., Pickford, D. B. and Hutchinson, T. H. (2004). Ecotoxicogenomics: The challenge of integrating genomics into aquatic and terrestrial ecotoxicology. *Aquat. Toxicol.* 67,143–154.

Sugawara, E. and Nikaido, H. (1992). Pore-forming activity of OmpA protein of *Escherichia coli*. *J. Biol. Chem.* 267,2507–2511.

Tao, H., Bausch, C., Richmond, C., Blattner, F. R. and Conway, T. (1999). Functional genomics: Expression analysis of *Escherichia coli* growing on minimal and rich media. *J. Bacteriol.* 181,6425–6440.

Thouand, G., Friant, P., Bois, F., Cartier, A., Maul, A. and Block, J. C. (1995). Bacterial inoculum density and probability of *para*-nitrophenol biodegradability test response. *Ecotoxicol. Environ. Saf.* 30,274–282.

Weber, A. and Jung, K. (2002). Profiling early osmostress-dependent gene expression in *Escherichia coli* using DNA macroarrays. *J. Bacteriol.* 184,5502–5507.

Zheng, M., Wang, X., Templeton, L. J., Smulski, D. R., LaRossa, R. A. and Storz, G. (2001). DNA microarray-mediated transcriptional profiling of the *Escherichia coli* response to hydrogen peroxide. *J. Bacteriol.* 183,4562–4570.

Systems toxicology: using the systems biology approach to assess chemical pollutants in the environment

Richard D. Handy*

School of Biological Sciences, University of Plymouth, Drake Circus, Plymouth, PL4 8AA, UK

Abstract. There are many complex problems in environmental toxicology that we have historically not been able to resolve in a satisfactory quantitative manner. These complexities include the effects of mixtures of pollutants, complex exposure profiles, or the complex responses of organisms or ecosystems over different timescales. The cell biology community, along with mathematicians developed the 'Systems Biology' concept. This is a modelling tool that was developed to understand and predict how complex biological process at the cellular, and sub-cellular level, work. It is also theoretically possible to apply this systems approach to toxicology, called 'Systems Toxicology'. This discipline is in its infancy. Historic concepts in the control of biological systems are outlined, and how these relate to the modern concept of systems biology. We then describe systems toxicology and its application to environmental pollution. System toxicology involves the input of data into computer modelling techniques, which use mostly differential equations, models of networks, or cellular automata theory. The input data can be biological information from organisms exposed to pollutants. These inputs could be data from the 'omics, or traditional biochemical or physiological effects data. The input data must also include environmental chemistry data sets and quantitative information on ecosystems so that geochemistry, toxicology, and ecology can be modelled together. The outputs could include complex descriptions of how organisms and ecosystems respond to chemicals or other pollutants and the inter-relationships with the many other environmental variables involved. The model outputs could be at the cellular level, organ, organism, or ecosystem level. Ecologically relevant outputs could be achieved ('systems ecotoxicology'), provided environmental variability is considered in the modelling. Systems toxicology is potentially a very powerful tool, but a number of practical issues remain to be resolved such as the creation and quality assurance of databases for environmental pollutants and their effects, as well as user-friendly software that uses ecological or ecotoxicological parameters and terminology.

Keywords: systems biology; systems toxicology; chemicals; pollution; control system; data modelling; risk assessment; ecotoxicology; toxicogenomics; cellular automata; Petri nets; networks; interactomes; modules; input; output; bioinformatics; environmental protection; software.

Corresponding author: Tel.: +44(0)1752-232900. Fax: +44(0)1752-232970.
E-mail: rhandy@plymouth.ac.uk (R.D. Handy).

ADVANCES IN EXPERIMENTAL BIOLOGY
VOLUME 02   ISSN 1872-2423
DOI: 10.1016/S1872-2423(08)00007-0

Introduction and aims

In the 1960s and early 1970s environmental pollution was largely assessed by using chemical detection methods to measure and monitor toxic substances in the environment. This approach consisted of setting environmental quality criteria, and defining concentrations of individual chemicals that should not be exceeded in the environment. For example, water quality standards for substances in river water or drinking water. The Control of Pollution Act in 1974 offered a basis on which the various agencies involved in environmental protection could enforce standards. The water quality standards, for example, were based on dividing the no observed adverse effect level (NOAEL) obtained from acute ecotoxicity tests (*e.g.*, *Daphnia magna*, algae, and fish tests) by a safety factor (*e.g.*, 100 or 1,000) to account for uncertainty and management issues such as the land use and consequences of a pollution incident. This quality standards approach has many practical advantages, but has always been criticised for being oversimplified, and lacking the complexity found in the environment, or lacking environmental realism. There are also concerns that laboratory test organisms are from limited genetic stock (or clones in the case of *D. magna*) and therefore lack biological realism in relation to the genetic diversity of wild populations. Consequently, a probabilistic risk-based approach is now used that takes into account the probable exposure in the environment, the types of organisms present, and their sensitivity to pollutants (*e.g.*, Brain *et al.*, 2006).

However, the problem of complexity remains. Different organisms can be exposed to mixtures of chemicals, and at varying concentrations in the environment, over variable timescales (*e.g.*, intermittent pollution; Ashauer *et al.*, 2006; Handy, 1994). Practical approaches have been sought to measure this complexity. For example, by the development of real-time monitoring devices to continuously measure the concentrations of chemical in the environment, or real-time bioassays (Dalzell *et al.*, 2002). Biological monitoring has also been suggested at different levels of biological organisation (from cells to animal behaviour, *e.g.*, review; Handy *et al.*, 2002), and the use of biomarkers of exposure/effect (review; Handy *et al.*, 2003). These approaches offer some insight into complexity. For example, by using a suite of biomarkers to unravel the toxic mechanisms and effects of a particular class of chemicals (*e.g.*, pesticides; Galloway and Handy, 2003). These approaches do not measure complexity directly in a single measurement because they rely on making biochemical or similar measurements individually in many different body systems or cell types to build up an overall picture of the effects of pollution on an organism. The one facet that links all these biological effects together is the control system(s) in the organisms. If we can study these systems and understand how they are influenced by pollutants, then perhaps we can use the system itself as a tool to measure complexity.

The topic of 'Systems Biology' has some historical routes in the study of the control of cellular and physiological processes, but for a new generation of molecular and computational cell biologists 'Systems Biology' is also an application or modelling tool to understand and predict how complex biological process at the cellular, and sub-cellular level, work. Systems biology might therefore be defined as the quantitative study of the control and operation of biological processes that uses computational tools to study the complexity of biological events, mainly at the cellular level. However, the systems biology concept could be applied to the control and function of an organ, or even at the organism level. Also, the systems biology approach does not necessarily need to be applied to the normal functioning of cells or body systems. It might also be applied to the study of dysfunction, such as the case for exposure to environmental pollutants. Thus the overall aim of this chapter is to ask whether or not the systems biology approach would be worth applying to toxicological problems, particularly to pollutants. This chapter outlines the historic concepts in the control of biological systems, and explains how the modern notion of systems biology evolved from these ideas. Then the idea of developing the systems biology approach as a tool for pollution and chemicals risk assessment, 'Systems Toxicology', is explored. Here we outline the advantages, disadvantages, and practical challenges of this new approach.

Control systems and the conception of systems biology

The notion of biological control systems

The notion that biological processes operate in an ordered way, with organisms responding to stimuli (an 'input') with an appropriate biological response (the 'output') that involves some kind of control system, is a very old idea (Fig. 1). These ideas probably originated in the field of comparative physiology in the latter part of the 19th century, and were part of the founding principles that August Krogh and others laid down for the study of physiological systems (Randall *et al.*, 2002). At that time, the context was the study of how body systems worked, and in particular, the control processes involved in making a functional

252

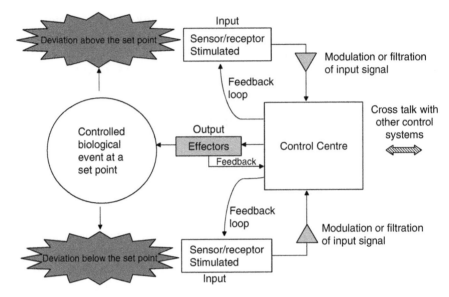

Fig. 1. An example of biological control system. Most biological systems operate from a set point that is controlled. Chemicals could damage sensors (input), the control system, the effectors (output devices), or the connections between them.

physiological system (*e.g.*, cardiovascular control of blood volume and blood pressure).

These early studies on physiological systems identified some key principles that are vitally important to understanding any biological system whether it be at the level of the whole organism, body system, cell, or sub-cellular compartment. It is worth understanding these founding principles of how biological systems are controlled before debating the modern context of systems biology. The following key points emerge from the comparative study of physiological control systems:

(i) Most biological systems operate from a 'set point' and the control system is designed to help maintain the desired biological parameter at that set point. So for example, the set point for resting blood glucose is $70-110\,\mathrm{mg\,dl}^{-1}$ in humans. We know that blood glucose levels may show transient changes when we eat, and the hormones insulin and glucagon are involved in ensuring that blood glucose is controlled within an acceptable level.

(ii) If there is a set point, then there is a control centre which monitors the set point. This control centre will initiate the output or biological response that corrects deviation from the set point. For example, in vertebrate physiology the brain often plays the role of the control centre. In the control of body temperature the hypothalamus is the thermostat that monitors sensory input from temperature receptors around the body and determines whether or not body temperature should be adjusted back to the set point ($37°C$) if the body starts to get too hot or too cold. Note that over long timescales (*e.g.*, seasons, years) set points can be altered or adjusted to new values – this is part of acclimation and adaptation to environmental change, or aging. Thus in the short term we may regard set points as fixed, but we should be aware that they can change as part of normal biological adaptation over longer timescales. Set points are therefore likely to change in response to adaptation, or tolerance of long-term low-level exposure to chemicals or other pollutants.

(iii) There needs to be a stimulus, or input, for the set point to be challenged. Importantly, the stimulus needs to be detected and this information passed to the control centre for processing. This stimulus can be a positive stimulus (*e.g.*, an increase or decrease in environmental temperature) that would cause the biological system to give a response (*e.g.*, adjustment of body temperature). The stimulus is positive in the sense that it makes the biological system respond. Alternatively, biological systems can detect the complete presence or absence of a stimulus (*e.g.*, taste receptors detecting salt in seawater vs. fresh drinking water), or respond to a change in the frequency, cycle, or pattern of the stimulus (*e.g.*, the pineal gland's involvement in the detection of day length). The latter is particularly important, since stimuli to biological systems are often complex and we should not focus our efforts on simply understanding mean values of parameters, but consider variability and intensity of stimuli or input. The stimulus need to be detected by some kind of sensor, this could be a sensory nerve in a physiological system, or a receptor on the cell membrane of a cell for example. However, it is also worth noting that the receptors also act as 'modulators' or filters of the input stimuli, *i.e.*, most sensors have a range at which they can detect stimuli and will not be able to detect every signal strength. Detectors can also be overloaded. Some sensors only fire a signal to the control system if change is detected and will not respond to a continuous or normal stimulus (this prevents the control system being overloaded with useless routine information).

(iv) There needs to be 'effectors' connected to the control centre which initiates the biological response or output from the control centre. So for example, in the case of a chemical burn on the hand, the 'effectors' are the muscles which draw the hand away from the painful stimuli.

(v) Finally, there needs to be some kind of feedback loop in the control system that ensures the output or biological response has been effective. These can take the form of either negative or positive feedback. These feedback loops need to re-measure the input or stimuli so that the control centre can determine whether or not further biological response or output is needed. Most normal biological processes work using negative feedback where the output results in a reduction of the original stimuli or a buffering of the stimuli so that further output or biological response is reduced or not needed. So for example, in the case of the chemical burn to the hand, the feedback loop in the CNS will be assessing whether or not pain to the injured area of skin has decreased following movement of the arm away from the chemical source. However, biological systems also sometimes use positive feedback. These positive feedback events are usually designed to control rapid or 'explosive' biological events. Classic examples include the sudden generation of an action potential in a nerve once the threshold voltage is reached, ovulation driven by a spike in circulating sex hormones, and the rapid proliferation of the endometrium in the uterus during pregnancy.

(vi) Redundancy in the control system. Most biological processes have more than one control loop, so that a biological set point can be partly or wholly controlled even though one part of the control system may be damaged. This may involve separate control systems for dealing with increases and decreases in parameters from the set point, or more than one control system (*e.g.*, neurological and hormonal control in a physiological system).

(vii) Communication between control systems. We must not forget that organisms are fully integrated, functional biological units, and to achieve this, the control systems for different organs, tissues, and cell types need to communicate with each other. This is the concept of 'networking' a series of control centres or processes together.

In summary (Fig. 1), a simple biological control system has inputs (stimuli) which are detected by receptors that may have some capacity to filter or modulate the input, and then pass this input information to a control centre. The control centre determines the output (biological response) which is mediated by effectors that deliver that response. The control centre monitors both set points in biological parameters and feedback following an output.

Why are biological control systems important in toxicology?

Toxicology is partly about the study of defects in biological systems and one can think of many examples where toxic substances could interfere with biological control systems (Fig. 1). This could be at the level of the receptor which detects the input. This could be chemical erosion of sensory nerve endings in the skin, or the blockade of a receptor on a cell membrane by a foreign compound (*e.g.*, mercury blockade of Na channels in excitable cells; Hoyle and Handy, 2005), or the inappropriate stimulation of a receptor by a foreign compound mimicking a normal biological signal (*e.g.*, endocrine disruption). Toxic chemicals might also interfere with the connections between the input receptor and the control centre. So for example, in a physiological control system this could be damage to a sensory nerve going to the brain, or in a cell to the second messenger systems (*e.g.*, cAMP cascade) that pass information from the cell membrane to the organelles and nucleus. The control centre could be damaged (*e.g.*, direct neurotoxicity of pesticides to the brain, genotoxicity and damage to the nucleus of a cell). Alternatively, the effector pathways could be damaged so that the output signal is not passed to the effectors (*e.g.*, transcription and translation of the genetic code to produce a response to cell injury). The effectors themselves could be damaged (*e.g.*, chemical blockade of muscle function in physiology, or the loss of antioxidants defences from a cell during intracellular chemical injury). It is also possible that the several parts of the system are damaged so that the feedback loop no longer works, and for example positive feedback is generated instead of the normal negative feedback (*e.g.*, immunotoxicity and hypersensitivity reactions to pesticides; Galloway and Handy, 2003), or that the feedback system does not work at all (*e.g.*, no response to a chemical burn with phenol because of its anaesthetics properties to nerve endings in the skin).

The modern concept of 'Systems Biology'

The modern concept and discipline known as 'Systems Biology' has evolved from molecular biology disciplines (genomics, bioinformatics), biochemistry, and cell biology. It also is multi-disciplinary in that mathematics, biophysics, and bioengineering are also involved. In essence modern systems biology takes the ideas of control systems above and studies them at the cellular level and in the context of the wealth of information we now have about the cell and cellular components (reviews, Kitano, 2002; Kremling and Saez-Rodriguez, 2007; Ivakhno, 2007; Wolkenhauer, 2001; Wolkenhauer and Hofmeyr, 2007). Systems biology aims to quantitatively understand biological control at the cellular level. If we can understand how a cell works then we can model the functions of the cell and potentially create an experimental and/or theoretical tool to predict the responses of the cell to stimuli. There are even suggestions that *in silico* models of cells, could ultimately be used to redesign or engineer an improved cell on the basis of the outputs of the modelling ('synthetic biology'; Barrett *et al.*, 2006). In the context of pollution, the main use of systems biology would be to predict the responses of the cell to chemicals. However from the viewpoint of hazard and risk assessment of chemicals to organisms or ecosystems, this cell biologist's definition of systems biology is a little restrictive. We expand the concept to higher levels of biological organisation such as organ function and body system functions that would be relevant to the toxicity of chemicals. For example, it would be useful to apply the systems biology approach to predict the effects of chemicals on heart or liver function, rather than just individual cell types within these organs (Noble, 2006).

The ultimate goal of systems biology for the cell biologist is to generate a computer model of cell functions (*e.g.*, Barrett *et al.*, 2006), or in the case of medical or toxicological sciences to expand this to models of organ/body system function (*e.g.*, Noble, 2006). Materi and Wishart (2007) identified several reasons why the development of such models is desirable.

(i) To generate data which is beyond the capabilities of current experimental techniques, or in the case of chemicals where it would be more ethically sound to generate the data by modelling.
(ii) Save time and money by doing experiments *in silico*.
(iii) Models may yield new (non-intuitive) information about how cells work.

(iv) It may be possible to identify missing components or processes in a cell using a computer model that are not easily identified by *in vivo* or *in vitro* experimental work.

(v) Enable complexity to be investigated. This might include the effects of multiple stimuli, or the role of networks of cells or cell systems.

Materi and Wishart (2007) argue that such models could be used in drug discovery, and in modelling drug toxicity such as distribution, metabolism, and pharmacokinetics to enable prediction of safe doses of drugs. Clearly, these ideas are also equally applicable to pollutants as toxic substances. In toxicology we already use predictive tools such as structure-activity relationships (SAR/QSARs) and decision trees applying the output of such tools in risk assessment (Doull *et al.*, 2007; Mazzatorta *et al.*, 2007). Doull *et al.* (2007) argue that this foundation of experience could be expanded to take advantage of computational biology and bioinformatics.

One of the main drivers for the establishment of the discipline of systems biology was the huge data sets derived from genomics and bioinformatics, and on the biochemical composition and physical structure of cellular components (Kremling and Saez-Rodriguez, 2007). This information has enabled some credible approaches to modelling of which there are two fundamental types: bottom-up or top-down approaches (Kremling and Saez-Rodriguez, 2007). The bottom-up approach essentially constructs the model from the basic components of the cell. So for example, it may model a biochemical pathway, and then network this together with other biochemical pathways, and spatial information about location/organelles involved within the cell. The approach is essentially modular in that each component is defined (*e.g.*, a simple biochemical pathway could be a modular unit) and the modules are put together to make networks, which are later merged to hopefully generate the overall model of cell function. Clearly, one difficulty with the bottom-up approach is that you need all the components (modules) to make the model work correctly. Of course, we do not know everything about the cell and there are bound to be missing components or modules. So, as an alternative, there is the top-down approach. The top-down approach considers experimental observation on biological responses (the outputs) and by using multi-variate statistical techniques such as cluster analysis to identify related components in the outputs. It is then possible based on the similarities and differences to construct networks.

Ideally, it would be good to employ both approaches at once, and this has been coined the 'middle-out' approach (Noble, 2002).

Terminology

Some of the phrases used in control systems such as 'input', 'output', and 'negative feedback' are outlined earlier. For convenience, some of the common terms and phrases used by the systems biology community are listed (Appendix 1). In general there seems to be reasonably good agreement about terminology between the engineers, chemists, and biologists working on systems biology. It would therefore seem appropriate for toxicologists/ecotoxicologists to adopt these terms.

Current applications of Systems Biology

Some selected examples of the current application of systems biology to the life sciences, with emphasis on biomedical/disease or toxicological applications, are outlined in Table 1. These examples illustrate several points. Firstly that the discipline is still in its infancy, and many of the studies are collecting data from real cells or organisms to do some preliminary modelling which results in either a modular component of a system model, or a simple network. In essence, the topic is at the proof of principle stage with relatively simple models, and we are far from having detailed models that could predict all the functions of a cell, organ or body system. However, even these relatively simple models are much more complex than anything we have been able to generate before and this is a major step forward in understanding the dynamics and complexity of biological processes.

The systems biology approach has been applied to biomedical sciences and drug discovery (Materi and Wishart, 2007; Moore *et al.*, 2007), and drug toxicity (Laaksonen *et al.*, 2006). Here the approaches tend to be for a top-down model where the biological response or outputs are correlated to generate simple networks. The tools used to generate the original data for the modelling tend to be genomic data obtained directly from experiments, or from databases; as well as proteomics or metabolomics to profile the drug metabolites or intended beneficial biochemical effects in the patient (*e.g.*, Laaksonen *et al.*, 2006). The ultimate goal is to predict the progression of disease and/or the effects of drug treatments, against the myriad of other environmental or biological factors that could influence these processes. Perhaps even to identify new ways of treating disease. These goals are still some way in the future, but

Table 1. Selected examples of the application of systems biology to cell biology, biomedical, or toxicological sciences.

Topic area	Type of model and test organism	Notes on the approach or results of the application	Reference
Biomedical	Gene expression profiling and correlation analysis. Mouse model of Fabry disease	Fabry disease manifests as vascular pathology and damage to the endothelial cells lining blood vessels. It is a genetic disorder resulting in the deficiency of the lysosomal enzyme, α-galactosidase A, resulting in lipid accumulation in the cells. The authors looked at gene expression (microarrays) made from knockout mice missing the relevant gene compared to normal mice. Using a top-down approach, correlation analysis of the gene expression pattern was used to suggest a simple network of key elements of the gene expression pattern. The next step would be to extrapolate the ideas into a more detailed model, and then logically define a new therapy for these patients	Moore *et al.* (2007)
Cell biology	Protein–protein interactions. Yeast cells	The model relies on the idea that functional responses of cells are ultimately defined by the dynamic interactions between proteins in the cell. These are physical interactions between proteins based on their structure and function and binding to macromolecules in the cell. The network of protein–protein interactions in the cell are called 'interactomes' and have been applied to mainly to yeast cells, *Saccharomyces cerevisiae* in this example. This model system is a yeast two hybrid (Y2H) that enables investigation of the role of protein activators in the expression of parts of the DNA	Cusick *et al.* (2005)

Table 1. (Continued)

Topic area	Type of model and test organism	Notes on the approach or results of the application	Reference
Cell biology	Mouse liver cell model of intracellular signalling pathways	Using primary hepatocyte cultures. The chapter collects experimental data on the activity of key enzymes in the cell (CYP3A, glutamine synthetase, glutathione transferase) and mRNA expression of cell signalling components and their level of phosphorylation. The data include time courses of responses and effects of varying the culture conditions on the liver cells. The next step could be to use the data in a bottom-up model of parts of liver cell function	Klingmuller et al. (2006)
Cell biology	In silico model of the kinetics of protein–protein interactions	One fundamental protein–protein interaction is the activation of proteins by phosphorylation, and their deactivation by dephosphorylation using kinase and phosphatase enzyme systems. Using data on the activation and deactivation rate constants, flux rates for reactions, and model network of protein activation–deactivation are created	Martelli et al. (2007)
Physiology	Modelling of immunity and inflammation	A bottom-up approach is applied to existing information in databases on a range of receptors, signalling pathways including various phosphoproteins, and various secretory proteins involved in immune responses. Genomic and proteomic data are used. A systems biology database (SBEAMS) and visualisation software (Cytoscape) was used to construct the modular components of the model, and then the networks. The model gives a foundation for understanding the complexity of the immune system	Aderem and Smith (2004)

Toxicology	Gene expression profiling and correlation analysis. Patients on high doses of statins	The logic of a top-down systems biology approach was used to analyse data from gene expression (microarray) from skeletal muscle and plasma lipid profiles (lipidomics) from patients on high doses of the cholesterol-reducing drugs called statins, compared to controls on placebo treatments. Discrimination analysis from the lipid profiling was combined with a partial least squares discrimination analysis of the gene expression profiles. The data were used to suggest biomarkers to monitor the toxicity of statins	Laaksonen et al. (2006)
Toxicology	Model of polycyclic aromatic hydrocarbon (PAH) pyrene metabolism in the bacterium, Mycobacterium vanbaalenii	An experimental approach was used to expose bacteria to PAH and metabolite profiles were measured along with protein profiling (proteomics) of the bacteria. Genome databases were also used to identify the genes involved in pyrene metabolism	Kim et al. (2007)
Toxicology	Gene expression profiling and correlation analysis. Modelling thermal injury in laboratory rats	An in silico model using data mined from burn injury studies on rats. Gene profiling identified a number of genes and transcription factors involved in thermal injury. Various cluster analysis methods were used to generate interaction networks of the genes involved in burn-induced inflammation	Yang et al. (2007)
Toxicology	Development of a model to predict the migration and location of individual tumour cells	Tumour cells were placed in 3D and 2D gels and allowed to migrate in response to the environmental conditions in the culture (level of nutrients for example). A computer model predicting the migration, clustering, and death of the cells was generated from the data. This may be used to identify the dynamics of the early stage of tumour formation based on the behaviour of the individual tumour cells	Mansury et al. (2002)

Table 1. (Continued)

Topic area	Type of model and test organism	Notes on the approach or results of the application	Reference
Toxicology	Gene expression and protein quantification. Nerve cells from rats and primates	Ketamine toxicity to receptors in the brain (NMDA receptor) and its application to the systems biology approach is explored. Data are collected on genomics, and key proteins are extracted from samples. Modelling is discussed but not applied to the data	Slikker *et al.* (2007)
Toxicology	Gene expression profiling in carp exposed to whole effluents	Common carp, *Cyprinus carpio*, are exposed to whole effluents for 21 days, microarrays are made and gene expression profiling conducted. The chapter discusses the data in the context of toxicogenomics, and advocates a systems biology approach, but no systems biology modelling is actually applied to the data	Moens *et al.* (2007)

some good attempts have been made at modelling at least some parts of cell function or suggesting diagnostic tools to monitor disease.

However, most of the literature is still focused on the fundamentals of cell biology and modelling parts of the behaviour of cells (*e.g.*, Cusick *et al.*, 2005; Klingmuller *et al.*, 2006; Martelli *et al.*, 2007). The choice of cells for modelling is varied but yeast cells and cells from the nematode *Caenorhabditis elegans* are particularly favoured because the genomes of these organisms have been sequenced (Cusick *et al.*, 2005). However for similar reasons, one might favour human cells given the available data from the Human Genome Project. Studies on the immune system and inflammation reactions are also featured (Aderem and Smith, 2004; Yang *et al.*, 2007). This is particularly important because of the central role immunity plays in health, but it is also of interest to the toxicologist.

Data modelling techniques

One feature that separates the discipline of systems biology from experiments that are simply reporting genomics, proteomics, metabolomics, or informatics is the use of computational methods that compare differences and similarities between multiple data sets, or use simple linear logic-based rules to generate networks. Central to the systems biology approach is the use of multiple data sets to quantitatively study complexity. Thus, for example, a scientific paper that reports some genomic data and also some metabolomic data (even if the individual data sets have been statistically analysed), but does not model the combined effects or differences of the data is not a paper on systems biology. Clearly, there are a large number of quality peer-reviewed publications that fit this description. No doubt it is just a question of time before these data sets are used in the systems biology approach.

The computational methods generally fall into two major themes, either multi-variate statistical methods for looking at differences/similarities in data, or binary or logic-based computational methods for generating networks. The latter essentially involves a series of simple decisions at each node in a network before moving onto the next node. These rules are used to build the computer-generated image of the network. A simple analogy to this approach might be the operation of a telephone system where switches are open or closed at each node in the network, or perhaps a decision tree used in taxonomy. The details of the equations and individual models are discussed elsewhere (Kremling and Saez-Rodriguez, 2007; Materi and Wishart, 2007) along with the software tools developed to provide a user interface for researchers wanting to process their data using a systems biology approach (Alves *et al.*, 2006; Kremling and Saez-Rodriguez, 2007). Table 2 summarises some of the computational methods that have been used in systems biology.

Ordinary differential equations (ODEs) are widely used in life sciences, including toxicology (*e.g.*, text books on biostatistics or quantitative

Table 2. Some computational methods used in systems biology.

Method or approach	Description	Notes
Ordinary differential equations (ODE)	Used to resolve a series of rate equations, and will generate graphs of the output	Already used in pharmacology and toxicology to describe kinetics. Well established and robust
Stochastic differential equations	More complex than ODE in that additional equations and random number generators are included	Works well when replication is low, and can be used for temporal modelling
Power law equations	Used mainly to simplify non-linear ODEs or to generate simple linear equations	Enables simplification, but the risk is loss of detail
Partial differential equations	Used to express temporal or spatial patterns. Gives a well-understood numeric output	The approach is sometimes used in epidemiology and may have some relevance to time/space-related incidence of chemical exposure
Petri nets	Often simple logic-based rules, to generate network connections	Simple non-mathematical idea, but assumes simple linear events
Cellular automata (CA) and dynamic cellular automata (DCA)	Similar to Petri nets in that logic-based rules are applied. Pairwise interactions based on Boolean logic are often used. DCA enables both space and time effects to be modelled	Used in biology for many years to describe cell–cell interactions

Source: Modified from Materi and Wishart (2007).

pharmacology) and are probably one of the most common computational methods used in systems biology (*e.g.*, Kitano, 2002; Noble, 2006).
One main disadvantage of the use of differential equations is the need for
input of precise values. So for example, to model the uptake and
metabolism of chemical then precise information on concentrations in
different tissues, reaction rate constants, diffusion rates, etc., would be
needed. Clearly, this precise level of detail is not always available for
chemicals.

Petri nets are more applicable to time-dependent processes such as
flow systems (and were originally used to model manufacturing processes
in industry). Although these have not been used in chemical toxicology, it
may be possible to apply the original use of Petri nets to modelling
chemical effluents from manufacturing (perhaps as part of direct toxicity
assessment (DTA)), as well as for deriving networks from metabolite
profiling (Materi and Wishart, 2007). Interestingly, Moens *et al.* (2007)
have applied genomics to whole effluents effects on fish.

Cellular automata (CA) are computer simulation tools which involve a
grid or lattice with the components (which could be cells, enzymes, or
reactants) located within the grid. The nearest neighbours sites in the grid
will have the biggest effect on each other. The advantage of the method is
that it is non-mathematical and works on simple decision rules, and
therefore suggests that the data does not need to be as precisely
quantified as in ODE.

Software packages

These have recently been summarised by Materi and Wishart (2007) from
the viewpoint of application to drug discovery and the reader is referred
to their excellent discussion of these packages. Some of the common
packages and web links are listed below, along with their main
computational method.

- Cell Designer, http://www.celldesigner.org/index.html (differential
  equation software)
- CellWare, http://www.cellware.org (differential equation software)
- Dynetica, http://www.duke.edu/~you/Dynetica_page.htm (differential
  equation software)
- E-Cell, http://www.e-cell.org/ (differential equation software)
- Gepasi, http://www.gepasi.org/ (differential equation software)
- SmartCell, http://smartcell.embl.de/ (differential equation software)

- Vcell, http://www.vcell.org (differential equation and partial differential equation software)
- Snoopy, http://www-dssz.informatik.tu-cottbus.de/index.html?/software/snoopy.html (Petri net software)
- CPN Tools, http://wiki.daimi.au.dk/cpntools/cpntools.wiki (Petri net software)
- CancerSim, http://www.cs.unm.edu/~forrest/software/cancersim/ (cellular automata software)
- Mcell, http://www.mcell.cnl.salk.edu/ (dynamic cellular automata software)
- SimCell, http://wishart.biology.ualberta.ca/SimCell/ (dynamic cellular automata software).

In addition to these software tools, some government departments and national level research programmes are starting to explore the use of 'Systems Toxicology'. Some recent examples (not exhaustive, website addresses valid at the time of writing, 24 November 2007) are listed below:

- The US Food and Drug Administration has recently established a "Systems Toxicology" division, directed by Dr Yvonne P. Dragan, http://www.fda.gov/nctr/science/divisions/systemstoxicology.htm
- The UK's Medical Research Council is establishing a systems toxicology research programme, http://143.210.176.81/Systems Toxicology/Home.aspx
- Link to a company called Genedata that are starting to produce software for applications in 'Systems Toxicology', http://www.genedata.com/solutions/systems_toxicology/index_eng.html.

The application of systems toxicology to pollutants

Systems toxicology is a newly emerging field, and the toxicological applications so far have been in biomedical toxicology looking at aspects of metabolism during disease states, or drug toxicity (Table 1). A few of these studies have incidentally used organic chemicals, which also happen to be pollutants, as tools in metabolic studies (e.g., polyaromatic hydrocarbons, PAHs; Kim et al., 2007). Unfortunately, examples where exposure to chemicals or the toxicity of pollutants has been the prime application for systems toxicology appear to be lacking. However, the toxicology community is very close to applying systems toxicology. There are detailed reports of how individual chemicals affect cells and the mode of action within cells (*e.g.*, cadmium; Martelli *et al.*, 2006), or details of metabolic pathways (*e.g.*, chloropene; Munter *et al.*, 2007). Several papers describing genomic data or other 'omics data argue that the next step is to apply computational systems biology to the data collected (*e.g.*, Moens *et al.*, 2007; Slikker *et al.*, 2007).

There are reviews on how the 'omics and other data on chemicals might be applied in a 'systems toxicology' version of the systems biology approach (Heijne *et al.*, 2005; Simon-Hettich *et al.*, 2006; Slikker *et al.*, 2007; Waters and Fostel, 2004; Waters and Yauk, 2007; Waters *et al.*, 2003). These reviews consider mainly pharmacology or biomedical toxicology. Nonetheless, we can also consider environmental toxicology and the effects of pollutants (Fig. 2). The first step is to consider what quantitative data needs to be used as the input to the modelling process. In the case of pollution, this could include data from the effects of pollutants on genomes, proteomes, or metabolomes ('omics data), but we must not forget more traditional sources of data that are of vital importance. For example, absorption, distribution, excretion, and metabolism (ADME) are fundamental to disposition and input modules could describe each of these facets for each chemical and/or organism. Suites of data describing biochemical or physiological variables in the organisms would also be useful. This could include enzyme activities in different organs and how they change (increase or decrease) in the presence of individual pollutants, or blood parameters such as plasma ion concentrations or haemoglobin levels, etc. However, we must be careful not to forget the environmental chemistry. Important inputs would be information on exposure profiles, concentrations and chemical speciation, abiotic factors in the environment such as pH, salinity, dissolved organic matter, dissolved oxygen, temperature, and so on, which are well known for their effects on toxicity. The environmental data could also include ecosystem information. This would be particularly important if the outputs were going to be used to define ecosystem effects, or be used to make management decisions about conservation of polluted ecosystems or habitat types. Clearly, much of the above input data is available, but at present is dispersed in the scientific literature and is not yet in database format. Nonetheless, the data could be prepared as input. The subsequent quantitative computer modelling could employ the main techniques used in systems biology (differential equations, Petri nets, cellular automata; Fig. 2) to generate the model outputs. Clearly, these outputs would ideally be information on complex processes that we have not been able to solve with traditional

268

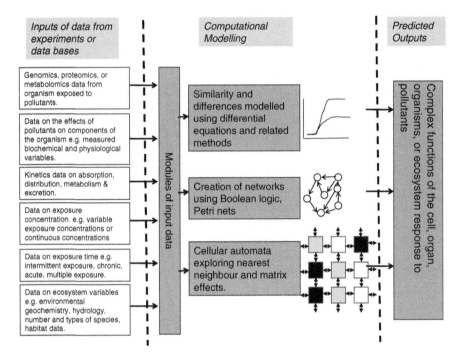

Fig. 2. The main steps in the systems toxicology approach. Biological and chemical data sets can be the data input. These data may need to be packaged into independent 'modules' of information or simple event (*e.g.*, uptake rate data) with some modules giving time information (*e.g.*, exposure profile) and spatial information (*e.g.*, target organ location in the organism, location of organism in relation to the pollutants). Some of the main modelling approaches are indicated which explore similarities/differences, relationships and effects of different modules or other components on the output. The output will be descriptions of complex events and complex biological responses to chemicals. In theory, the outputs could be a description of how a cell, organ, organism, or even an ecosystem, responds to pollutants. The outputs could be re-introduced as inputs to refine the model further.

methods such as the effects of mixtures or complex exposure patterns. However, these outputs can also be at different levels of biological organisation from the cell right through to the ecosystem level.

Ensuring ecological considerations in systems ecotoxicology

Systems toxicology clearly has its origins in cell and molecular toxicology, and yet successful pollution control measures are often monitored at the ecosystem level in terms of biodiversity, habit type and quality, population growth rates of the organisms in the ecosystem, or the reproductive success of sensitive or rare species in the ecosystem (*e.g.*, Schäfer *et al.*, 2007). The scientific community needs to ensure that the 'eco' is included in the application of systems toxicology to environmental pollution, *i.e.*, that we are doing 'Systems Ecotoxicology'. While data on environmental chemistry (pH, salinity, pollutant concentrations, etc.) can be input to a computer model, this alone will not add ecological relevance. For example, we can easily explain water quality effects on metal toxicity at the organism level, and make organism level models (*e.g.*, biotic ligand models; Hollis *et al.*, 2000), but our goal for systems ecotoxicology is to make a systems-based ecologically relevant model that also explains the effects of pollution. A systems model should also require input data that are relevant to the higher tiers of the ecological risk assessment process such as data on species sensitivity distributions, biodiversity, and quantitative data that describes habitat type (*e.g.*, soil types in an agricultural landscape, percentage of tree cover, and species of trees in a forested area) as well as population dynamics. In addition, if exposure is going to be linked to ecosystem effect, then detailed information on the fate and behaviour of the pollutants in the ecosystem will need to be linked with the precise location(s) of the organisms in the ecosystem (*e.g.*, GPS positions), their ecological niche and function in the ecosystem, as well as temporal/spatial changes in ecosystem quality. In theory, all of these data could be used as input to a computer model. Furthermore, much of these data may already exist. For example, ecological risk assessments and catchment management plans (*e.g.*, Schäfer *et al.*, 2007) might include much of the exposure data for an ecosystem, while conservation bodies (*e.g.*, in the UK, English Nature) and statutory ecological monitoring programmes could provide habitat and population data. It is just a question of data management, data quality assurance, and the willingness of different agencies to provide compatible data sets. Of course, there will always be the need to collect new up-to-date information on the specific ecosystem being protected, but this applies to any approach, not just systems toxicology.

The central question remains as to whether systems toxicology and related approaches can yield ecological relevant information that cannot be more easily obtained with traditional ecological methods. Genomics and the use of microarray technology can sometimes identify environmental problems that are associated with specific genes. Thomas and Klaper (2004) argue that gene hunting can identify specific traits (*e.g.*, genes controlling food preferences in pest insects on crops), and this information might then be used to manage the specific environmental problem associated with that gene. Procaccini *et al.* (2007) suggest that monitoring genetic diversity in wild populations is crucial, since genetic diversity is fundamental to population growth and survival. Gene chips can also be part of a package of techniques used to provide the supporting data for risk assessment (Robbens *et al.*, 2007). Alternatively, Kammenga *et al.* (2007) indicate caution because genome-wide expression profiles can be greatly influenced by confounding factors in the environment, and microarray technology is often validated in rather standard and fixed conditions in the laboratory. The latter point highlights a general concern about methodology for microarrays and the 'omics in that responses are often measured from pooled individuals, and yet it is the inter-individual variability that provides the phenotype plasticity that enables some organisms in a population to survive pollution (*i.e.*, sensitive and tolerant organisms within a single population). It is therefore important to consider this in statistical analysis of microarray data (Kammenga *et al.*, 2007), so that inter-individual variability can be linked to population survival.

However, the above examples of using microarray technology to tackle environmental issues are not necessarily adding ecological relevance to a systems ecotoxicology approach. Pollino *et al.* (2007) showed that Bayesian theory can be used to 'parameterise' ecological information for modelling, and give the example of modelling a fisheries catchment. Sulkava *et al.* (2007) also indicate that variability in environmental monitoring data could be used in ecological informatics software to detect long-term trends or patterns in pollution. This illustrates that we do have tools to add ecologically relevant information. Clearly, the scientific community is on the very threshold of using systems toxicology with data on the biological structures and toxic effects of pollutants, and there seems to be no theoretical reasons why this approach could not be applied with strong consideration of the ecology.

Practical benefits for the regulatory agencies involved in the control of pollution

At the beginning of this new field, it is worth asking what practical benefits would systems toxicology bring to the assessment of pollution. What are the complex problems in risk assessment of chemicals or pollution control that we cannot resolve, but might be moved forward by the systems toxicology approach? Second, even if we could identify aspects where systems toxicology would be useful, would this be scientifically more robust, more cost-effective, or practicable than existing risk assessment approaches?

If the issue of complexity is considered, there are many aspects in the assessment of pollutants/chemicals that show complexity that are not easily resolved with current hazard or risk assessment methods (Table 3). These include information on environmental concentrations (fate and behaviour of chemicals) in terms of spatial or temporal patterns of distribution, or the effects of a myriad of abiotic (*e.g.*, pH, salinity, hardness, type of organic matter) and biotic factors (age, sex, target organs, adaptation, acclimation, exposure history, etc.) on bioavailability. The toxicity of mixtures and intermittent or pulsed exposure to chemicals is also an area of complexity where systems biology might be applied. For example, consider the issue of toxicity of mixtures of chemicals. Apart from the DTA or whole effluent toxicity assessment approach which tests the effects of real industrial effluents on organisms (Hutchings *et al.*, 2004; Wharfe *et al.*, 2004), regulatory testing has not really resolved this long-standing ecotoxicological problem of mixtures/ temporal exposure regimes. However, there are large amounts of information on the toxicity of individual chemicals, and some traditional experimental data on mixtures in the literature (*e.g.*, McCarty and Borgert, 2006) or pulsed exposures to chemicals (Ashauer *et al.*, 2006; Handy, 1994). If some genomic, proteomic, or metabolomic data can be generated from organisms exposed to mixtures of chemicals, then perhaps it could be possible to apply the systems biology approach to build a network describing the interactions of the components of the mixture. Clearly these are only ideas, but it does illustrate that there are complex and long-standing problems in the assessment of chemical pollutants that a fresh approach such as systems toxicology might help to resolve.

Barriers to the application of systems toxicology

There are a number of practical and technical issues to resolve before systems biology/toxicology can be applied to some of the complex problems outlined in Table 3. These include:

(i) Proof of principle experiments with different chemicals or pollutants. Before there is significant investment of resource in this new area for environmental protection. It would be useful to see some key research published (experimental work at the

Table 3. Potential issues where a systems toxicology approach could help assess the complex effects of pollutants.

Problem	How systems biology/toxicology could help
1. Predicting chemical concentrations and bioavailability at specific locations in the environment	Chemical concentrations are a function of numerous abiotic factors such as the type of environmental matrix, volume flow of water through the site, ligands present, pH, temperature, particle sizes and adsorption processes, etc. Also biotic factors such as processing of chemicals by microorganisms, biodiversity and habitat type, etc. Perhaps differential equation software can be used to network the effect of these many parameters on chemical concentrations
2. Spatial resolution of chemicals in single habitat types	This may be particularly useful for agrochemicals applied over large areas of single crops, or for the fate and behaviour of chemicals in different types of crops in adjacent fields. DCA-based software could be used to describe the interactions between plots of land and give detailed topographical maps of chemical concentration over large areas of agriculture land
3. Mixtures of chemicals in the environment	Similar to problem 1 above, except that different chemicals rather than biotic and abiotic factors would be the main focus of the analysis
4. Absorption, distribution, and metabolism of chemicals by organisms	All of the current applications of systems biology currently used in drug discovery could equally be applied to chemicals
5. Toxicity of mixtures	Current regulatory toxicity tests are generally not designed for mixtures. Data from single exposure tests and from genomic studies on multiple exposure may be merged to network mixture effects on organisms

Table 3. (*Continued*)

Problem	How systems biology/toxicology could help
6. Toxic effects of intermittent or variable pollution exposure	Time-dependent changes in toxicity or exposure concentration, and combination of both might be modelled with Petri net or other dynamic software tools
7. Epidemiology of acutely toxic chemicals or radiations and public health	The spread of contamination by person-to-person contact in densely populated areas could be modelled using DCA. At a different level of biological organisation this could equally be applied to the spread of chemicals between cells within an organ

bench), which demonstrates the use of the systems toxicology approach for specific groups of chemicals, such as toxic metals, pesticides, and persistent organic pollutants. This would focus on the computational biology of multiple data sets generated in such experiments. This would build confidence in the application for chemicals, before dealing with the issues outlined below.

(ii) Data quality in chemical and toxicological databases. Differential equations can be sensitive to small errors in the inputed data. It is therefore vital that there is some unified quality assurance procedure on the content of such databases at national level, and across the European Union, to ensure that the original data is correct; and that information in different databases are compatible to do multi-variate analysis. For the establishment of databases on toxicogenomics and related 'omics for toxicology, the chemicals risk assessment community and ecotoxicologists, should start by adopting the existing standards for such 'omic databases. In particular for the standardisation, reporting and management of data from toxicogenomics (Baken *et al.*, 2007). It would also be useful to develop bioinformatic databases for pathological or toxic states. For example, all the existing protein structure databases rely mainly on modelling the structure of the protein from the known or predicted amino acid sequence of the normal protein (*e.g.*, Swissprot, http://www. ebi.ac.uk/swissprot/). A database would be useful for denatured or damaged proteins, so that this information can be put into the systems biology approach for toxicological conditions. This would of course require some solution chemistry parameters (*e.g.*, charge screening of proteins with ions) to add to such databases, but such parameters are largely ignored at present.

(iii) Allow some time for the scientific community to build up the relevant genomic, proteomic, and metabolomic data sets. These tools have only recently been applied to the toxicity of substances, with the exception of genomics (*e.g.*, Craig *et al.*, 2006; Munter *et al.*, 2007). The published data for the toxicity of chemicals are not nearly as diverse as those from the cell biology community, and more bench work on the 'omics using chemical pollutants is needed to provide the data to be used in systems toxicology. These data sets should include molecular information on the inter-individual variability of test organisms (*e.g.*, polymorphisms in mRNA expression), since this variability is central to both the evolution of resistance to chemicals, and the survival of wild populations.

(iv) Custom software tools for chemicals. The current software tools outlined above are mainly designed for cell biologists or molecular biologists. The user interface on some of these tools should be redeveloped using the relevant terminology and menus appropriate to toxicology and chemistry. At least one group has recognised this issue, in terms of ensuring data quality input into software tools for chemical toxicology (Linden *et al.*, 2007).

Conclusions and recommendations

Overall systems biology, and its application to toxicology as systems toxicology is a scientific discipline in its infancy that has not been widely applied to the assessment of chemicals or the effects of environmental pollution. However we can at least identify several complex pollution problems that, despite decades of research, have not been satisfactorily resolved using traditional chemical or biological methods. These tend to be topic areas where complexity is the key problem to moving forward, such as temporal or spatial patterns of exposure, mixtures and intermittent exposure, or fine resolution of exposure or multiple toxic effects. There is also considerable scope for developing new (non-intuitive) predictive tools for existing chemicals. Systems toxicology is therefore worth some investment of time and resources by Government departments, national research councils, and European agencies involved in chemicals risk assessment, etc. Systems biology/toxicology should be regarded as an additional tool, not as a replacement for current risk and hazard assessment procedures. The key practical issues to overcome are outlined above, and the specific recommendations below help to partly resolve some of these problems:

(i) Proof of principle studies should be conducted with different groups of pollutants or chemicals. These studies should include original work at the bench so that data quality for the whole experiment and subsequent application of systems toxicology can be monitored and managed. Chemicals could include examples of toxic metals, pesticides, and persistent organic pollutants such as PAHs. These should ideally be conducted with organisms where the genomes have been sequenced, but that are also relevant test organisms for toxicology. Similarly, well-characterised environmental matrices should be used for experiments that model fate and behaviour in the environment in combination with toxic effects.

(ii) Computational software should be developed, or evolved from the cell biology applications, for the chemistry and toxicology community. This would be a prerequisite of the wider application of systems biology in chemical toxicology in the long term.

(iii) The amount of 'omics data, and data available to input into the systems biology approach specifically for chemicals is relatively small. This will grow as toxicologists and chemists in academia or industry do research in this new field. However, funding is needed for scientific networks or workshops to bring these scientists together. Importantly this should involve computer scientists with an information technology capability to build new databases.

(iv) A review of the technical quality and quality assurance of existing data sets on chemicals should be sought, so that minimum standards can be set for the use of these data in systems toxicology. The molecular toxicologists and cell biologists working in this field may not know about all the data sets available on chemicals and awareness of these data sets, and how to access them, will need to be in place.

276

Appendix 1: Glossary of terms

Bioinformatics	The management and analysis of biological data using advanced computational techniques
Boolean logic	The mathematics of logic, first described by George Boole in the 19th century. Boolean logic is an algebra of two components (true or false), and is used in the construction of decision trees and is the basis of binary code (0, 1) used in computers
Bottom-up model	An approach in computational modelling where the model is constructed from the basic components of the biological process (the modules) and the effects of inputs to each module. Hence, the bottom-up approach tends to use modules to construct networks, which are in turn used to construct overall functions in cells
Cellular automata (CA)	A computational approach used in systems biology using pairwise logic to define interactions between neighbouring bodies. Software tools show a grid or lattice with the components (which could be anything defined by the user such as cells, enzymes, or reactants) located within the grid. The software calculates the effect of neighbouring components on each other so that a pattern across the grid emerges. The approach can also include a temporal component, and the latter is called *Dynamic Cellular Automata*
Control centre	Inputs need to be processed into outputs in biological systems. A control centre or processing centre is needed to do this. Control needs to be defined in computational models, but examples of biological structures acting as control centres might be the nucleus of a cell, or the brain in physiological control. Control centres monitor the inputs and outputs using various types of feedback loops
Genomics	The science of studying DNA sequences and properties of entire genomes. Genomics is used to investigate the responses of genomes to toxicants and is called *toxicogenomics*
Input	A term used in computational modelling to describe the information or starting data put into a part of a model, or generically for the whole model. In systems biology the input is a biological stimulus such as an electrical or chemical signal from another cell, extracellular fluid (*e.g.*, chemical in the blood), or the external environment
Interactomes	A term used in computational modelling to describe a collection of interactions or a network of interactions. This term is often used in the context of proteins to describe collections of protein–protein interactions within cells
Metabolomics	The study of patterns, and pattern recognition in metabolism. Metabolomics often involves the simultaneous profiling of

Appendix 1. (*Continued*)

	numerous metabolites in a single biological sample. Names are sometimes given to different groups of metabolites. So for example, the profiling of lipid metabolites is sometimes called 'Lipidomics'
Middle-out model	A combination of both the bottom-up and top-down approaches to computational modelling
Module	The basic units of a computational model may consist of modules. The module is a defined biological event that can be characterised, such as a simple biochemical pathway, a simple signalling pathway, a structure in a cell, or a module could identify a location or periodicity of an event (*i.e.*, modules defining the space and time of a simple biological event, rather than the event itself)
Network	A term used in computational modelling to describe the inter-relationships between a series of modules. For example, a network could describe the overall biological function of a collection of biochemical reactions or the combined effects of several signalling pathways in a cell
Output	A term used in computational modelling to describe the information or data generated from part of a model such as the output from a network or module, or generically for the whole model. In systems biology the output is a biological response. This could be something simple such as a reaction product or enzyme induction in a module describing the output from a biochemical reaction series, or the output from a network of events in the cell (functionality of more complex processes in the cell, *e.g.*, cell volume control), or the overall behaviour of a cell
Petri nets	A computational method in systems biology based on simple logic-based rules that are used to generate network connections. Petri nets can be graphically presented as bipartite graph or plot with nodes represented by circles, and transitions between nodes by rectangles
Proteomics	The study of patterns of protein expression, and the properties or complex interactions between multiple groups of proteins
Receptor or detector	The biological structure detecting the input signal. This could be a molecular receptor on a cell membrane for example, or a sensory nerve ending in a physiological control system
Systems biology	The quantitative study of the control and operation of biological processes that uses computational tools to study the complexity of biological events, mainly at the cellular level. However, the systems biology concept could be applied to the control and function of an organ, or even an organism

Appendix 1. (*Continued*)

Systems toxicology	Similar to systems biology, except the biological processes are toxicological events
Top-down model	An approach in computational modelling where observations are made on outputs (the biological responses) of cells, organs, or physiological processes. The approach uses statistical techniques to identify related and dissimilar components in the output and therefore, by logical deduction, to construct at network that would explain the observed output

References

Aderem, A. and Smith, K. D. (2004). A system approach to dissecting immunity and inflammation. *Semin. Immunol.* 16,55–67.

Alves, R., Antunes, F. and Salvador, A. (2006). Tools for kinetic modeling of biochemical networks. *Nat. Biotechnol.* 24,667–672.

Ashauer, R., Boxall, A. and Brown, C. (2006). Predicting effects on aquatic organisms from fluctuating or pulsed exposure to pesticides. *Environ. Toxicol. Chem.* 25,1899–1912.

Baken, K. A., Vandebriel, R. J., Pennings, J. L. A., Kleinjans, J. C. and van Loveren, H. (2007). Toxicogenomics in the assessment of immunotoxicity. *Methods* 41,132–141.

Barrett, C. L., Kim, T. Y., Kim, H. U., Palsson, B. O. and Lee, S. Y. (2006). Systems biology as a foundation for genome-scale synthetic biology. *Curr. Opin. Biotechnol.* 17,488–492.

Brain, R. A., Sanderson, H., Sibley, P. K. and Solomon, K. R. (2006). Probabilistic ecological hazard assessment: Evaluating pharmaceutical effects on acquatic higher plants as an example. *Ecotoxicology and Environmental Safety* 64,128–135.

Craig, A., Sidaway, J., Holmes, E., Orton, T., Jackson, D., Rowlinson, R., Nickson, J., Tonge, R., Wilson, I. and Nicholson, J. (2006). Systems toxicology: Integrated genomic, proteomic and metabonomic analysis of methapyrilene induced hepatotoxicity in the rat. *J. Proteome Res.* 5,1586–1601.

Cusick, M. E., Klitgord, N., Vidal, M. and Hill, D. E. (2005). Interactome: Gateway into systems biology. *Hum. Mol. Genet.* 14,R171–R181.

Dalzell, D. J. B., Alte, S., Aspichueta, E., de la Sota, A., Etxebarria, J., Gutierrez, M., Hoffmann, C. C., Sales, D., Obst, U. and Christofi, N. (2002). A comparison of five rapid direct toxicity assessment methods to determine toxicity of pollutants to activated sludge. *Chemosphere* 47,535–545.

Doull, J., Borzelleca, J. F., Becker, R., Daston, G., DeSesso, J., Fan, A., Fenner-Crisp, P., Holsapple, M., Holson, J., Llewellyn, G. C., MacGregor, J., Seed, J., Walls, I., Woo, Y. T. and Olin, S. (2007). Framework for use of toxicity screening tools in context-based decision-making. *Food Chem. Toxicol.* 45,759–796.

Galloway, T. and Handy, R. (2003). Immunotoxicity of organophosphorus pesticides. *Ecotoxicology* 12,345–363.

Handy, R. D. (1994). Intermittent exposure to aquatic pollutants-assessment, toxicity and sub-lethal responses in fish and invertebrates. *Comp. Biochem. Physiol. C* 107C,171–184.

Handy, R. D., Galloway, T. S. and Depledge, M. H. (2003). A proposal for the use of biomarkers for the assessment of chronic pollution and in regulatory toxicology. *Ecotoxicology* 12,331–343.

Handy, R. D., Jha, A. N. and Depledge, M. H. (2002). Biomarker approaches for ecotoxicological biomonitoring at different levels of biological organisation. In *Handbook of Environmental Monitoring* (eds F. Burden, I. McKelvie, U. Förstner and A. Guenther), pp. 9.1–9.32, McGraw-Hill, New York.

Heijne, W. H. M., Kienhuis, A. S., van Ommen, B., Stierum, R. H. and Groten, J. P. (2005). Systems toxicology: Applications of toxicogenomics, transcriptomics, proteomics and metabolomics in toxicology. *Expert Rev. Proteomics* 2, 767–780.

Hollis, L., McGeer, J. C., McDonald, D. G. and Wood, C. M. (2000). Effects of long term sublethal Cd exposure in rainbow trout during soft water exposure: Implications for biotic ligand modelling. *Aquat. Toxicol.* 51,93–105.

Hoyle, I. and Handy, R. D. (2005). Dose-dependent inorganic mercury absorption by isolated perfused intestine of rainbow trout, *Oncorhynchus mykiss*, involves both amiloride-sensitive and energy-dependent pathways. *Aquat. Toxicol.* 72, 147–159.

Hutchings, M., Johnson, I., Hayes, E., Girling, A. E., Thain, J., Thomas, K., Benstead, R., Whale, G., Wordon, J., Maddox, R. and Chown, P. (2004). Toxicity reduction evaluation, toxicity identification evaluation and toxicity tracking in direct toxicity assessment. *Ecotoxicology* 13,475–484.

Ivakhno, S. (2007). From functional genomics to systems biology – meeting report based on the presentations at the 3rd EMBL Biennial Symposium 2006 (Heidelberg, Germany). *FEBS J.* 274,2439–2448.

Kammenga, J. E., Herman, M. A., Ouborg, N. J., Johnson, L. and Breitling, R. (2007). Microarray challenges in ecology. *Trends Ecol. Evol.* 22,273–279.

Kim, S. J., Kweon, O., Jones, R. C., Freeman, J. P., Edmondson, R. D. and Cerniglia, C. E. (2007). Complete and integrated pyrene degradation pathway in *Mycobacterium vanbaalenii PYR-1* based on systems biology. *J. Bacteriol.* 189,464–472.

Kitano, H. (2002). Systems biology: A brief overview. *Science* 295,1662–1664.

Klingmuller, U., Bauer, A., Bohl, S., Nickel, P. J., Breitkopf, K., Dooley, S., Zellmer, S., Kern, C., Merfort, I., Sparna, T., Donauer, J., Walz, G., Geyer, M., Kreutz, C., Hermes, M., Gotschel, F., Hecht, A., Walter, D., Egger, L., Neubert, K., Borner, C., Brulport, M., Schormann, W., Sauer, C., Baumann, F., Preiss, R., MacNelly, S., Godoy, P., Wiercinska, E., Ciuclan, L., Edelmann, J., Zeilinger, K., Heinrich, M., Zanger, U. M., Gebhardt, R., Maiwald, T., Heinrich, R., Timmer, J., von Weizsacker, F. and Hengstler, J. G. (2006). Primary mouse hepatocytes for systems biology approaches: A standardized in vitro system for modelling of signal transduction pathways. *IEE Proc. Syst. Biol.* 153,433–447.

Kremling, A. and Saez-Rodriguez, J. (2007). Systems biology – an engineering perspective. *J. Biotechnol.* 129,329–351.

280

Laaksonen, R., Katajamaa, M., Päivä, H., Sysi-Aho, M., Saarinen, L., Päivi, J., Lütjohann, D., Smet, J., Van Coster, R., Seppänen-Laakso, T., Lehtimäki, T., Soini, J. and Orešič, M. (2006). A systems biology strategy reveals biological pathways and plasma biomarker candidates for potentially toxic statin-induced changes in muscle. *PLoS ONE* 1(1),e97. doi: 10.1371/journal.pone.0000097

Linden, R., Sartori, S., Kellermann, E. and Souto, A. A. (2007). Substance identification in systematic toxicological analysis using a computer system for chromatographic parameter calculation and database retrieval. *Quim. Nova* 30,468–475.

Mansury, Y., Kimura, M., Lobo, J. and Deisboeck, T. S. (2002). Emerging patterns in tumor systems: Simulating the dynamics of multicellular clusters with an agent-based spatial agglomeration model. *J. Theor. Biol.* 219,343–370.

Martelli, A., Rousselet, E., Dycke, C., Bouron, A. and Moulis, J. M. (2006). Cadmium toxicity in animal cells by interference with essential metals. *Biochimie* 88,1807–1814.

Martelli, C., Giansanti, A., Arisi, I. and Rosato, V. (2007). Asymptotic states and topological structure of an activation–deactivation chemical network. *J. Theor. Biol.* 245,423–432.

Materi, W. and Wishart, D. S. (2007). Computational systems biology in drug discovery and development: Methods and applications. *Drug Discov. Today* 12,295–303.

Mazzatorta, P., Tran, L. A., Schilter, B. and Grigorov, M. (2007). Integration of structure-activity relationship and artificial intelligence systems to improve in silico prediction of Ames test mutagenicity. *J. Chem. Inf. Model.* 47,34–38.

McCarty, L. S. and Borgert, C. J. (2006). Review of the toxicity of chemical mixtures: Theory, policy, and regulatory practice. *Regul. Toxicol. Pharmacol.* 45,119–143.

Moens, L. N., Smolders, R., van der Ven, K., van Remortel, P., Del-Favero, J. and De Coen, W. M. (2007). Effluent impact assessment using microarray-based analysis in common carp: A systems toxicology approach. *Chemosphere* 67,2293–2304.

Moore, D. F., Gelderman, M. P., Ferreira, P. A., Fuhrmann, S. R., Yi, H. Q., Elkahloun, A., Lix, L. M., Brady, R. O., Schiffmann, R. and Goldin, E. (2007). Genomic abnormalities of the murine model of Fabry disease after disease-related perturbation, a systems biology approach. *Proc. Natl. Acad. Sci. USA* 104,8065–8070.

Munter, T., Cottrell, L., Ghai, R., Golding, B. T. and Watson, W. P. (2007). The metabolism and molecular toxicology of chloroprene. *Chem. Biol. Interact.* 166,323–331.

Noble, D. (2002). The rise of computational biology. *Nat. Rev. Mol. Cell. Biol.* 3,460–463.

Noble, D. (2006). Systems biology and the heart. *Biosystems* 83,75–80.

Pollino, C. A., Woodberry, O., Nicholson, A., Korb, K. and Hart, B. T. (2007). Parameterisation and evaluation of a Bayesian network for use in an ecological risk assessment. *Environ. Modell. Software* 22,1140–1152.

Procaccini, G., Olsen, J. L. and Reusch, T. B. H. (2007). Contribution of genetics and genomics to seagrass biology and conservation. *J. Exp. Mar. Biol. Ecol.* 350,234–259.

Randall, D., Burggren, W. and French, K. (2002). *Eckert's Animal Physiology*, 5th Edition, W. H. Freeman, New York.

Robbens, J., van der Ven, K., Maras, M., Blust, R. and De Coen, W. (2007). Ecotoxicological risk assessment using DNA chips and cellular reporters. *Trends Biotechnol.* 25,460–466.

281

Schäfer, R. B., Caquet, T., Siimes, K., Mueller, R., Lagadic, L. and Liess, M. (2007). Effects of pesticides on community structure and ecosystem functions in agricultural streams of three biogeographical regions in Europe. *Sci. Total Environ.* 382,272–285.

Simon-Hettich, B., Rothfuss, A. and Steger-Hartmann, T. (2006). Use of computer-assisted prediction of toxic effects of chemical substances. *Toxicology* 224,156–162.

Slikker, W., Paule, M. G., Wright, L. K. M., Patterson, T. A. and Wang, C. (2007). Systems biology approaches for toxicology. *J. Appl. Toxicol.* 27,201–217.

Sulkava, M., Luyssaert, S., Rautio, P., Janssens, I. A. and Hollmen, J. (2007). Modelling the effects of varying data quality on trend detection in environmental monitoring. *Ecol. Inform.* 2,167–176.

Thomas, M. A. and Klaper, R. (2004). Genomics for the ecological toolbox. *Trends Ecol. Evol.* 19,439–444.

Waters, M., Boorman, G., Bushel, P., Cunningham, M., Irwin, R., Merrick, A., Olden, K., Paules, R., Selkirk, J., Stasiewicz, S., Weis, B., Van Houten, B., Walker, N. and Tennant, R. (2003). Systems toxicology and the chemical effects in biological systems (CEBS) knowledge base. *Environ. Health Perspect.* 111,811–824.

Waters, M. D. and Fostel, J. M. (2004). Toxicogenomics and systems toxicology: Aims and prospects. *Nat. Rev. Genet.* 5,936–948.

Waters, M. and Yauk, C. (2007). Consensus recommendations to promote and advance predictive systems toxicology and toxicogenomics. *Environ. Mol. Mutagen.* 48,400–403.

Wharfe, J., Tinsley, D. and Crane, M. (2004). Managing complex mixtures of chemicals – a forward look from the regulators' perspective. *Ecotoxicology* 13,485–492.

Wolkenhauer, O. (2001). Systems biology: The reincarnation of systems theory applied in biology? *Brief. Bioinform.* 2,258–270.

Wolkenhauer, O. and Hofmeyr, J. H. S. (2007). An abstract cell model that describes the self-organization of cell function in living systems. *J. Theor. Biol.* 246,461–476.

Yang, E., Maguire, T., Yarmush, M. L., Berthiaume, F. and Androulakis, I. P. (2007). Bioinformatics analysis of the early inflammatory response in a rat thermal injury model. *BMC Bioinformatics* 8(10). Available at http://www.biomedcentral.com/1471-2105/8/10

Index of authors

Gates, J.L. 89
Gavaghan, C.L. 32, 44, 47, 151
Ge, W. 4
Gebhardt, R. 260, 263
Gedamu, L. 91
Geisler, R. 84, 108
Gelderman, M.P. 258–259
Gensberg, K. 33, 48, 84, 144
George, E. 101
George, S.G. 84, 92, 95, 104, 108
Georgiev, O. 139, 145–146
Gerrie, E.R. 84–85, 95, 104, 108, 111
Gerritsen, A. 87
Gerwick, L.G. 51, 96
Gevaert, K. 27
Geyer, M. 260, 263
Ghai, R. 267, 274
Gharbi, K. 201
Giansanti, A. 260, 263
Gibb, J.O.T. 150
Gibbs, A.R. 84–85
Giesy, J.P. 87, 141
Gilbert, D.G. 173, 202
Gilek, M. 33
Gill, M. 5
Gillespie, J.W. 94
Gimeno, S. 141
Ginsburg, G.S. 6
Girling, A.E. 271
Givan, S.A. 34, 45, 96
Glaholt, S.P. 173–174, 179, 195–196
Glasner, J.D. 223, 230
Gleason, D.F. 31
Glenisson, P. 111, 225
Glover, C.N. 79, 84, 95, 169
Gocayne, J.D. 188
Godoy, P. 260, 263
Godwin, B.C. 82, 152
Goeden, M.A. 223, 230
Goetz, A.K. 4
Goksoyr, A. 31, 33, 89, 141
Goksøyr, A. 86, 89

Goldin, E. 258–259
Golding, B.T. 267, 274
Gomes, X.V. 82, 152
Gomez-Ariza, J.L. 32, 35
Gomiero, A. 31, 46
Gong, Z. 35, 44, 88–89
Gonzalez, H.O. 34, 95
Gonzalez, P. 93
Gonzalez-Valero, J. 88
Goodsaid, F.M. 4–5, 83, 108–109
Goralski, K. 32
Gordon, D.A. 54
Gorenstein, M.V. 27
Gornati, R. 35
Gorr, T.A. 189
Gotschel, F. 260, 263
Gottardo, N.G. 94
Gottschalg, E. 101
Goulden, C.E. 168
Govoroun, M. 90, 93
Gracey, A.Y. 54, 196
Graham, D.W. 54
Gramatica, P. 55
Graney, R.L. 26
Gray, L.E. 89
Greaves, P. 101
Greeley, M.S. 35, 79, 84, 95, 103
Green, C.F. 5
Greenberg, B.M. 32
Gregor, J. 223, 230
Greven, H. 139
Greytak, S.R. 89
Griffin, J.L. 151
Grigorov, M. 257
Grinwis, G.C.M. 87
Grissom, S. 101
Grizzle, J.M. 77
Groskinsky, B.L. 178
Gross, P.S. 31
Gross, T.G. 33, 50, 95, 100
Gross, T.S. 93
Grosvik, B.E. 31, 33

302

Lyons, C.E. 88
Lyssimachou, A. 93

Ma, T.W. 87
Ma, Y. 4
Maack, G. 88, 90, 94, 96, 101, 111
Macdonald, N. 39
MacGregor, J.T. 29, 79, 257
MacIntosh, A. 92
Maciver, F. 177
Mackay, T.F.C. 199
MacNelly, S. 260, 263
Madden, T.L. 203
Maddix, S. 77, 87, 93
Maddox, R. 271
Mader, S. 82
Maeda, H. 13–14
Maggini, M. 8
Maggioli, J. 5, 56, 58, 60
Magnuson, S.R. 4
Maguire, T. 261, 263
Maiwald, T. 260, 263
Major, H. 151
Makhijani, V.B. 82, 152
Makynen, E.A. 88–89
Male, R. 90, 93, 103, 141
Malek, R.L. 4
Malléa, M. 241, 243
Mange, A. 148
Mangone, M. 202–203
Manly, K.F. 177
Mann, M. 27
Mansury, Y. 261
Maples, N.L. 34, 48
Maqsodi, B. 4
Marabini, L. 87
Maras, M. 35, 84, 96, 106, 108, 110, 173–174, 195–197, 270
Marcel, R. 141
Marchand, J. 95
Marchant, G.E. 11
Marcino, J. 87

Marcovich, D. 88
Margulies, M. 82, 152
Marino, F. 140
Marinone, M.C. 169
Marion, M. 87
Markowitz, V. 111, 225
Maronpot, R. 101
Marra, M.A. 84
Marro, A. 86
Marsano, F. 31
Marsh, J.W. 169
Martelli, A. 267
Martelli, C. 260, 263
Martin, A. 50
Martin, K. 96, 179
Martin, M.T. 4
Martinez, A. 237
Marton, M.J. 195
Martyniuk, C.J. 84–85, 95, 104, 108, 111
Massabuau, J.C. 93
Masson, R. 32
Materi, W. 256–258, 263–265
Matese, J.C. 109, 111, 225
Mathavan, S. 35, 44
Mathieu, A. 86
Matic, I. 230
Matin, A. 241
Matsubara, T. 88–89
Matsumura, N. 77
Matsuo, M. 199
Matsushima, O. 141
Matthews, J.B. 87, 141
Matthews, S.B. 169
Matthiessen, P. 89
Mattingly, C.J. 39, 57, 183
Mau, B. 223, 230
Maudelonde, T. 148
Maul, A. 222
Maule, A.G. 32
Maund, G. 174, 195–197

Subject index

320

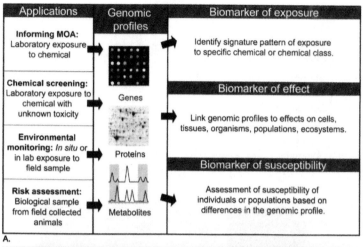

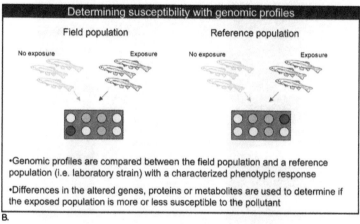

Plate 2.1. The potential applications of ecotoxicogenomics. (A) The pattern of gene expression, protein expression, or metabolite levels provides genomic profiles that can act as biomarkers of exposure, effect, and susceptibility. These genomic biomarkers can be applied to the following applications: informing mode of action (MOA), chemical screening, environmental monitoring, and risk assessment. (B) Illustration of how genomic profiles can be used to determine the susceptibility of a wild field population. Both a field population and a reference population are exposed to the pollutant of interest; controls that are not exposed are also used in both populations. Genomic profiles are established for both populations using a variety of genomic methods (microarrays are shown here as an example technique). The differences between the two profiles are investigated to determine if any of the differences correspond to biomarkers of effect that might determine whether the populations are being affected in the same way.

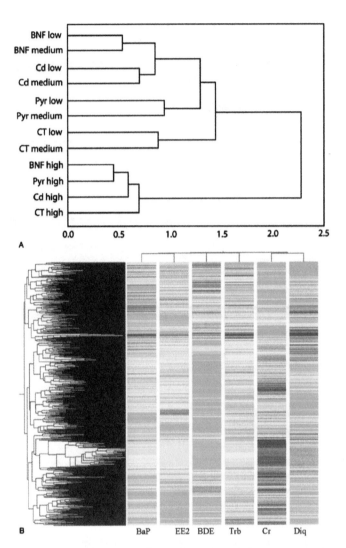

Plate 2.4. Dose- and toxicant-specific effects in rainbow trout (*Oncorhynchus mykiss*, Walbaum) toxicogenomics studies. (A) Pearson correlation clustering of expression profiles in this study of exposures of trout fry show specific profiles at low doses of chemicals, while overlapping expression profiles at higher doses suggest non-specific effects. β-naphthoflavone (BNF), cadmium (Cd^{2+}), carbon tetrachloride (CT), pyrene (Pyr). Adapted from Fig. 1 of Koskinen *et al.* (2004) and reprinted with permission from Elsevier. (B) Pearson correlation clustering of the expression profile of differentially expressed genes in rainbow trout exposed to model contaminants (*y* axis) vs. contaminants (*x* axis) show contaminant-specific profiles. The gene tree is coloured as a gradient with respect to expression level, with red denoting five-fold induction, yellow denoting no change (fold change of 1), and green denoting five-fold reduction in expression levels. Brominated diphenylether-47 (BDE), benzopyrene (BaP), chromium (Cr), Diquat (Diq), ethinylestradiol (EE2), and trenbolone (Trb). Reprinted from Fig. 8 of Hook *et al.* (2006b), with permission from Elsevier.

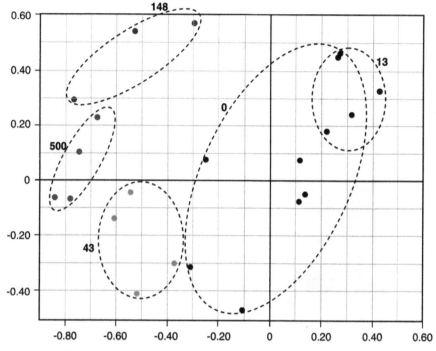

x-axis: PCA component 1 (39.17% variance)
y-axis: PCA component 2 (14.36% variance)

Plate 4.2. Principal components analysis (PCA) of mean polishing per chip and
per gene normalised cDNA microarray data for earthworms exposed to Cd at
the shown concentrations of 13, 43, 148 and 500 µg g^{-1}. Each biological replicate
consists of a cDNA sample generated by pooling tissues from three adult worms
from a single box exposed for 28 days at 12°C and under a 16 h:8 h light:dark
photoperiod.

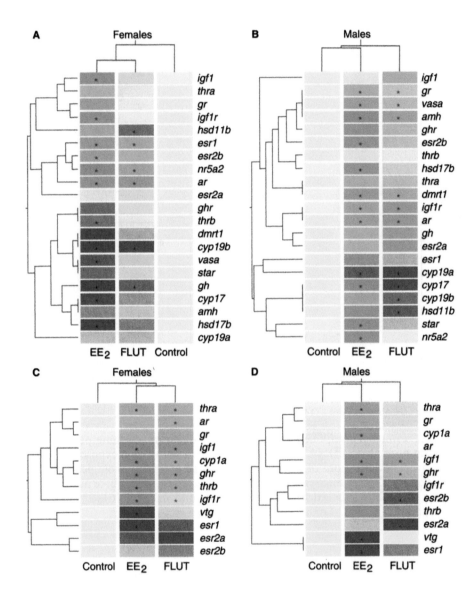

◀————————————————

Plate 3.1. Comparison of the effects of exposure to the oestrogen ethinyloestradiol (EE$_2$; at 10 ng l^{-1}) or the antiandrogen flutamide (FLUT; at 320 µg l^{-1}) for 21 days on the gene expression profiles of the gonad (A and B) and liver (C and D) in adult male and female fathead minnow (*Pimephales promelas*). Gene expression data were generated through real-time RT-PCR. Clustering of both genes and conditions were performed using Pearson's correlation as a similarity measure. The gene trees are displayed horizontally and the condition trees are displayed vertically. Yellow colouration represents a relative expression of 1 (when compared with mean expression of the control group), green colouration represents downregulation and red colouration represents upregulation of gene expression relative to the control group. Each treatment group consisted of eight male and eight female fish and each fish was analysed in triplicate. Statistically significant differences in gene expression between control and EE$_2$-treated fish and control and flutamide-treated fish, are denoted by an asterix ($P < 0.05$; one-way ANOVA, followed by Dunn's post hoc test). Gene acronyms are as follows: *esr1, oestrogen receptor 1; esr2a, oestrogen receptor 2a; esr2b, oestrogen receptor 2b; ar, androgen receptor; gh, growth hormone; ghr, growth hormone receptor; igf1, insulin-like growth factor 1; igf1r, insulin-like growth factor 1 receptor; thra, thyroid hormone receptor alpha; thrb, thyroid hormone receptor beta; gr, glucocorticoid receptor; cyp17, cytochrome P450 17; cyp19a, cytochrome P450 19a; cyp19b, cytochrome P450 19b; star, steroidogenic acute regulatory protein; hsd11b, hydroxysteroid 11-beta dehydrogenase; hsd17b, hydroxysteroid 17-beta dehydrogenase; vtg, vitellogenin; amh, antiMüllerian hormone; vasa, vasa homologue; dmrt1, doublesex and mab-3-related transcription factor 1; nr5a2, nuclear receptor subfamily 5 group A member 2.*

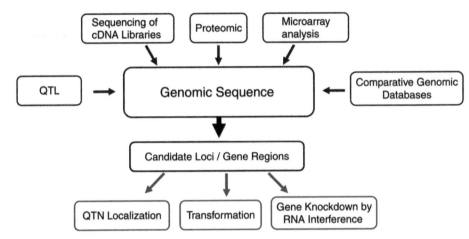

Plate 5.3. Strategies for applying the *Daphnia* genomic tool box for functional analysis of toxicologically relevant genes and gene networks. A series of tools (blue) are available to probe the genome (black), via candidate gene, global expression profiling or genetic mapping approaches. These are validated using standard molecular biological approaches (red).

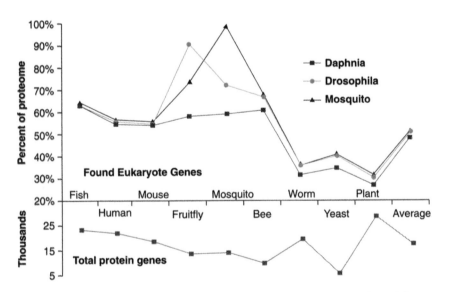

Plate 5.5. Per cent of full gene sets of nine eukaryote genomes found in new genomes *Daphnia pulex, Drosophila virilis*, with out-group *Anopheles gambia*. Lower line shows count of protein gene sets. *D. virilis* has 90% similarity to model fruitfly *D. melanogaster* and *A. gambiae* has 100% similarity to itself.

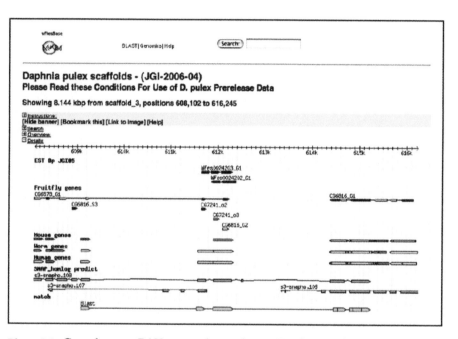

Plate 5.6. Cytochrome P450 gene located on *Daphnia pulex* genome. This GBrowse map view at wFleaBase is returned from a BLAST search for mouse gene MGI:88607 (GenBank:NP_067257) and matches well the predicted *Daphnia* gene s3-snapho.108. Two *Daphnia* EST matches part of this gene and homologous genes from fruitfly, mouse, worm and human match.

Printed and bound by CPI Group (UK) Ltd, Croydon, CR0 4YY

03/10/2024

01040431-0008